D0983959

Inspection and Gaging

SIXTH EDITION

Inspection and Gaging

A training manual and reference work that discusses the place of inspection in industry; describes the types of automatic and manual gaging and measuring devices employed; shows the proper techniques of using inspection equipment; and outlines the various duties of inspection personnel.

Clifford W. Kennedy (Deceased)

Edward G. Hoffman

Steven D. Bond

INDUSTRIAL PRESS INC.
200 Madison Avenue, New York, NY 10016

Library of Congress-in Cataloging-in-Publication Data

Kennedy, Clifford W., b. 1895.
  Inspection and gaging: a training manual and reference work that discusses the place of inspection in industry . . . / Clifford W. Kennedy, Edward G. Hoffman, Steven D. Bond. – – 6th ed.
    p.   cm.
  Includes index.
  ISBN 0-8311-1149-6:
  1. Engineering inspection.   2. Gages.   I. Hoffman, Edward G.
II. Title.
TS156.2.K46 1987
620'.004'4– –dc19   87–17393
                                                                          CIP

INDUSTRIAL PRESS INC.

200 Madison Avenue
New York, NY 10016–4078

First Printing

Inspection and Gaging, Sixth Edition

2 4 6 8 9 7 5 3

Preface to Sixth Edition

Since the publication of the First Edition of *Inspection and Gaging*, many changes have taken place in the technologies of measuring devices and gages. This Sixth Edition will introduce the reader to these new tools and techniques, which are grounded in the ever-changing realm of the physical sciences.

The role of inspection and gaging as it relates today's industry has also evolved since the First Edition was published. Inspection and measurement are no longer only concerned with accepting and rejecting parts, but also with supplying information, which should be used to improve quality and increase competitive performance. In spite of these changes, the foundation for operating the most complex of modern inspection tools rests on basic principles of inspection and gaging that have been in use for many years. The ultimate effectiveness of all tried-and-true measurement operations always involves the use of a tool by a human.

People serving the inspection or measurement role in industry today perform tasks and functions that are different from their earlier counterparts, not so much in what they do, or even in how they operate a measuring device, but more in the use of techniques that make their measurement data more useful. Information is the goal, and balancing the dynamics of quality, reliability, cost, and product variation is the game. In this sense, inspection data are useful only if they are understood and used to solve problems and to reduce product and process variation.

Today, inspection involves measurements—the quantifaction of a variable—and judgments—some less tangible and more qualitative than others. The ability to make these judgments effectively is the first step toward modern inspection and gaging. Expressing these judgments and measurements clearly and succinctly is the next step. The use of this information for quality and process improvements is the final step.

Inspection and Gaging, Sixth Edition, provides a background to time-honored measurement and gaging techniques, and is useful when comparing different approaches to inspection requirements. This edition then carries forward these basic principles into today's inspection environment. It answers the following questions: What changes have occurred in inspection tools and techniques, and what new methods are available? How are the new techniques performed, and how do they compare? What methods are becoming industry standards, and how do they supercede the more traditional methods of manual measurement?

This Sixth Edition carries on in the First Edition's original style and intent—as an up-to-date guide to inspection and gaging that blends the technical disciplines of inspection with the needs of managers, supervisors, teachers, shop workers, engineers, technicians, and all others concerned with reducing critical variations. *Inspection and Gaging*, Sixth Edition, will serve as a guide to effective variation control, by outlining the choices available in the field of inspection and gaging, and how those choices can be applied to the requirements of modern industry.

Preface to First Edition

This book, or the idea of writing it, was war born. During the critical and hectic period from Pearl Harbor on, industry strained every fibre to supply war material fast enough. From the most minute instrument parts to the massive pieces of completed ordnance, this vast array of manufactured product had to be inspected and re-inspected throughout its various stages of production. Most of the inspection burden fell on the manufacturers and contractors.

Thousands of workers of both sexes were hired and assigned to inspection duties. Practically none of this horde knew anything about inspection methods and techniques. They came, most of them, from other fields of endeavor — housewives, store clerks, insurance salesmen; too few had had factory experience of any sort and those who knew anything at all about a machine, a bench, or even a wrench were diverted hurriedly to the hungry maw of production.

Many weary hours were spent by the few who possessed some industrial inspection experience, in teaching, demonstrating, and coaching. Many of us would have given the proverbial shirt, could he have presented the neophyte with a book explaining at least the elementary principles of inspection operations, to expedite the process of learning.

Today, new people are continually being inducted into industrial inspection departments in a new push to build up the sinews of defense. Even those who are not so completely uninitiated in manufacturing processes and machining methods, usually have only a vague or erroneous conception of the equipment and techniques of inspection and gaging. Then, it is not difficult to observe that in many plants so-called trained and experienced inspectors and their supervisors have obviously missed many of the essential principles somewhere along the line.

It seemed likely, therefore, that a book such as this is hoped to be, would prove as valuable for today's inspection forces for instruction and reference as it would have been during the early days of World War II.

No attempt has been made to include within the covers of one book every last ramification of inspection, gaging, testing, and quality control as it exists in the present huge diversity of American industrial production. This would be impossible. It did seem possible, however, to present, explain and illustrate many basic principles and procedures which form the common denominator for most inspection requirements. It is believed that such basic information will provide a sound foundation on which the individual inspector can build additional knowledge and skills related to his specific needs. If such a foundation has been successfully laid in this text, the book will have well served its intended purpose.

Providence, Rhode Island   CLIFFORD W. KENNEDY
July, 1951

Acknowledgments

Acknowledgment is made, with thanks, to the following companies and organizations who have contributed more or less directly to this effort in the form of photographs, diagrams, quotations, examples and case histories. Their gracious grants of permission made the author's task easier but, of greater importance, made the contents of the book much more valuable to the reader.

American Bosch Company
American Machinist
American Standards Association
Ames, B. C. Co.
Bausch & Lomb Optical Co.
The Bendix Corporation, Micrometrical Division
Brown & Sharpe Manufacturing Co.
Brush Development Co.
Bryant Chucking Grinder Co.
Carl Zeiss, Inc.
Crane Company
Dearborn Gage Co.
Eastman Kodak Co.
Electrolux, Inc.
Engis Equipment Co.
Farrand Controls, Inc.
Farrand Optical Co., Inc.
Federal Products Corp., a unit of Esterline Corp.
Fellows Gear Shaper Co.
Gaertner Scientific Co.
General Electric Company
Gillette Safety Razor Company
Greenfield Tap & Die Company
Hamilton Standard Propeller Division
Hanson-Whitney Company
Hunter Spring Company
International Business Machines Co.
Johnson Gage Company

Jones & Lamson Machine Co.
Link Division, General Precision, Inc.
Lufkin Rule Company
Machinery
Metals and Controls, Inc.
Moore Special Tool Co.
Norton Company
Optical Gaging Products, Inc.
OPTOmechanisms, Inc.
Colt Industries, Pratt & Whitney Machine Tool Div.
Portage Double Quick, Inc.
Remington Arms Company
Ryan Aeronautical Company
Saco-Lowell Company
Scherr, George Co., Inc.
Automation & Measurement Division of The Bendix
    Corporation (Formerly Sheffield Corporation)
Sperry Company
Standard Gage Co.
Starrett, L. S. Co.
Sylvania Electric Products Corp.
Syracuse University Press
Taft-Pierce Manufacturing Co.
Van Keuren Company
Wagner Electric Company
Wilson Mechanical Instrument Co.

Any attempt to list the many individuals, friends and associates, the engineers, chief inspectors and quality managers throughout the country who have helped supply other material and experience for this book would pose a publisher's problem akin to Who's Who. Our gratitude for their help is none the less sincere though we fail to print an honor roll.

C. W. K.

Contents

CHAPTER 1

The Need and Function of Inspection in Industry

Our five senses are basically instruments for inspection based on self-preservation, curiosity, or enjoyment. We look to the left (most of us who are still living) to see if it's safe to cross the street, and listen at railroad crossings. Our noses tell us when there is a gas leak, and our taste buds warn us before we eat something that will sicken or poison us. We feel in the dark for the light switch before bumping into the furniture.

Babies put everything into their mouths; it is their way of comparing the new and mysterious with what they know, their manner of inspecting. We see a sunset, smell a rose, hear music, taste candy, and stroke the soft silkiness of a dog's ear.

Either for protection or gratification we are inspecting all day. The dress is inspected that has just been brought back from the cleaners; you try the brakes on your car after the garage has finished adjusting them. Sighting along the green from ball to cup before that critical putt is an inspection. Many inspections are so commonplace that we are not conscious of making them until something is awry. You look at the top of your dresser every morning of your life and never really notice it until one of the familiar objects is removed or something is substituted for it.

The dictionary defines an inspection as the critical examination of something, and the inspector in industry is supposed, by the very nature of this title, to examine products more closely than workers performing other tasks on them. But in the broader industrial sense, an inspection is a critical examination directed to some predetermined purpose. You might inspect a store show window with great care and comment only on a lack of symmetry or some equally irrelevant condition, but the head window dresser would instantly detect the absence of an item of merchandise or the wrong price tag. A factory inspector would disregard the oil, dirt, and chips on lathe-turned pieces, knowing they are to be washed and nickel-plated later, and observe the shoulder burr that would prevent ready assembly in mating parts.

In the broader industrial sense, inspection, if it were necessary to

1

define it in a very few words, might be called *the function of comparing or determining the conformance of product to specifications or requirements*.

Because inspection is so varied and related to many different applications, the underlying *purpose* of an inspection is what sets one inspection measurement apart from another. The purpose of inspection may be to inspect in order to ascertain which products pass—what is acceptable—or inspection may be instituted to see how many products fail—what products may be unacceptable. More specifically, inspections are measurements, visual assessments, tests, or some relevant evaluation of a product, process, or the act of making a product.

All inspections strive to isolate and assess the *relevant* detail—the detail or quality or characteristic of a situation (or product) that is relevant or important. The purpose, then, of all inspections is always related to a good understanding of relevant detail, of requirements or specifications, and of the proper assessment (and dispositioning) of that product with its attending detail.

Inspection detail is simply the measurement of relevant variation—whatever the measurement involved, variation may eventually reach a point of being excessive. When this happens, the *purpose* of inspection is clear: to play a part in the reduction of excessive variation.

The Study of Inspection

The subject of inspection in manufacturing plants may be approached over one of several routes. For the purpose to be served here, one road should have been traveled, at least a short stretch of it. Whoever undertakes the study of industrial inspection should also pursue a general knowledge of present day industrial methods. Lengthy interruptions to explain common manufacturing operations and terms will be avoided in this text.

The mention of shop inspection may immediately signify to many an ability to measure. It calls up visions of micrometers, height gages, surface plates, indicators, and the host of measuring devices developed for the use of the shop inspector. Measurement is an important part of inspection; practice and proficiency in efficiently securing accurate measurements are essential. All the emphasis, however, should not be placed on learning this phase of inspection.

For others, the term implies visual inspection, or the act of looking at parts and pieces and classifying the work by eye as satisfactory or rejectable. Visual skill and good judgment in determining the quality of shop products are assets the successful inspector must develop.

Different products, processes, and types of manufacture, of course, alter details of the viewpoint of inspection. The foundry inspector thinks of such things as castings, blisters, core alignment, and snagging. In the paper mill or weave room, inspection means the survey of broad areas of product or tearing off a swatch for more detailed analysis. Inspection frequently includes testing, as for viscosity or tensile strength; it may involve reading electrical instruments. In a silver shop the inspector requires an expert knowledge of finishes, while the candy inspector, many times, is a taster. A unique inspector is the person in the match factory who strikes and blows out 3,000 matches a day, samples of the day's production, to observe how they act in the various safety tests he or she, conducts. These are details of procedures and techniques that are learned on the job.

Inspection also may be studied from the point of view of its function and responsibilities in the modern industrial organization as compared with, for instance, production, engineering, and purchasing. This viewpoint is valuable since so many inspectors fail to see the forest for the trees. They become snarled in details and petty wrangles, often self-righteous when their smallest decisions are questioned, and forget that other people and departments are also interested in their assignments and inspection results. There is a basic framework of inspection applicable to practically every industry, which is a part of a total organization's responsibility for putting out quality products, and inside which the particular local pattern of inspection details is arranged.

The role of quality inspection and inspectors is rapidly changing and expanding in modern industry. Effective inspection groups are well aware of their relationships to other groups or parties within the manufacturing plant. Measurement and assessment is indeed the most important role that inspectors have to perform, but extensive communication of results, as well as analysis and action on those results, especially when shared among inspection forces and other groups, is what makes inspection invaluable.

Inspection in the Small Shop and in the Large Plant

Consider a small shop employing, say, less than 10 workers. The owner is at once boss, sales manager, treasurer, and foreman of this "one-man" business. At almost any time he can examine each operation in detail, inspecting the fabricated parts or completed units of the product. He knows what his customers think. He can readily check on the quality of materials or parts he buys. Furthermore, each employee shares the owner's concern and knows instantly when shabby workmanship appears in the shop. In such a group, anything resembling formal inspection is unlikely.

At the other extreme there are, for instance, the huge automobile plants where many thousands of people work on three shifts. Other thousands of dealers and salesmen distribute the cars and repair parts throughout the world. The "boss" of such an organization has only occasional opportunity to inspect comprehensively the finished product. He can practically never examine critically an individual operation. Yet the quality of each car concerns him more, perhaps, than almost any other problem, for he knows the customer is every bit as critical of the big plant's product as he is of the merchandise offered by the one-man business.

Whatever the scope of the inspection organization, the basic principles remain the same. Inspection details measure and assess relevant characteristics of a product or process in terms of requirements and specifications derived from a desire to meet expectations about the product's level of quality and performance, however those details are measured.

Manufactured products are expected to meet a level of quality and performance through a system of standardization, and the purpose of standardization is to minimize variation within a product and its components and to keep this variation under control. But standardization always produces variation; the questions are: How much variation is acceptable, and how does that translate into product requirements? These questions have to be answered by information from the inspection department.

When a shop grows larger, more people working in it must assume some of the responsibilities and anxieties of the owners of the business. They must perform more of the duties the "boss" still wishes he had time for. So, to an extent, an inspector represents the boss when it comes to seeing that the manufactured products conform to the desired standards. Standards represent the epitome of quality requirements. To understand and administer a desired standard is to define the standard, and apply the method of measuring or representing the standard.

The job of inspection is to ensure the ongoing maintenance of standards and requirements, and this means continuing measurement in order to control and reduce the variation that threatens the integrity of the standards and requirements.

A comprehensive treatment of shop inspection should include a study of instrumentation—gages, meters, special apparatus, visual aids, and their selection, manipulation, and maintenance. It should include a study of the more commonly used industrial inspection routines, the planning of an effective day's work, the keeping of essential records, and the elimination of unnecessary activities. An inspector, in many cases, should be well grounded in statistical quality control techniques.

If he has a flair for mechanics, he is that much better equipped. He needs business sense, too, in connection with his duties, especially an instinct for detecting unprofitable routines. He should be as wary of adopting superfluous inspection procedures as a good merchant is of merchandise that won't sell. Finally, there is what is popularly known as the human angle. The inspector who can get along with people, influence them, get them to accept his decisions and like them, is possessed of a most valuable asset.

The Function of Inspection in Industry

A good way to look at the function of inspection is to compare the industrial organization with government. In government (American variety) there are the legislative, executive, and judicial branches. An industrial organization has engineering, production, and inspection departments. Where one or two legs of the governmental triumvirate are combined or missing, there is apt to be dictatorship. Dictatorships in an industrial organization may, like their political counterparts, come into existence because of the weakness of one or two of its functional branches.

Basically, whatever a factory produces is made because the customer wants it. Customers are to industry what the "people" are to the politicians. Hence, in industry, what the people — the customers — want is determined largely by the sales and engineering departments. Theoretically, in most plants, the final responsibility for exactly what is to be made, and frequently how it is to be made, rests on the shoulders of the engineers. They are the legislators.

Then, theoretically as well as practically, the production department carries out engineering directions. It makes the specified products, producing them in the quantities required by the sales department, and many times following the manufacturing procedures in detail prescribed by the engineering department.

It is the function of the inspection department to balance the engineering, support, and manufacturing departments by designing and maintaining a system that assesses the quality levels of the work that is done, and the products that are made, according to objectified standards of measurement criteria. (See Fig. 1.)

The basis of inspection lies in understanding customer requirements, and the translation of those requirements into objectified measurements that can be carried out in the manufacturing arena.

Customer requirements, from a producer's point of view, balance a level of quality and reliability within the product's performance with *cost*. A customer expects to get what he pays for, and that's the bottom

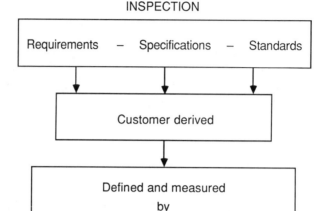

Figure 1

line, a line that cannot be leveled without accurate data from the inspection group.

Quality levels are customer-derived. Inspection systems support customer-derived quality levels, and the improvement of those levels of variation is the contribution of inspection working with other groups and departments.

Some customers buy inexpensive automobiles and other customers will purchase expensive automobiles, and each customer should receive a level of quality that they pay for. The detail of inspection for each particular level of customer satisfaction will vary. Balancing cost considerations with customer expectations, and the achievement and maintenance of a good match, is the goal.

Multiple Functions of the Inspection Departement

The products are made and are ready to be sold. Has the law been transgressed? Here is where, in the more progressive shops, the inspection department enters. Like the judiciary in government it has, in a way, many functions. It must simultaneously interpret the law and also decide whether or not the decrees have been satisfactorily met. To this degree, it stands in judgment between engineering and production. But, operated properly, inspection performs another service. Wise course decisions are based not only on the law but also on the viewpoints, feelings, and rights of the people. The inspection department, in a plant, represents the customer. This responsibility should be kept in mind in deciding troublesome shop family disputes.

In the same vein, the inspector should never try to make the laws — originate, change, elaborate, or establish specifications — but, as with a court, his wise interpretations of the law, of constitutionality, if you want to put it that way, will in the end effect desirable changes in the applications of the laws.

The quality balance of acceptable versus unacceptable detail in a product, component, or even a manufacturing process is indeed the responsibility of the inspector. He must understand and interpret according to objective criteria or specifications in an inspection. In a sense, then, the inspector must act as critic, to assess relevant detail, and reject, accept, or communicate about that detail. It is also the responsibility of the inspector to *help fix problems* that are uncovered through inspection. Helping to fix problems means reducing excessive variation. This level of change in a component, product, or process, from acceptable to excessive variation, is the responsibility of the inspector to assess. The definition of relevant (inspection) criteria or detail is the first step. The definition should contain an objectified measurement by inspection — a method to apply the criteria or standards. Then the inspections take place — are collected, and the observed variation is judged to be acceptable or unacceptable according to the stated requirements.

If the variation is deemed acceptable, the inspection is either continued to ensure on-going compliance with the requirements, or the inspection sequence is over. If the measured variation is unacceptable, then other variables such as time or personnel are factored into the results and considered as a possible cause of excess variation. In other words, the data are analyzed. Out of this analysis comes opportunities for improvement, and problems are fixed.

Cost, of course, plays an important role in balancing the quality scale — are the changes, improvements, or reductions in variation cost-effective? Management groups in manufacturing systems are chartered with this important and delicate decision. What is important to the customer, and how much will they pay for these requirements and performance levels — this is the stuff of relevant inspection detail.

Part of this inspection cycle, which involves awareness, assessment, reporting, and acting, includes the *Education* of other production or manufacturing personnel. Ultimately, responsibility for some amount of inspection (and *most* of the *level* of *quality*) of parts, components, and the finished product lies with the person who *does the work*. An adequate translation of customer requirements through specifications and product criteria *must* be available to production employees. In modern manufacturing environments, inspectors assist in this task.

Therefore, the work of inspection personnel is much more broad than that of the direct measurement itself. Although the act of inspection measurement is the most critical element of inspection, it is simply, in the final analysis, a cog in the wheel of product or process improvement cycles. (See Fig. 2.)

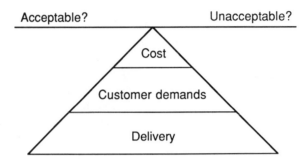

Fig.2. Quality balance.

Applying another viewpoint, it might be said that the inspector's job includes a management function and responsibility. He is assigned to do what the plant manager, the purchasing agent, the chief engineer, and the production manager themselves should do, in theory, and perhaps would like to do if they had time. This assumption should offer no reason for the inspector to puff up with importance. It has been suggested solely to emphasize a degree of *responsibility* that the good inspector should assume. Because the inspector does take over the responsibilities of quality and performance, the sales manager can devote that much more attention and diligence to where and how the product can be sold, the purchasing agent to better and better sources of supply, the engineer to design and methods, and the production worker to the actual manufacturing operation.

In manufacturing, the inspection act itself entails the comparison of a product with the specification or some standard. It includes determining whether or not the part, product or batch is free from faults, blemishes, or defects, that it is made to prescribed tolerances, sizes, color, taste, or texture. A portion of manufacturing inspection frequently overlooked is the responsibility for making sure that the correct materials have been used and especially that all of the processing operations have been performed.

In order to institute these extensive levels of responsibility for manufacturing quality and performance, an inspection system must be far-reaching within the organization and have excellent communicative abilities. Assemblers or operators must understand critical inspection detail and standards. Supervisors must likewise understand these criteria directly. All support groups — engineers, materials handlers, and others — must be aware of customer requirements appropriate to their particular level of activity. Finally, management must understand the results of inspection systems, for they must ultimately pay for most improvements in variation. Consequently, to be successful in the manufacturing environment, problem-solving activity that is focused or centered around data has to be communicated to multidisciplinary groups. The inspector is important to such an effort, because the inspector supplies data — facts that describe variation in real terms. From these facts come an assessment of the acceptability of the measured variation, and a search for causes and solutions in the effort to reduce variation in the competitive arena.

Therefore, inspection must constantly interact with the process and systems it is measuring on a continuous basis, and at many levels, in order to ensure that the results of inspections are fully understood, and integrated within a system that strives for constant improvement.

Need for Inspection

Let one man make one thing, another man make a second thing, and a third put the two together to form a subassembly or product, and you have a basis for inspection. When different people make parts there are likely to be as many different standards of workmanship. Few instances occur in present day manufacturing where a product is made complete from raw materials by a single worker. The tendency is to sub-divide a job into a series of separate operations. Operations are also frequently performed on products at widely separated locations and time intervals. Parts are routed, for example, to milling machines and then upstairs to be drilled. Frequently, they are sent out of the plant for plating, say, or for hardening.

Years ago, if a man wanted a screwdriver, he cut off a piece of steel, heated, hammered, forged, and ground the blade. He turned a bit of hardwood for a handle or whittled it out with a draw shave. Nowadays the stock is cut to lengths by machine in one place, the blades are drop forged in dies, tempered in hardening furnaces, and then ground by another group. The handles are turned in the wood-working department by one man and the holes in them bored by another. They go over to the paint shop to be dipped and baked. The ferrules are blanked and drawn on presses somewhere else. Finally the screwdrivers are assembled. A dozen workers, or more, have had some part in making a single screwdriver.

Variation is now notorious in manufacturing systems that rely on interchangeability and specialization. As specialization is perfected, requirements and associated competitive factors become more demanding; the need for finer discrimination and less variation is on the increase.

More direct measurements are demanded of processes and products. These direct measurements must be available speedily, economically, and accurately, and most often the job of obtaining these measurements falls to trained and reliable inspectors. Such measurements of critical parts and dimensions must be as free of subjectivity as possible, so that they can be accurately assessed and acted upon. Cost improvements eventually become the final bastion of the competitive position as the marketplace becomes more demanding. The organization that can measure and adapt based on these measurements will be the most successful economically.

In addition — or as a consequence of specialization — the individual worker is interested almost entirely in his own task and cares little about the effect of operations preceding or succeeding his. Repetitiveness, monotony, even drudgery, enter in. Too often in modern industry the operator is less an artisan and more of an automaton.

Engineers and methods experts are employed to devise machines, techniques, and routines for faster, simpler, and cheaper production. Incentive plans are added under which the worker's pay envelope depends considerably more on the quantity of units he completes than on the quality.

As this scenario progresses, individual operators or assemblers on the manufacturing line may lose touch with customer requirements; interpretations of these requirements in terms of criteria and standards for parts or finished assemblies must be available to inspectors and assemblers alike. "Self-inspection," or inspection of process work by the people who perform the work, is essential in maintaining an acceptable level of quality.

Follow-up to these self-inspection data is a job for inspection personnel. Inspection data are not simply a scoreboard for what happened, but a springboard for action. Active feedback and response, based on a cross-analysis of data among inspectors and line workers, can yield a highly accurate and timely assessment of data. Regular inspection personnel become data and information handlers, and more information allows them to be better at the job of inspection.

The collective responsibility for quality of several operators never adds up to the care displayed by one man, himself performing all the operations; the growth of a business and an increase in the number of workers dilute individual responsibility. So the "vitamin tablets" of inspection must be added to the industrial diet in order to ensure healthy products.

In the larger organizations, the continuous pressure of technical improvements in methods, tooling, and equipment, the effects of time and motion study, and the turnover in help which seems an inevitable consequence of expansion — these and similar conditions all result in the production of increasing quantities of defects, scrap, and rework unless steps are taken to check them. Systematic, routine inspection seems to be the best means. Cost reports universally indicate savings in scrap and rework that are more than substantial enough to pay for an inspection setup when it is properly installed.

Some plants employing, say, from 3 to 50 people seldom consider or adopt formal inspection procedures until the combination of continuing growth and complaints from customers makes a routine examination of its products necessary. At first, something like a formal inspection operation may be initiated on the products just before they are packed for shipping. Occasionally, too, a need arises for examining parts, materials, or subassemblies that are purchased for more or less direct use in the products the small factory is making. So an inspector is hired or someone is assigned to that routine. In some such manner the inspection procedure gets a start and grows to be one of the manufacturing routines.

Final Inspection Is not Enough

Very shortly after the final inspection of finished products is instituted, it is discovered that no product is better than the components that make it up. Then inspection is extended. The various parts are to be inspected before they are assembled. It is not only more economical to discover defective units in the early stages of manufacture, but many times it is impossible to test or inspect an assembled product in such a manner as to detect defective workmanship buried inside the article.

It is cheaper, for example, to test a radio condenser as it is made than to tear down the whole set later in order to replace a dud. A motor's field can be given a high voltage breakdown test as it is made, a type of test that cannot be as safely applied after the field coils are in the stator and the armature is assembled. Just one warped or loose wrist pin connection inside the piston in one of the cylinders of a $15,000 automobile can completely ruin the owner's pleasure in that car. Furthermore, the troublesome wrist pin may not make itself evident either at the normal factory test or when the dealer demonstrates the car on a short run around the city. The defect may not "speak out" until later at 70 mph or when the car is toiling up a mile high mountain pass. In most modern industry there is need for a detective agency, some portion of the organization specializing in and responsible for uncovering actual or potential hidden defectiveness.

Another economic reason for inspection as a part of the manufacturing routine is the fact that later operations so many times fail to correct previous defectiveness. Suppose a small motor bearing is drilled as shown at *b* in Fig. 3, rather than correctly as the dotted lines *a* indicate. The reamer, in the next operation, will follow the originally drilled hole no matter how solidly the reamer and bearing are chucked or held. After a motor is assembled with such an out-of-line bearing, there will be vibration, heat, and rapid wear, at least, if the armature shaft doesn't actually freeze in the bearing.

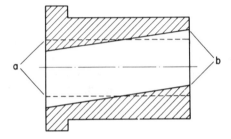

Figure 3

Inspections paralleling certain production operations help to prevent cumulative defectiveness in products. Weaving faults may be blamed on the loom, but frequently the trouble can be traced back to the spinning frames, to carding, or even to some carelessness in scrubbing the original sheared wool. You recall the jingle: "For the want of a nail . . . the battle was lost." Modern industry increasingly relies on inspectors to detect omissions in the process that accumulate error on error like a snowball rolled down hill.

As the snowball rolls downhill and costs go up, process feedback is all-important to detecting problems early on and avoiding an accumulation of scrap, rework, and related overhead costs. Processes are comprised of four essential ingredients; materials that are used, procedures or methods of manufacture, tools or machines that help do the work, and the people who do the work. Solving quality problems, and cutting associated costs, involve an understanding of the way these factors interrelate to cause a problem of excessive variation, and to solve a problem of excessive variation. Inspections that detect variation before costs get out of hand are those inspections that can stop such variations from continuing.

Product assessment is important, because this *is* the result of the work that is being done. Customer requirements are obviously linked easily to product evaluation and measurement. But process evaluation — that which makes the product — is also important, and can be more timely in terms of supplying information about the causes of variation, excessive or otherwise.

Modern manufacturing environments are supplied inspection data on both the process and the product, and then assessments are made from both sources. Change or corrective action from processes is much more direct than causes and solutions extracted from product variation. Such a responsive system requires a clearly proven and understood system of linking product variation to process measurements and change. This system is usually cooperatively derived from engineering support groups and manufacturing inspection personnel.

Mass Production Requires Interchangeability

Most present day products are put together using assembly-line methods. Take a factory that makes electric flat-irons. You will see a long bench or conveyor at which operators sit or stand. On and above the bench you will see sole plates, among other things, and lengths of cord, nichrome heating units, plastic handles, electrical connector parts as well as trays of screws and bolts. It makes no difference that some of the parts were made three months ago and others just yesterday, or that plastic parts were molded by a concern half way across the continent. They must all fit together accurately — and quickly.

Our whole system of producing large quantities of products at low prices hinges very considerably on interchangeable parts. In the flat-iron assembly line, it makes no difference which sole plate an operator picks up to fit with a heating unit and other parts selected entirely at random, the assembly goes together because the parts are interchangeable. The same idea shows up in the weave room where a truck full of spools of yarn are moved up to the loom. It makes no difference whether the loom fixer racks up a particular beam of yarn on the right-hand side of the loom or the left or in the middle so long as the yarn is uniform enough to make the spools truly interchangeable. Complete interchangeability of parts is economically essential in mass production of fabricated products.

The automobile epitomizes interchangeability for most of us. At the same time it brings up the subject of repair parts. Whether it is a king pin, clutch plate, or door handle, we expect the purchased repair part to replace accurately the worn or broken article.

Interchangeable parts, then, reduce assembly costs by eliminating nagging and costly delays in selecting, filing, and fitting. Interchangeable parts permit quick, cheap, and satisfactory repairs. But the most important factor in connection with truly interchangeable and accurately fitting parts, perhaps the one most frequently overlooked by factory people, is the fact that they make better, longer wearing, and more reliable products when assembled together.

The engineering department is charged with the responsibility for designing parts that will be interchangeable; it establishes tolerances that make this possible. Frequently, it helps in developing the processes and methods to ensure interchangeability. The production department then assumes the duty of producing the parts as specified; in the end it is responsible for *making* them interchangeable. Finally, the inspection department must determine the fact that the parts are interchangeable, or not.

100 Per Cent Inspection Doesn't Produce 100 Per Cent Results

It should be remembered by anyone who has anything to do with 100 per cent inspections, either visual or manual, that 100 per cent is not 100 per cent in results. If defectives are present in a batch, even the most expert inspector will miss some of them at least once in a while. Where only a scattering of defectives exists in a large lot, the situation is akin to hunting for a needle in a haystack, and if there is an abundance of defectives to be culled out, it is like brushing white dog hairs off a blue serge suit.

There are several reasons for human failure in 100 per cent inspections. The greatest enemy perhaps is monotony. A certain human attribute seems necessary, some sort of patience or peculiar stamina, to withstand the relative boredom of the repetitive series of inspection motions. Coupled with that kind of stamina should be also a degree of conscientiousness. If an inspector feels he hasn't those qualities, if 100 per cent inspection makes him nervous, bored, and impatient, if a don't care attitude develops, he should doubtless seek a transfer to other work.

Even from the best, the most alert inspectors, fatigue steals efficiency. You may be conscientious, you may wish to work your best, but somehow in spite of yourself, after a while, the mind quits making decisions, the optic nerve evidently "numbs" and the retina becomes "paralyzed," even though the hands obediently continue with the required motions.

A third factor affecting inspection efficiency is ineptitude. Some people, otherwise completely normal, intelligent, conscientious and able, simply cannot inspect and do it well. An inspector should seek a transfer away when, after a reasonable apprentice period, he appears to be truly inept or awkward doing inspection work. (See Fig. 4.)

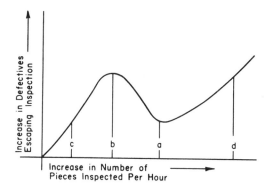

Figure 4

Setting up 100 per cent inspection also exerts a peculiar "psychology" on manufacturing departments. Knowing that, if they do make substandard items, inspection will comb it out, production people inevitably slacken their attention to quality and zealously concentrate on quantity, on making new and ever better production records, forgetting that it is the *net good* product that pays the dividends.

Sometimes an attempt is made to justify 100 per cent inspection expense on the basis that from it reports are secured as to how much substandard product is actually manufactured. But, somehow, the reports seldom do much good. In the first place, under the best conditions, it takes time to complete 100 per cent inspections and get the reports, and the time lag is usually nearly fatal. To tell an operator on Tuesday that he made 8 per cent scrap last Friday has little effect. Yesterday is gone, tomorrow has not yet arrived; most of us shrug off the importance of today.

No doubt the best cure is prevention. Applying this axiom to manufacturing means engineering the process that is least likely to produce substandard products. If the blue print calls for ± .0005-inch tolerances, it is nearly useless to manufacture such parts on a machine that cannot hold closer than ± .002 inch. However, the best intentioned, the best designed, and the best made equipment and machines "get off the beam." It is good manufacturing economics, for instance, to secure the maximum production from the equipment, machine, or process, the most production with the least possible number of interruptions or shutdowns for adjustments, repairs, or replacements whether it means sharpening a tool, dressing a wheel, reclothing a card, or changing a set of rolls. But such practice, economically sound as it may be, does produce marginal if not substandard or actually defective products — at least a small percentage right at the time of adjustment.

Under the conditions, then, the man best located not only to detect the trouble but to inspect, and to sift out defective units is the operator. Let the necessary gaging, testing, and visual inspection operations — the time for them — be added to his operating cycle. Why handle each piece all over again a second time at some later and distant inspection? Another axiom comes up here: the closer to the source of manufacture the decision is made in regard to conformance, the better will be the over all quality.

Statistical quality control (SQC) can be used to alleviate the burden of 100 per cent inspection. It is always wise to consider the shortcomings of any system along with its favorable points. The inspector many times faces the problem not only of technology — what is the best gage, apparatus, or instrument to use — but also of economics. He should question constantly. Is this inspection necessary? Could or should the operator do it? Must every unit be examined — 100 per cent inspection — or will sampling serve as well? In general, the more an inspector deliberately attempts to work himself out of a job, not in the sense of slacking or soldiering or buck-passing, of course, the sooner he will be promoted to a better job.

Special Duties of the Inspector

In the routing of pans, boxes, and trucks of parts from machine to machine, a batch of work may be carelessly or inadvertently detoured and miss an operation. Certain small cam sections in an intricate weaving machine are to be shaped, milled, and drilled. Then they are to be hardened before a final grinding operation. Suppose it is discovered after they are hardened that the drilling operation has been somehow overlooked. There is not only the nuisance and cost of rerouting the work back to drilling but the pieces must first be annealed. Then they must be hardened again after the omitted drilling job has been performed. By that time they may be warped or ruined beyond repair. Add to this the probable expense of someone's lost time waiting for the pieces to reach his operation. An inspector can serve profitably as a sort of truant officer to prevent such occurrences.

Ensuring that assembled parts do indeed result in high or acceptable levels of quality and competitive levels of reliability or performance also involves work from the inspection or Quality Assurance

(QA) Department. Final testing, in a functional mode, as well as final inspection, is critical to most manufacturing operations. A system that has been "tried out" or stressed to some extent can lead to a more reliable product (if the feedback from the testing is used to fix problems).

On-going reliability testing, or testing designed to determine where failures occur in the lifetime of a functioning product, is also essential to maintaining high quality. Conversely, preproduction testing is also generally an accepted responsibility of the QA department. Such testing helps determine the producibility of a product, and assess the viabilities of a design, by anticipating problems before they occur in an actual manufacturing operation. This type of "up-front" process work also allows effective inspection stations to be set up throughout the factory line, because the critical points or variation are known ahead of time.

Personnel in these systems with definitive and specialized purposes may have special roles as inspectors, both in taking the measurements and in accumulating, analyzing, and translating the data to other interested support groups. But again, the principle remains the same; one of criteria or measurement definition, the operation of the criteria, the measurement itself, the reporting sequence, and, finally, the analysis and response cycles.

The Inspection Department's Second Responsibility

If the inspection department should be said to have only two responsibilities, the second one, in addition to ensuring measurable conformance, would be that of judging appearance. This includes the questions of workmanship and standards, to be taken up in detail subsequently.

From mouse trap to mansion, the choice of which product — yours or your competitor's — the customer will take may well hinge on its appearance either at first glance or after prolonged scrutiny. Standards of appearance falter under the repetitiveness and pressure of incentive-paced production and someone must stand guard over them. The worker, absorbed in his tools, overlooks appearance. We fail to notice how shabby our everyday shoes have become until we go in to buy a new pair. The artist stands back from his picture but someone else can more readily point out where the coloring is dull or the shading too deep. Impartial self-criticism is difficult.

In industry the inspector is not only trained as a specialist in the early detection of many substandard details which the production group, naturally more intent on quantity, ignore or overlook, but he is equipped with special apparatus for the purpose. He may use a microscope, for example, a special type of light, a color comparator, or chemical analysis apparatus. He has gages, perhaps, that magnify almost intangible changes of dimension on indicators.

The judging of "appearances" in quality can be quite deceptive, and also quite important to the often-vague issue of customer requirements. But, the fact is a large proportion of inspection decisions are based on these so-called *attribute* decisions — choices that characterize the quality level of a product as "good" or "bad," and not both, and not halfway in between. (See Fig. 5.) Of course, these attribute decisions can directly affect the function of a product, as well as appearance, and are most often used as a quick way to assess the quality "level" of a product or process.

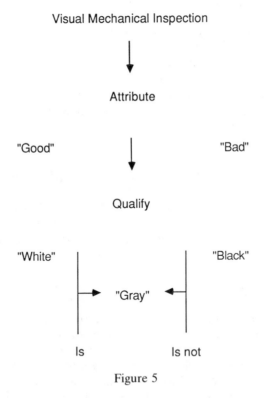

Figure 5

Again, a balance of quality or performance attributes is sought in relation to the customer's needs and requirements. The pass/fail criteria must make sense in the marketplace. This is a part of the very definition of attribute measurements. If this makes sense, the attribute must be objectively described, and a method for measuring the characteristic is devised. Then "appearance" becomes a natural and profitable part of the inspection routine.

Classifications of Inspection

Industrially, inspections have come to be typed or classified in terms that are descriptive of the sort of work performed or the location of the inspection in the process or shop.

According to the general sort of labor involved, we have,

Manual Inspection
Visual Mechanical Inspection
Test Inspection
Mechanized or Automatic Inspection

Generally descriptive of the area in which the inspection is performed, there are,

Process Inspection
Batch Inspection
Final Inspection
Receiving Inspection
Tool and Gage Inspection

Further subdivisions of inspection work and relevant terminology, common in industry, will be taken up as each major division is discussed. Inspection procedures in the average factory embrace many of the classifications previously listed in various combinations, although different local names may be used for the same general type of inspection.

Inspection personnel are also classified under a variety of names. There are sorters or detailers — and inspectors, of course. As the work involved becomes more technical and administrative, the expression inspection or quality engineer appears. Lately, we are hearing about quality analysts and quality control engineers. In the upper brackets of inspection department supervision there are inspection foremen, chief inspectors, quality superintendents, and quality managers.

As the lists above imply, inspection work is often specialized. One group of inspectors may do only visual work day in and day out, becoming fast, adept, and expert. Others are trained in gaging, testing, or handling special apparatus. Specialization in inspection follows the products and operations. One man may know all the details of receiving inspection but not much of what to look for in the finished product his factory makes and ships. A foundry inspector could not readily pass on the conformance of wooden handle grips made over in the wood turning shop.

Inspections are often combined with production operations. The inspector may, for instance, screw on a nameplate or stamp a serial number as part of his routine. Calibrating and indexing are often assigned to inspection groups. The inspector may be asked or required to keep records, not only his own, but for stock, production control, or cost accounting. Under some systems, a worker cannot be paid accurately until his work is certified in some fashion and the inspector, many times is in the best position to say that all required operations have been performed.

Extra Duties of the Inspector

Inspection covers a miscellany of other industrial items. Inspectors are sent into shipping departments to sample and by various means discover shipping errors or sloppy, ill-secured cartons and crooked labels. Where the external appearance of the package is a strong factor in its eye appeal on the store counter or shelf, formal inspection is frequently required. To hold the inspection department responsible for checking all tools, dies, and jigs is a fairly common present-day practice, this work frequently including the testing and calibrating of gages, measuring instruments, and test apparatus.

Inspectors are assigned to trouble shooting, and sometimes actually to expediting. Where products are failing at a certain operation, an inspector is assigned to trace the causes of the trouble, the trail often leading him back not only through a series of previous operations but perhaps to the raw materials themselves. In a similar vein, if certain parts fail to show up for subsequent operations in accordance with the production control schedule, an inspector may have to take journeys farther back into the shop to find out why and report on the reasons for the delays.

Parts, goods, and products may become damaged in transit through a plant. In one shop, many hundreds of carefully ground and lapped disks were ruined daily for assembly purposes because everyone handling them tossed them carelessly into tote boxes or onto benches. The parts became badly nicked, scratched, and dented. Inspectors were employed to attempt to police or control the careful handling of parts. In an ammunition plant inspectors were assigned to examining empty tote boxes. Powder from loaded ammunition components occasionally spilled out and, unless the boxes were given careful daily washings, dangerous accumulations of explosives could occur.

In his shop area, an inspector may also be asked to report on general conditions (which nevertheless affect the quality of the products) such as the maintenance of machines and equipment, the lighting or the housekeeping.

Where scrap or junk has been made or where parts have been rejected simply on the basis of being somewhat over the specification, an experienced inspector is frequently present to aid in devising rework operations that may in one manner or another salvage losses. A few progressive plants call in an experienced inspector or inspection supervisor to take part in product design and process or methods planning because of his peculiar and unique shop knowledge. An inspector many times can foretell why or where a planned part, product, or operation will fail the desired specifications.

Good inspection exerts a worthwhile though intangible, perhaps, moral influence on an operation, a department, or a whole shop. It is a common saying in manufacturing that, in the interests of operating economy, the inspection force should be reduced to a minimum, of course — but never take away the last inspector. The fact that there are traffic officers restrains many of us from driving sprees and disobeying traffic rules even though for long periods we don't see one "inspecting" traffic.

In one form or another inspection has become an integral and essential division of manufacturing routine. It will be found on most factory organization charts along with sales, engineering, production, purchasing, stores, maintenance, tool room, or first aid, and the inspector shares in the cooperative effort that makes up a successful manufacturing enterprise.

CHAPTER 2

How Specifications Aid the Inspector

If the primary duty of an inspector is to compare conformance with the specification, then the natural question arises: What is the specification?

Then a mother sends her boy to the store for a spool of black thread, he and the storekeeper are in trouble if he brings home blue, green, or brown. Black thread was specified. Life is crammed full of specifications. Working hours are from 8 A.M. to 5 P.M. with an hour at noon. Specifications! A railroad timetable is a set of specifications; so is a concert program. Or a cook book. The number you dial on your phone or repeat to the operator is an exact specification. Almost any advertisement you read in a newspaper or magazine contains specifications, direct or implied, and you are disappointed when the purchased merchandise fails to live up to your impression of those specifications. Catalogs, as contrasted to advertisements, are much more specific.

Many specifications have been set by tradition. They have become common standards. Do you recite the specifications for muslin, a pair of shoe strings, or ten-penny nails as you casually purchase them? Other specifications are natural chemical compounds or reside in formulas that have been handed down from century to century. We buy sugar, salt, or ale without any particular reference to specifications except, perhaps, that we are attracted by some particular brand name or feel that a certain supplier more successfully keeps impurities from his product.

There are definite specifications back of practically everything you use, from a safety pin to a skyscraper. One reason for a high standard of living such as exists in this country today is our ability to specify our wants pretty exactly. And get them!

The Blueprint as a Specification

In the majority of manufacturing plants, the most common form of specification is the blueprint. On it the engineer usually describes the particular part or product in minute detail. The blueprint conveys the process engineer's and the shop man's ideas of how a part should be made; it includes the designer's ideas so that the part will fit and function properly in the assembly; many times it represents years of experience and trial and error; directly or indirectly it indicates what management and the customer think the part or product should be like.

The blueprint *is* the primary link between the designer(s) of a product and the people who make the product. This document is the final word in relevant detail, and serves as the gospel of requirements for what the customer expects in terms of conformance to the expected form, fit, and function.

In the physical sense, a blueprint is a developed photograph. Rather than light and shadow affecting sensitized paper, however, the blueprint is made by having strong light shine through the original drawing (or tracing) and the chemical change leaves white lines on a blue surface. Other types of sensitive paper will produce a black print (black lines on white paper) or a brown print. Some stress has been given to this detail because the blueprint, black print, or brown print is the exact mechanical, the legal duplicate of the original drawing. It saves ruining an original drawing by using it on the shop floor. Some shops, too, issue brown prints in place of the traditional or routine blueprints in order to emphasize certain critical, special, or important work.

Blueprints portray a huge variety of products, from battleships to bullets, from an Empire State Building to an inkwell. There are assembly prints and detail prints. Their whole purpose, of course, is to transmit the idea of what somebody wants to those who are to make it, and in so doing, they vary through a tremendous range of complexity. In the manufacturing plant, they may show an extremely simple part or they may show a complicated assembly of parts.

The Importance of Process/Operation Sheets

The substantial detail contained in blueprints is not generally needed for production or in-process inspection. Blueprints are used to describe individual parts or components, and assemblies containing multiple components. In the manufacture of parts, or in the use of manufactured parts for fabricating some other product, blueprints are mainly used in the inspection area or laboratory itself. Setups, time, and expertise needed to assess all requirements on a print must be done on a limited or sample basis, to gain information, rather than an inspection as a check of quality or conformance.

The on-going or in-process inspections commonly carried out in a manufacturing house deal with only a very small part of the dimensional detail and other information held on a blueprint. A particular operation on the assembly line needs to check only a few items — perhaps only one thing. Maybe the dimensional conformation of a part is already ensured by the time it gets to the line. In such a case, the process or operation inspection sheet may call for a qualitative assessment by the line worker on the functionality of the assembly. The operation sheet describes such a functional check, in detailed terms using words, numbers, and pictures (and sometimes other descriptive aids) so that the factory worker can adequately perform a self-inspection.

Process or operation sheets are usually written by the Quality or Manufacturing Engineer, and will sometimes be included as part of the procedure for building or assembling at a given work station. A check or self-inspection of the part will be conducted as the part comes *to* the station, to ensure that "goodness" of quality is present as parts arrive at the work station. Another checklist or operational inspection sequence is performed as the work of the given station is *completed*, *before* the workpiece moves on to the next station.

A blueprint is generally *not* useful in such an inspection operation. It is up to the Process or Quality engineer to identify the *relevant* detail or variables that need to be inspected for at a station, and include a method for doing so on the process sheet. This detail reflects what is important *at* a particular operation or assembly level, not all of what the blueprint calls for. This assessment will generally be linked to a functional assessment of the part in question as well.

The inspection may include a direct dimensional assessment — resulting in *numbers* that can be charted on an X bar and R chart, and perhaps this quantitative inspection is conducted on a sample basis, to indicate the level and range of parts relative to the (blueprint's) dimensional requirements. But even more often, the inspection calls for a qualitative inspection — is it "good" or is it "bad" — according to stated and sometimes functional criteria. What is the rate of this goodness — is it only one part, three in a row, 20% of production? At a certain point action must be taken. This information is then fed back to supervisors or support engineers who evaluate the situation, and act to make corrections if they are needed.

Quite a lot of talent and consideration to cost is needed to extract the critical or relevant detail from a blueprint or performance specification, and write the suitable inspection into the middle of a manufacturing process. The inspection must be understood and carried out the same way by everyone, in an efficient and cost-effective manner.

Nonconformance to specifications and requirements becomes more costly as the nonconformance is compounded through the production line. When is an inspection likely to uncover the nonconformance, and how soon is too late to find and fix a problem? Inspection detail must be important to the requirements of the specification and the customer, or the activity is wasteful.

Process inspection sheets must be highly qualified and understandable in a standardized manner, like any inspection. A considerable amount of detail, usually by written word, is needed to make such documents effective.

Reading Blueprints

If the blueprint is the most common form of manufacturing specification, the inspector naturally must know how to read one. To conserve time and space for the purpose to be served here, it will be assumed that the reader has a working knowledge, at least, of blueprints — that he has had some practice in reading them. More important to the immediate subject are certain elements in connection with drawings and blueprints that pertain especially to the inspector's interpretation of the specification illustrated by the blueprint.

Blueprints and other guidelines for inspection such as operation sheets and procedures contain a level of detail relating to customer requirements in terms of attributes characteristics (pass/fail) or variables information or both. Blueprints generally serve up a wealth of detail concerning variables data or criteria — real numbers that form a direct measurement or continuous description of the assembly or product.

Numbers most accurately describe a situation or product because they offer a tangible method of ascribing criteria to a physical object. "The object is long or short" can quickly be detailed or clarified if an observer is able to directly measure the object in feet and inches. When this is possible, a more descriptive unit of measurement is available to the inspector, and furthermore, the inspector can break down this unit into smaller and smaller increments (continuous data) if the need arises for even less variation out of the customer's requirements.

Shop or manufacturing procedures and blueprints contain both attribute and variable demands or criteria on the product. Some are explicit and some are implied, and some applicable standards are more important than others in terms of form, fit, and function, or customer requirements. Knowing how to sort out these details, and adequately assess and report on the "quality" level of a part or assembly is the job of the inspector.

The translation of blueprints and other technical documents detailing customer requirements into operations sheets or procedures is invaluable in inspection operations. Vendor relations change, and customer requirements are often restated, both in concrete terms and in order of importance for the inspector. If inspectors have the benefit of support engineering groups who can write an inspection as an operation and keep them in touch with these changes, accuracy and timeliness of the resulting data can be improved. (See Fig. 1.)

Operationalize Inspection!

Define what is to be measured.

Define how measurements are to be taken.

Construct form to list all required measurement results.

Decide who will collect data for how long?

Arrange appropriate sample.

Report!

Respond!

Figure 1

Communication of criteria and relevancy is increased if such operation sheets are used, and they are handy as checklists and training aids for new inspectors and supervisors who are required to keep up with a wealth of detail.

Therefore, inspection criteria, essentially specifications or requirements for a product or even a situation, are not born in a vacuum. The *need* for a certain level and range of requirements grows out of a natural and progressive demand for a constantly improving situation or quality level.

In an industrial setting, and even in the everyday world, that level of requirement or specification will vary with the context — a piece part will be closely scrutinized when it is first introduced into a manufacturing process, every dimension may be checked, and even life testing may be run on the part. Once it has proven itself, especially over time, and through the process, where other parts and other quality requirements of those parts coexist simultaneously, the inspection demands for the part will lessen and increase for the assembly.

The quest for improving quality involves ever more demanding requirements, and cost competition to attain them. The effort also involves cooperation among many different levels and degrees of definition of requirements, and so also inspection routines. In plotting the level, the rate, and the course of an inspection, the purpose and the relationship of the purpose to the end goal — the most critical requirements — must always be understood, or the critical relationship among quality, cost, and customer satisfaction will not be attained.

Being Sure to Get the Correct Print

The number one question for the inspector who picks up a blueprint is whether or not it is the official or authorized print for the particular purpose. In most shops the question can be answered by asking another one: Is this the very latest print?

One logical answer might be: Refer to the print the man working at the machine is following. Surely he would not be allowed to use so much time, material, and equipment making a quantity of incorrect parts. While the answer is logical, it is not always, unfortunately, the correct one. Many factories have set up systems with iron clad rules under which it would seem impossible for production operators, tool makers, and machinists to continue work with anything but the very latest blueprint information. Yet they do — time after time — and it becomes an inspector's duty to see that both he and the operator are provided with correct specifications.

In fact, it is a primary responsibility of the inspector to communicate this need to his support groups and engineers, particularly to manufacturing and quality engineering staffs. The inspector must know if the information relative to the detail of his inspection criteria is current and relevant. Pursuing the wrong or outdated detail in an inspection operation can be more dangerous than no inspection. The system that supports the inspector must constantly be assessed and questioned by the inspection staff, as surely as the inspectors must do the direct inspections.

There are a number of reasons for this sort of trouble — carelessness, habit, and laziness leading the list. The engineering department may alter a specification, for example, and duly send out copies of revised prints, but the latest issues may rest on a foreman's desk for days among unread "shop mail."

Checking for the Latest Revision

On the shop floor, revised blueprints almost never look a bit different than the superseded copies. Revision notices, blueprint dates, and authorized signatures usually appear in somewhat obscure boxes in the formal margin or border of the print. Hence, the use of superseded prints is somewhat excusable.

Knowing and using (as well as assessing and correcting) the system that provides inspection detail is an important part of an inspector's job. Extensive product and process criteria are utilized increasingly in many functions in modern manufacturing systems, and the availability of that detail is critical to the efficient operation of a manufacturing unit. Assemblers, material handlers, management, and engineering groups and supervisors all have occasional and even on-going need for inspection criteria. The maintenance and communication of these criteria is a direct function of the inspection force.

Not only may the revision notice be somewhat obscure, but the very specification change itself may be so slight, yet important, as to escape attention. Observe print A in Fig. 2; then look at the revised copy, B in Fig. 2. It takes more than a casual glance to see that one of the taped holes has been changed from $\frac{1}{4}$-20 thread to $\frac{1}{4}$-28 thread, the engineering department having decided for some good reason to use a fine thread.

One system for issuing revised prints is by messenger directly from the engineering department, who not only personally distributes the latest copies, but also simultaneously hunts for and himself destroys the superseded copies. Where the factory system is inefficient, the next best step is for the inspector to assume responsibility for securing and using the latest revised prints himself.

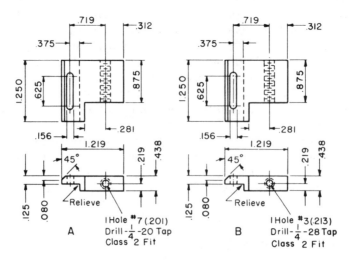

Fig. 2. (A) Original drawing of a small part. (B) Revised drawing showing changed thread size.

For the inspector, ignorance of the law is no excuse. If he is to measure conformance, he always must be sure of the specifications.

Checking for the Correct Part

The second and similar step, where blueprints are to be relied on in connection with an inspection, is to be sure that the print for the *correct part* is at hand. Usually the particular unit is clearly defined on a print by a prominent part number. Even though considerable prominence is usually given the part number on a blueprint, it is still necessary to warn the inspector to double check it before proceeding with his work. As an example of what is meant, part numbers 123K-B83 and 128K-B83 or some such combination as 123X-B38 can all be readily mistaken for one another.

In addition, the parts themselves may be closely alike and look the same to all outward appearances from a quick glance at their blueprints. In the case of the part illustrated in Fig. 2, for instance, it may be decided, because of the need of supplying old model repair parts, to give part B (Fig. 2) another part number rather than to consider part B a revision of print A (Fig. 2) and the two part numbers might have been cataloged as #1248 and #1249.

It is important, as a part of the data derived from an effective inspection sequence, to be able to identify *where* the part or product has come from, perhaps *who* has dealt with the part, *what* was done to the part, and other detail relevant to the accurate inspection and reporting on the part. Lot number, batch number, assembler number, dates, and various codes are all critical to dealing with the *results* of inspection data. Just as it is important to set the *right* part for an inspection sequence, it is crucial to be able to attach a considerable amount of detail to what has happened to the part or product within the manufacturing sequence, should corrective action become necessary.

Vendors or suppliers, incoming or receiving inspection, the process itself, test sequences, the customer, and field service are all extensions of the total system that serves the customer in a manufacturing effort. Inspections are undertaken throughout this system, and the results or data from many of these inspections is critical to the performance or control of the entire system. One part depends on the other, and the communication of inspection measurements is necessary to improve parts or all of the system.

Practically all shops have some form of shop or production orders. The part number of the product to be made is used in the orders. The inspector verifies the use of the proper blueprint by referring to the manufacturing order or memorandum for the correct part number.

Many blueprints show the dimensions and specifications of more than one part on a sheet. Sometimes a dozen different small parts, with their part numbers or letters, appear on a single blueprint. On the other hand, the blueprint may portray a single outline with a table of dimensions or specifications for different sizes of the particular kind of part as shown in Fig. 3.

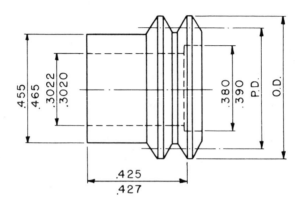

PART NO.	MODEL LETTER	RANGE	O. D.	P. D.	WIRE SIZE	WIRE MEASURE
AL-522	A	4-6P	.7274	.6553	.09623	.79963
AL-529	B	7-10P	.6553	.6120	.05774	.69862
AL-523	C	11-13P	.6303	.5970	.04441	.66358
AL-524	D	14-20P	.6011	.5795	.02887	.62280
AL-525	E	22-30P	.5831	.5687	.01924	.59750
AL-526	F	32-40P	.5741	.5632	.01443	.58489
AL-534	G	44-56P	.5663	.5586	.01031	.57406
AL-544	H	64-80P	.5605	.5551	.00722	.56594

Fig. 3. Drawing with tabulated data showing parts having same shape but different dimensions.

Except for the fact that the sort of mistake, the type of carelessness, mentioned above happens to be quite commonplace on the average factory floor or in an inspection crib, it would seem at first glance almost ridiculous to need to warn an inspector (a) to know what part number or symbol or class of part or product he is about to inspect; (b) to verify the part number from the manufacturing order, and (c) to be sure he has the latest authorized blueprint or specifications.

Making Sure of Mark-overs and Sketches

Very often an inspector will run across a blueprint on which some dimension, tolerance, or instruction has been changed — marked over — in, say, pencil, crayon, or ink. Many shops permit this practice. Most organizations do have a definite routine for making authorized and official blueprint changes, but it may take a matter of hours, if not days, to get the drawing changed and new blueprints out on the manufacturing floor. At the same time it may seem necessary to those in charge to execute a desired change immediately. Hence the expedient of marking over prints on the factory floor. The inspector need only question the authority of the red pencil revision to make sure that it is not already superseded by still another pencilling over. Such informal changes should be signed or initialed and dated by the engineering or manufacturing executive authorizing them. To the inspector, a "mark-over" without signature or initial should be about as valid as an unsigned bank check because, once in a while, unauthorized factory personnel attempt blueprint changes as a sort of emergency expedient.

Inspection serves as a gate between engineering and the people who do the work — whether that gate opens from the engineer to the assembler or from the assembler or process to the engineer and other support groups. When a process change or specification change is made, it will likely impact the production floor. Initial and on-going feedback to and from the person who ordered the change is essential, and inspection must provide that information.

Sometimes the system or method for that feedback is very formal, as in an experiment or modeling program where the results of a change are sampled; inspectors may be intimately involved with such cross-discipline experimentation. Other times that system is less apparent, and the entire responsibility for ensuring that the feedback occurs rests informally with the inspector.

The same frame of mind and watchfulness should be displayed toward freehand sketches and the like, which are not uncommon on factory floors. (Even a sheet torn from a catalogue has been known to serve as an official print.) All such informal specifications should be properly initialed and dated by someone in position to assume responsibility.

Throughout his career, the inspector is faced with the problem of determining correct specifications, of trying to find out exactly what is wanted, and of keeping abreast of changing conditions. Like any good judge, he must read the law continually. And he must be sure of the official character of any document he deals with.

Is the Material Correct?

In his examination of a blueprint, preliminary to inspecting the corresponding manufactured part, the inspector should next make note of the type of material called for: cast iron, brass, machinery steel, aluminum, or whatever it may be. It probably seems peculiar to put the question of material high on the list of things to be looked for when checking any product against its blueprint, but one oversight not uncommon in manufacturing, ridiculous or impossible as it may seem, occurs when parts or products are made of the wrong material. The blueprint of a casting, for instance, may call for bronze. But a chain of errors, starting perhaps in the engineering or purchasing department, may cause the pieces to come in as cast iron. A sudden change in long standing specifications frequently starts the trouble. Both the machine operator and the inspector, accustomed to the cast iron, fail to notice the new specification. A similar error occurs at times in connection with screw machine parts made from rod. The stockroom may have inadvertently supplied stainless steel, for example, rather than the specified drill rod.

The amount of detail carried on blueprints or inspection instructional documents is critical. It is wasteful to inspect for irrelevant detail, and no one can assume that inspections will occur if criteria are not clearly present and explained. Therefore, the inspector must take responsibility for asking questions that bridge the intents of engineers with those who carry out the work.

Inspection of the material of a part is like many other types of modern inspection — difficult or impossible to quantify. New tools and inspection devices are invented every day to quantify detail such as material surface finish or color. But the need to qualify relevant attributes such as these is on-going, and the money and time for expensive equipment is not always there. Therefore, the means of making these assessments must be as fully documented and detailed in a procedural form — by the responsible engineer. If three shifts with three operators per shift are going to call a "good" or "bad" surface finish on a part, the method and criteria need to be clear, or chaos and confusion will be rampant. Blueprints often do an inadequate job of specifying these "gray" areas (gray *is* a specific mix of black and white), and they need to be rewritten and interpreted in operational terms that everyone can understand and adhere to.

Order of Making Measurements or Observations

As the experienced inspector studies a new blueprint he is probably looking for the more important dimensions and tolerances. In other words, he seeks out the items and elements in connection with the manufacture of the part or product he is about to inspect that will require the greatest attention and care on his part. Usually the tolerances and notes supplied by the engineer or draftsman give him good clues. His own manufacturing experience and his particular knowledge of the product guide him, too, in selecting first those "characteristics" of the part or product he should perhaps pay closer attention to in the course of his inspection. What is meant here can be simply illustrated by referring to Fig. 4.

Without knowing anything specific about the particular part or product of which Fig. 4 is a partial drawing, it is reasonable to assume some sort of axle or shaft with a shoulder flange against which something — a wheel, a pulley, a crank arm perhaps — is to be tightly clamped. Anyway, in measuring and checking a batch of such pieces, the inspector is cautioned by the very tolerances themselves, as shown on the print, that certain elements of the shaft require greater care in manufacture and closer observation at inspection.

Evidently the designer was concerned with the .748-inch diameter. It cannot be oversize at all; and undersize by not more than .0002-inch. In addition he has asked for an extra fine surface finish as shown by the symbol f. The inspector's mind immediately turns to consideration of the available high precision measuring instrument he can use to check this diameter with. On the other hand, he can use micrometers or vernier calipers to measure the diameters of the flange and the long part of the shaft, which the print shows as 1.875-inches and 1.125-inches, respectively. In these latter dimensions, an error in manufacturing of even .010-inch probably would be tolerated and an error of a thousandth or two in the way the inspector made the measurement would cause no harm.

The inspector's second choice of a "critical" dimension on the part shown in Fig. 4 would be the length of the precision turned and ground

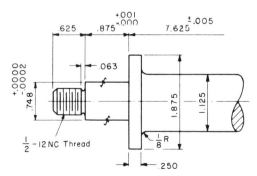

Fig. 4. Drawing a piece-part showing some of the dimensions with which the inspector is concerned.

shaft section — marked as $.875''^{+.001''}_{-.000''}$ on the print. He could check the .625-inch length of threaded section and the .250-inch thickness of the flange with a steel rule, perhaps, and secure sufficient accuracy for the purpose of inspection. But he must be sure to arrange for an accurate measurement setup, and sufficiently precise instruments when he verifies the shaft segment length of $.875''^{+.001''}_{-.000''}$

If the several sections of the shaft shown in Fig. 4 are to be classed in order of measurement importance, the inspector would then, as the third mental step when he looks at the blueprint, select the threaded end of the shaft and plan to secure a suitable $\frac{1}{2}''$-12 NC thread gage.

If for no other purpose than to recommend reasonable orderliness in going about an inspection assignment, the selection of information from a blueprint has been classified in the following order of importance:

(a) Be sure the blueprint is for the correct part number.
(b) Verify it against the shop order.
(c) Be sure the blueprint is the latest, authorized, official revised version.
(d) Verify the official character of all blueprint markovers, of all freehand sketches and the like appearing on the manufacturing floor.
(e) Check for the kind of material specified, on the blueprint and the shop order and at the operation.
(f) Catalog somewhat in order of importance the dimensions and specifications which the print shows may require extra attention at the inspection.

Blueprint Notes Should Be Read

It is wise next — adding another letter, (g), to the above list — to read all notes the draftsman may have put on the drawing. Such additional specifications, warnings, and information may not outrank, in importance at the inspection, some of the regular dimensions to be verified. Nevertheless, a definite place in the routine of reading a print should be assigned to notes and special information, otherwise the inspector may neglect a special piece of information. Good habits are established as easily as bad ones.

Often, some of the most important detail held on blueprints — important to the use of the part in manufacturing — are listed in *words* in the notes or addendums. The reason for this is clear — these criteria may be difficult or impossible to express in numbers — they may not be (readily) quantifiable. Therefore, these notes often require a special method or at least an understanding based on a thorough acquaintance with the "gray area" of attributes inspection.

The notes will not say "the part must be clean," perhaps some obscure government specification will be referenced, and the laboratory may or may not be able to carry out the inspection. In any case, attention to this detail often determines the functional performance of the part.

An example of the effect of blueprint notes is shown in Fig. 5. Among the several draftsman's notes is one which directs the machinist to spot face the casting where a .375-inch hole is drilled and reamed through it, an operation just about as important for the use of the product as the hole itself, but one which the operator may readily neglect. On this same print is another note which orders two surfaces to be machined square *but not painted*. The paint shop might overlook this specification and fail to mask these two surfaces prior to spray painting. In like manner, the paint shop could forget to mask the "spot face" mentioned above.*

The draftsman's notes on Fig. 5 also tell the machinist the exact size tap drill to use in making a tapped hole. If too small a tap drill is selected, the tap that is subsequently used may bind, even though the

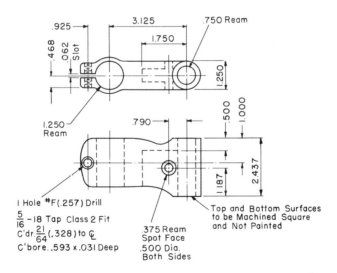

Fig. 5. Drawing that shows how notes are used to supplement dimensions.

*Masking is a term used in paint shops. It means that surfaces that should not be painted will be covered with metal or wood templates, strips of paper or tape, or greased. Holes are frequently filled with wood plugs. So protected, the part or product may be quickly sprayed, dipped, or brushed. The masks are designed to be readily removed after the paint is applied. Wood, cardboard, and mental masks are used over and over again.

machinist succeeds in threading the hole without breaking the tap, and, in addition, the mating screw will quite likely bind, strip, or not enter at all at assembly. When the theory and elements of screw threads are studied in a chapter farther on, the reader will not only appreciate the draftsman's good sense in specifying tap drill sizes, but also, more clearly understand why an inspection should include a check on the size of tap drill used by the machinist.

The notes on the print shown in Fig. 5 emphasize, too, the value, to the inspector, of knowing where and how the particular part is to be used in the assembly or on the finished product. If, for example, he knows that a shaft will go in the .375-inch reamed hole and that the shaft is to be turned by a handle attached to it, the hub of which will normally rub against the side surface of the casting, the reason for the smooth spot face called for becomes apparent. The value of knowing the subsequent use of any part or product about to be inspected will be emphasized again in the later sections discussing the subject of standards.

Another form of draftsman's notes concerns surfaces that are to be finished, or machined, in contrast to those that are to be left as the foundry mold, the forge die or the steel mill roll formed them. The symbol f is used to mark finished surfaces.

The practice of calling for finished surfaces with the symbol f has developed so that now it has more to do with the actual grade of surface finish. The f (see A, Fig. 6) may mean a certain degree of fine finish and an ff an even better degree, as at B in Fig. 6. Recently, other and more exact symbols for surface finish have been adopted, to keep pace with improving methods for obtaining finer surfaces, as illustrated at C in Fig. 6. The meaning of this type of designation and the technicalities of surface finish, or surface roughness, are discussed more fully in a later section on surface standards.

Perhaps these several examples will impress the inspector with the necessity for having a definite step in his routine, a particular time during the reading of a blueprint, in which he will mentally emphasize draftsmen's notes. Prior knowledge of all specifications enables a more expert inspection of the product.

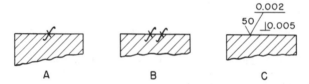

Fig. 6. Two ways of denoting degrees of surface finish. Early method: (A) Fine finish; (B) Finer finish than at (A). (C) Improved method gives specific data.

Sequence of Operations Should be Noted

The next regular chore on the list of inspection do's in connection with reading a blueprint is to know or determine the particular operations to be performed on the part or product. Only one of the total number of operations indicated is to be performed, ordinarily, at any one machine or at any one point in the process. Sometimes several operations are carried out at one location, but not all of them; occasionally the part shown on a print can be completely fabricated on a single machine. Some shops do furnish prints of semifinished parts, but in most factories the blueprint describes the completed part, and operation sheets, or their equivalent, are relied on to detail the several operations or subdivisions of the total job. (Operation sheets are described in a section farther on.)

To illustrate the inspector's interpretation of a blueprint in terms of the individual operations performed on the part, refer to Fig. 7. Here is the sketch of a short shaft that is to be turned, in which a keyway is to be milled, and through which a hole is to be drilled. If an inspection takes place at the lathe, the inspector would measure dimensions A, B, C, D, E, and F. After the pieces leave the milling machine he would verify G, H, and I, and at the drill press he would check the hole diameter K, making certain also that the hole has been drilled on center $L - L$ — checking dimension J also — and perpendicular to center line $M - M$.

In other words, an inspector must usually select from a blueprint the dimensions and notes pertinent only to the machine, operation, or station in the process where the particular work is being performed.

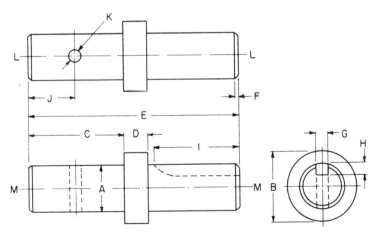

Fig. 7. Dimensions A, B, C, D, E, and F are measured after lathe operation; dimensions G, H, and I after milling; diameter K, dimension J after drilling.

Even the most detailed incoming or laboratory-type, full-blown (dimensional and qualitative) inspection usually contains a (process) instruction or sequential inspection sheet to guide the inspector. The order and significance of individual inspections will help in the efficiency of the overall inspection routine and ensure that the important points are identified. The skill and time needed to define such a system, and maintain it, is worth its cost in payback through time and accuracy of the overall inspection.

Watch for Conditions Implicit in the Drawing

When a draftsman draws a picture of a part, there are certain things he expects will be obtained when the part or product is made, even though his drawing does not explicitly cover such details with actual dimensions or notes. For example, if he calls for a simple tongued piece similar to that sketched at A in Fig. 8 he expects it to be closely like A and not in any one or several of the different shapes that can actually emerge from a milling machine, as shown in exaggerated fashion at B – of Fig. 8.

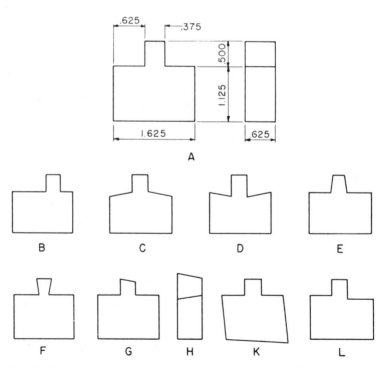

Fig. 8. (A) Drawing giving dimensions and shape of part. (B), (C), (D), (E), (F), (G), (H), (K), and (L) show possible variations from (A) due to improper machining.

Where a horizontal and vertical line intersect on a drawing the designer intends a 90-degree angle although he makes no special note of it on his drawing. If two lines are parallel, the two surfaces represented are supposed to come out parallel from the machining operation. A hole is supposed to be bored straight and not on a slant; it is supposed to be round and not oval.

As an example of the possibilities an inspector needs to be on the alert for, observe the several types of holes pictured in Fig. 9. In this case the drawing called for a round, straight hole of diameter A. Any drilling or boring operation has a tendency, however, to produce one of the forms listed in Fig. 9. Although in each case the dimension A is met, at one point at least, it is evident that the other difficulties pictured could make trouble at, say, assembly.

Thus, in addition to checking the part for conformance to *explicit* dimensions and instructions given on a blueprint, the inspector must watch for conditions *implicit* on the print. Common terms for such items, in addition to those listed in Fig. 9, are squareness, parallelism, waviness, warp (distortion), sprung, kinked, eccentric, run-out, and out-of-line.

Lack of Surface Parallelism

Shaper, planer, milling machine, and grinding operations produce inspection problems in parallelism — waviness and warp, for example. An effort has been made to picture such conditions, in exaggerated

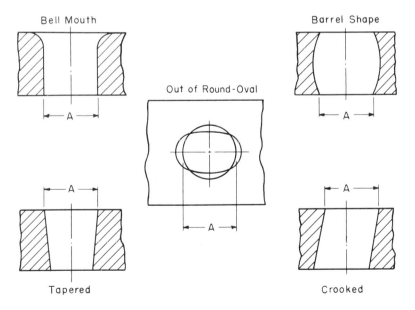

Fig. 9. Possible variations of a supposedly round, straight hole.

form, in Fig. 10 where the blueprint requirement of perfect rectangularity is illustrated at A but where lack of parallelism, waviness, and warp show at B, C, and D, respectively.

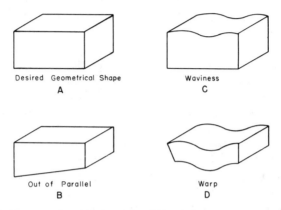

Desired Geometrical Shape
A

Waviness
C

Out of Parallel
B

Warp
D

Fig. 10. (A) Rectangular block. (B), (C), (D) Variations due to improper machining.

Undesirable Shaft Taper

Taper is a form of lack of parallelism. In some instances front taper is spoken of, or back taper, especially the latter. Taper (front taper) and back taper are illustrated in the views of a shaft section in Fig. 11. Here again, dimension d has been met at least at one point on the shaft sections at B and C.

If a shaft like that at B in Fig. 11 were assembled in a hole in a mating part, it could bind in the hole (because of the wedging effect of a front taper) or, if the hole were large enough, it could make an extra

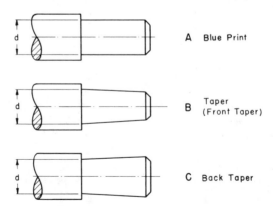

A   Blue Print

B   Taper (Front Taper)

C   Back Taper

Fig. 11. (A) Shaft required. (B) (C) Variations due to improper machining.

loose fit. The back taper condition of C (Fig. 11) causes even more trouble in the assembled product. When the shaft with back taper is assembled in a hole, especially if the hole is on the small side for size, the fit is tight and the assembler instinctively forces the shaft in. The large or bull end of the back-tapered shaft section then broaches the hole, in effect, and a loose fit between shaft and hole is inevitable. All this, even though the shaft was intrinsically made to the correct diameter.

Concentricity, Run-out, and Eccentricity

The terms concentricity, run-out, and eccentricity are used more or less interchangeably in the shop and on blueprints, with run-out the favorite shop term. Where two holes or shaft sections are not on the same centerline, they are not concentric. While the shop calls the condition run-out, the draftsman may put a note on the drawing calling for "concentricity within .002 inch." However, the actual, measurable error is eccentricity — the lack of concentricity.

Part A of Fig. 12 illustrates a coupling piece as the designer wanted it and B in Fig. 12 shows a few of the possible combinations in regard to eccentricity that may be found in the part when it emerges from the machining operations. Note also in Fig. 12B the out-of-line and slanting set screw hole.

Lack of concentricity (or eccentricity) as a problem is not confined necessarily to metal cutting or machining operations. In manufacturing rubber or plastic insulated electrical wire by the mile, for instance, one of the problems at the insulation extrusion equipment is to maintain the

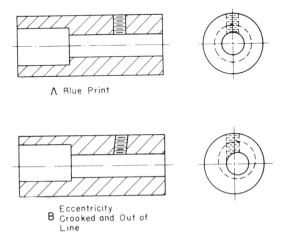

A Blue Print

Eccentricity
B Crooked and Out of
Line

Fig. 12. (A) Tapped hole called for on blueprint. (B) Crooked and misaligned tapped hole actually produced.

centrality of the copper wire within its insulation coating. If the wire gets off center, leaving a thin section of insulation on one side and an unnecessarily thick layer opposite, the finished insulated wire will not withstand the voltage test required because of the lack of sufficient insulation on one side of the wire.

Unsatisfactory Hole and Shaft Conditions

In Fig. 9, several conditions which may appear in the bored or drilled holes were illustrated. In much the same manner shafts, bushings, and all turned and cylindrical ground work may leave an operation as anything but a perfect geometric cylinder. The presumed cylinder may be tapered, barrel shaped, like an hour glass, or oval. See Fig. 13 for illustrations, in exaggerated form, of these conditions. They may be imperceptible to the naked eye or ordinary measuring instruments, but one of them, or some combination of them, exist on any piece that is turned or cylinder ground and can be discovered where sufficiently fine, accurate measuring equipment is used.

The experienced shop inspector is especially alert to what he usually calls taper and out-of-round. In other words, he may not analyze the absolute condition in terms of hour glass effect, say, or ovality or eccentricity. Conversationally, the expressions taper and out-of-round cover all conditions. But he knows that, irrespective of

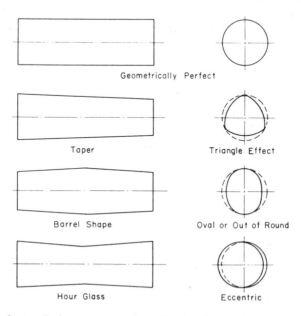

Fig. 13. Perfect cylinder and variations in the shape of a cylinder caused by improper grinding.

the exact analysis of such a general condition or the specific term to be applied, it always exists to some degree after any turning, boring, or cylindrical grinding operation — internal or external, and that it makes trouble at assembly or causes wear, vibration, or interference where it becomes excessive on component parts.

The "Triangle Effect"

A rather special condition, in the class of ovality or eccentricity yet unlike them, is the so-called "triangle effect" also illustrated in Fig. 13. The triangle effect is almost inevitable in centerless grinding operations. It frequently gets by an inspector because special measuring setups are required to detect it. A shaft or cylindrical part is held in a centerless grinder in cutting position between the grinding wheel itself and a revolving guide or pressure wheel and a stationary guide blade or rail, as shown diagrammatically in Fig. 14. Unless, or even though, the guide wheel and guide blade are very carefully set, there is a momentary, yet continuous condition where the grinding wheel is cutting or reducing the diameter of the workpiece on its side — at A in Fig.

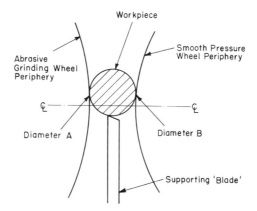

Fig. 14. Schematic diagram of centerless grinding operation.

14 — while, in effect, a larger diameter of the workpiece is momentarily rolling against the revolving guide wheel — diameter B in Fig. 14. The result is a tendency toward the triangle effect, if not the actual condition. The so-called "triangle effect" may result in six or nine points — two or three triangles may be set up in effect, or under certain grinding conditions five or even seven points may be present. The skilled operator or tool setter can so nearly compensate for the triangle effect that it cannot be detected (practically), but it may appear at any minute because the grinder adjustments may vary a trifle from temperature changes, vibration, and the like.

Unwanted Burrs and Fillets

When the draftsman shows a square edge or corner on the print as in the case of the simple step block shown in Fig. 15 part A, he expects it to be sharp, clean, and square. Otherwise, he would have put a note on the blueprint to "break all edges" (meaning to file, grind, or emery belt a slight chamfer on an otherwise very sharp corner) or he would have provided for radii or definite bevels by means of notes and dimensions on the print.

At the machine, however, most usually a burr is formed as indicated at B in Fig. 15. Some of the metal is crowded or pushed out over the normal edge in the form of an overhanging lip, frequently razor sharp, by the cutting tool or grinding wheel. Burrs are not wanted on manufactured parts.

A close relative of a burr is the fillet. Here is the case where the square sharp edge of the cutting tool or wheel becomes rounded off (or the machinist begins to withdraw the tool just at the corner), thus leaving a tiny bead of unwanted metal in the corner as shown at C in Fig. 15.

Experience with local conditions in his own shop will point out a number of similar product conditions to the inspector which are not specifically warned of on the blueprint but which nevertheless prevent conformance of the part to specifications. In the average factory the responsibility for detecting many of the troubles described and implied in the preceding paragraphs seems to gravitate to the inspector. For some reason, operators and tool setters direct so much attention and energy to maintaining dimensions and the requirements in specific blueprint notes — at the same time keeping an eye on the production quota — that they overlook other essentials that are also plainly pictured on the blueprint. The obvious is easily ignored.

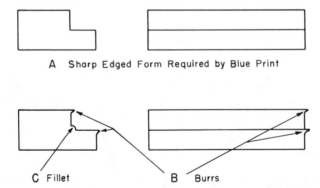

A   Sharp Edged Form Required by Blue Print

C   Fillet   B   Burrs

Fig. 15. Although blueprint calls for sharp corners, burrs and fillet may be produced by machining.

It is important that the inspector *report* accurately the results of measurements and observations such as the presence of undesirable attributes such as burrs or fillets. It is also critical that the inspector knows *what is important*, and perhaps what criteria are more relevant than other criteria.

For instance, depending on the purpose of an inspection, a burr may be so prohibitive to the acceptance of the part in question that complete detailed measurement of the rest of the dimensions is not necessary. It may be important to set feedback to the process that is passing on the burred condition, and then so on to continuing detailed assessment. It is a question that can only be answered by an inspection group that is well tuned in to the products and processes that they are measuring. (See Fig. 16.)

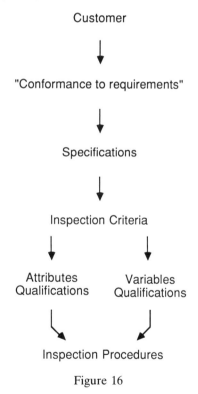

Figure 16

Operation Sheets as Specifications

Many factories use shop orders, operation sheets, process records, or similar devices for specifying the exact operations to be done and their order of accomplishment. Such a breakdown enables, among other things, a timing of each job and is an important element in production control, wage incentive rates, and scrap reduction. An operation sheet also serves an educational purpose by directing new workers in the most economical way to proceed in manufacturing the product. The process record or an operation sheet is the official record of methods. Usually it travels with the production order to the job and is in turn accompanied by the required blueprints.

A typical commercial operation sheet is displayed in Fig. 17. This particular operation sheet applies to a part having a contour of the type outlined in Fig. 8. A glance at this operation sheet discloses its value to the inspector. In a preceding paragraph a preparation routine was suggested — certain orderly steps the inspector should adopt in securing information from the blueprint concerning the product he is about to inspect. One more step could well be added to that list: look for an operation sheet or a process record.

From this operation sheet, Fig. 17, the inspector can immediately

PART NUMBER BS - 28	.375" X .875" COLD ROLLED STEEL					
OPERATION	TOOLS	MACHINE	MCH. NO.	SPEED	FEED	JIG OR FIXTURE
CUT OFF TO 1.125 (1 at a time)	Abrasive Whl	Abrasive Cutter		Set	Hand	Clamp
GRIND (4) SIDES TO REMOVE BURRS	80 Gr. Blt.	Belt Grind	4	Set	Hand	
STRADDLE MILL TO 1.062 (14 pcs)	.375 X 4.000" side cutters	Miller	4	2-92	522' min	T-1765
FILE 4 EDGES-2 SIDES (to burr)	#6 File	Bench			Hand	
STRADDLE MILL SPRING SLOTS Top (17 Pcs.).250" wide File edge (2) before removing Reverse Straddle Mill Fixture Bottom (17 pcs) .250" Wide File edge (2) before removing	2-.250" Cut	Miller	4	2-92	.644 min	BS-25 T1
GRIND ALL OVER (To Burr) Lines to run lengthwise	100 grit belt	Belt Grind			Hand	
MILL STEP (.062" X .313") 5 Pcs. File Burrs	.375" Side	Miller	29	105	.750 min	Mill Vise Parallel T-1765
DRILL 5 HOLES 4 Spring Holes .250" Deep 1 Contact Hole .375" Deep	.093" Drill .118" Drill	Drill Pr.	18	3-580	Hand	T- 1763
DEGREASE	Wire Basket	Degreaser			Hand	
TAP 4 SPRING HOLES	4/48 Tap	Drill Pr. w/Att.	17	1-480	Hand	Mach.Vise
TAP 1 HOLE	6/40 Tap	" " "	"	" "	"	" "
DEGREASE	Wire Basket	Degreaser			Hand	

Fig. 17. Typical operation sheet for a part similar to that in Fig. 8.

pick conditions to be checked that he might not have thought of if only the blueprint had been examined. It may be necessary for him to verify the stock size (.375-inch × .875-inch cold rolled steel, as shown on the heading of the operation sheet). If by accident the operators have been furnished with stock too thin or too narrow, the parts they make may not "clean up," that is, finish to the required dimensions, in subsequent operations.

Observe on the operation sheet the burring and grinding operations specified to free the parts from burrs — items that did not appear at all on the blueprint. The inspector may find it necessary to verify all drill and tap sizes, making certain that those which the operators are using correspond to the operation sheet instructions and that both are consistent with the corresponding blueprint notes or that they will produce the size and kind of tapped hole the blueprint specifies.

Note, also, the sixth item on the operation sheet, Fig. 17, which specifies that belt grinding marks or lines are to run lengthwise of the piece, a condition probably essential to the ultimate appearance of the finished product in which the piece will be assembled.

Burring operations are specified in items 4, 5, and 6, for instance, so that the parts will fit accurately in the drill jig specified for item 8 (jig #T-1763) — another example of a factor not mentioned on the blueprint of the finished part.

The inspector is aided, too, when an operation sheet calls for the removal of burrs. If he were to measure the length of an unburred piece, for instance, he would get a false dimension — measuring not the correct length of the body of the piece but the false length produced by the overhanging burrs. Remember that a burr or edge overhang so slight as to be scarcely visible or tangible even to the fingers, can still cause an error of several tenths of a thousandth, if not a full thousandth, in the measurement secured. If no routine deburring operation is called for, an inspector must make sure to free the pieces he is measuring from burrs by filing or with emery cloth.

It should be clear, now, that an operation sheet or process record is essential information for the inspector. Many plants, however, do not have operation sheets. They depend on the native skill and "know-how" of the machinists or their supervision or upon word-of-mouth instruction, memory, and tradition. Then, too, the manufacture of many products does not respond naturally to operation-sheet control. Foundries, plastic molding, wood working, punch and forming press shops, and wire, brass, and paper mills, to name but a few of many places, may not rely on the specific operation sheet or process record.

When no official operation sheet prevails, an inspector must, of course, gain the equivalent information as rapidly as possible from experience, from observation, and by asking questions. He cannot know too much, too soon, about the unexpected effects of the process on a product he is assigned to inspect and judge. It might be wise for him, especially where he is new on a job, to write out, in effect, his own operation sheet.

Written Specifications

In many types of industry, such as rubber, plastics, paper, fabrics, wire, and sheet metal, blueprints and operation sheets may not be used much, if at all, as has been said. There may be a form of process record. More likely, however, a set of formal, written specifications are used — manufacturing specifications. Government procurement agencies, of course, make frequent use of specifications. In fact, nearly every article, staple, or device required by federal government departments is covered by detailed specifications.

Specifications may occasionally be found where blueprints would be expected. This is especially true in the manufacture of more or less staple articles. A sample specification covering the purchase of ordinary wood screws is quoted below:

> Wood screws shall be furnished in three types; namely, flat, oval, or round head as particularly specified.
>
> Wood screws shall be made of steel or brass as particularly specified.
>
> The length of all screws shall be measured from the largest diameter of bearing surface of the head to the extreme end of the point measured parallel to the axis of the screw.
>
> The diameter shall be measured on the body of the screw under the head.
>
> Standard screws shall be furnished with gimlet points.
>
> Wood screws shall be furnished plain, uncoated, unless blued, nickel plated, or other special finish is particularly specified.

Specifications are ordinarily so peculiar, particular, or unique in or to some class of industry that no extended discussion concerning them is possible here. The points to be emphasized are that specifications are used in industry and the inspector must be sure to know about them; in studying specifications, he must make sure of observing and remembering essential items in them.

It is said that the fifth cause attributed to process variation (besides materials, methods, manpower, and machines) is inspection itself. The first aspect that should be investigated after the reporting of discrepent (to the requirements) inspection results is the manner and method of the inspection itself. If the inspection can be said to be objective and carried out in a consistent and repeatable fashion, the data from that inspection are said to possess integrity, and can be trusted.

Purchase Orders

Perhaps the most common industrial form of written specifications is the purchase order. In many instances the customer's purchase order, or a copy of it, is the only information the shop has from which to manufacture parts or products. As a simple example, a woodworking plant might receive a purchase order for "one thousand $\frac{1}{4}$-inch dowel-pin rods, birch, three feet long."

More frequently, however, the purchase order requires something more complex than pieces of hardwood dowel. It may refer to a catalog part or number. It may be accompanied by a blueprint or a more explicit and voluminous set of written specifications.

The wording of a purchase order can be valuable to an inspector because, many times, all of the information appearing on it is not transmitted through to the shop or the information is interpreted in a manner satisfactory for manufacturing but not for inspection.

For example, an order is placed with a certain plant for a large quantity of compression springs to be used in footbutton automobile starter switches. This plant's sales and engineering departments had studied the customer's requirements at first hand, in addition to reading the customer's purchase order with its accompanying blueprint. Then the engineers drew up a blueprint of the spring for use on their own shop floor. The print indicated the wire size to be used, the temper of the wire, the number of turns and coils and their spacing or pitch, the diameter of the wound spring, the total length of the required spring, the type of ends, and a number of other mechanical details. In other words, the print gave the shop mechanics every piece of information they needed in order to make the springs exactly as ordered.

When the springs were presented for inspection, they could have been readily checked for dimension — wire size, coil diameter and pitch, and the like — and approved for shipment as being according to blueprint. In this case, however, the inspector was not fully satisfied. He wanted, correctly, to examine the springs from the viewpoint of the customer. Technical dimensional details were all right so far as they went, but he knew that performance, after all, was the thing that counted. What did the customer expect the springs to *do?* Examination of the purchase order brought to light a specification paragraph that the factory's engineers had not copied directly onto the blueprint. They felt that they had accurately translated this information for the shop in the form of wire size, coil diameter, pitch, etc., and that the springs, so made, would perform in accordance with the following paragraph.

When compressed to 1-inch length, spring is to exert $11\frac{1}{2}$ oz. – 15 oz. pressure. Closed coil length — 15/64 inch. End coil must not close before spring is compressed to $\frac{1}{2}$.inch.

This was the definite information the inspector needed in order to make a comprehensive test of the conformance of the springs ordered.

One point to be emphasized here is that an inspector should keep always in mind the function, the performance, of the product, part or detail whose conformance to a print or specifications he is measuring. In order to do so intelligently, he may need to seek more information than appears on the relatively bald specification of a blueprint.

In these days of self-inspection, with less dependence on the formal and repetitive laboratory inspections, the inspector may indeed be someone less schooled in the quality pursuits than a professional inspector. This being the case, inspectors are more extensively screened from blueprints, purchase orders, and other documents that may contain general or specific conformance requirements.

Instead, inspections are usually defined and ordered through a support engineering group that can translate *all* the requirements into a system that defines the critical criteria, as well as a method and order for the inspection. It is customary today not to allow much interpretation of the inspection criteria within the direct inspection circles themselves — in other words, the detail of the inspection guideline is intended to remove doubt and uncertainty about the order and significance of assessments.

This does not mean that inspections are not questioned — such feedback from the front lines is necessary in order to maintain and improve a relevant inspection system. But it does mean that once the inspection sequence and criteria are set and defined, and defined well, everyone should inspect in the same fashion.

Other Types of Specifications

Specifications appear on factory floors also in the forms of contracts, shop orders, letters, memoranda, and sketches. A contract is a formal form, to put it that way, of a purchase order or letter. Usually the particular specifications included in a contract are quite detailed and complete. The contract form is used in place of a purchase order to cover unusual contingencies such as special delivery dates or penalties for delivery delays, to describe special customer acceptance procedures or details of payments. Where huge quantities of product are being purchased, extra manufacturing facility investments may be involved, many months may be needed to complete the order — these are some of the reasons a contract sometimes supersedes the ordinary purchase order.

As for sketches, verbal orders, memoranda, and similar informal types of shop orders (which are nevertheless specifications), the inspector needs first to question the authority or official character of such instruments, as has been already intimated, and, secondly, to make sure he has all the information required to make a thorough inspection of the product.

In the larger organizations, seeking some of the sort of information described or implied up to this point is technically more a function and responsibility of the inspection department, itself, and not of the individual inspector. The latter cannot readily visit engineering, sales, and purchasing departments, or the laboratories, nor can he travel to customers' plants to secure the detailed knowledge he needs. Nevertheless, local restrictions should not prevent his persisting in some manner. Continued, intelligent study in his own area will always help. He can and should, of course, approach his own supervision for additional needed information. The danger is that of familiarity; that he may take too much for granted; that he feels he knows all about the part or product he is inspecting. Remember, always, that the requirements may have been altered since the last time the particular part was made. With catalog items, regular production, and staples especially, the inspector's error is usually that of presuming more knowledge than is actually possessed. Facts form sharper tools than opinions or assumptions. Don't be lazy; dig out all the information needed to make a comprehensive inspection.

Inspection Specifications

There is a growing tendency in many shops to issue detailed instruction sheets to inspection personnel. Such specifications, slanted to the needs of the inspection department, help materially to cover the sort of lapses implied in the preceding paragraph. They are valuable, also, as check lists, of a sort, preventing inspectors many times from forgetting essential inspection steps. an example of an inspection planning sheet is shown in Fig. 18.

Additionally, these sorts of checklists, guides, or tools are helpful to a variety of groups and functions, especially to the workers on the production line who are engaged in "self-inspection." It is not intended to make inspectors of operators or assemblers, but rather to identify and provide these workers with a level of control over their work that is

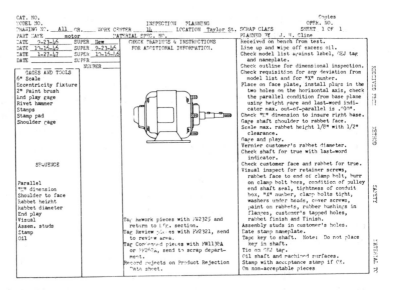

| CAT. NO. | | | | | | Copies |
| MODEL NO. | | | INSPECTION PLANNING | | | OPER. NO. |

Fig. 18. Example of an inspection planning sheet which is used to guide the inspector.

appropriate to whatever they can control, that is, the work that they do directly.

A complete and detailed inspection of a part or assembly is not in order here, but rather an assessment of work that would lead a worker to understand a problem before it multiplies and results in added value to the product in terms of scrap or rework.

These guidelines can take many forms — pictures, lists, diagrams, samples, and even training sessions — as long as they are current and understandable, and applicable, to the particular audience.

Care must be taken not to introduce irrelevant detail into these procedures or guidelines — many modern manufacturing procedures are now written with built-in criteria at critical points, in the case of both attribute and variables data. Many frustrations and problems encountered by inspections can be solved by inspection groups if they simply "spread the work around".

CHAPTER 3

Tolerances and Allowances

Modern industry has developed on the basis of interchangeable manufacturing. Interchangeable manufacturing means the production of parts to such degree of accuracy as is necessary to permit the assembly and proper functioning of the parts without further machining or fitting, although the individual parts may have been made at different times or in different manufacturing plants. In other words, the parts are theoretically, at least, interchangeable.

Ideal interchangeable mating parts would be those without any kind of dimensional variation, that is, they would be exactly the size called for on the blueprint or specification. In actual practice, however, there are factors that make it impossible to meet this ideal condition. Some of these factors are:

(1) The machines that are used to produce the parts have inherent inaccuracies built into them and therefore cannot produce perfect parts.

(2) In setting up the machine, that is, adjusting the tools used in the machine, the operator cannot make perfect settings or will at least contribute to the accrued variation. Inspection guidelines or details can be helpful in this respect, and again, inspection groups need to be able to provide the proper level of detail to the people who produce and handle the product.

(3) Variations in the properties of the material being machined introduce errors.

(4) The prohibitive cost of attempting to overcome entirely the first three factors favors making the parts as inaccurate as is tolerable, that is, just good enough to do their intended job and no better. The balance among factors of cost and quality is, to say the least, expensive.

Limits of accuracy are needed, therefore, for the various parts and also for groups of assembled parts; then manufacturing and gaging equipment can be used to obtain and check the established limits. In production and manufacturing, then, variation is a way of life. The inspector must understand the amount or level of variation acceptable to the customer. Limits or tolerances define the boundaries of this variation, and the more clearly these limits can be defined, the better they can be used and administered.

Basic Requirements — Basic Dimensions

A basic dimension is the theoretical or nominal size, which, for practical reasons, is only approximated; or, it is the dimension that would be obtained if perfection were possible and did not result in increased manufacturing costs. However, since perfection is impossible and also unnecessary, so far as the dimensions of machine parts are concerned, it is general practice to give a base or *basic dimension* and then indicate by supplementary "tolerance" dimensions just how much the actual dimension can vary from the basic without causing trouble; or, to put it another way, how much *inaccuracy* is allowable without causing a part to fit or function improperly.

Anyone at all familiar with mechanical devices knows that there are wide variations in the accuracy required for different parts. Some, for example, might need to be within 0.001 or 0.002 inch, or less, of a given or basic size, whereas other parts might function perfectly if within, say, 0.010 to 0.020 inch of this basic dimension. To illustrate further, suppose that a hole (which is to receive a stud) requires a diameter of about 1.250 inch (see Fig. 1); but assume that the actual diameter may vary from this 1.250 inch size as much as 0.005 inch oversize without too much play between the hole and stud. In this case, 1.250 inch is the basic dimension or the dimension aimed at. In producing such holes,

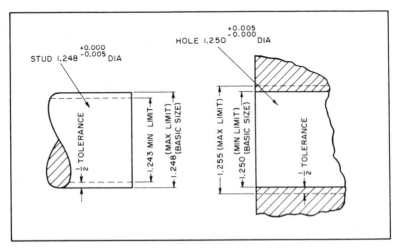

Fig. 1. Graphic illustration of the meaning of the terms limit and tolerance.

some might happen to have a diameter of exactly 1.250 inch, whereas the diameters of other holes would be up to the maximum of 1.255 inch and yet serve equally well. Since, as a general rule, approximate dimensions are easier to obtain and maintain than very accurate ones, unnecessary accuracy is avoided unless it is obtained without extra cost or effort. The same principles hold true for attributes data — a surface scratch on a component cabinet is assessed or judged to be unacceptable to a customer — the inspection criterion calls for "no scratches," and the inspection call is a pass/fail or go/no go option. However, such scratches can also be held to a range or "tolerance" of acceptance. (See Fig. 2.)

Attributes Scale:

"Qualification of a Qualification"

Criteria: "Reject for scratches."

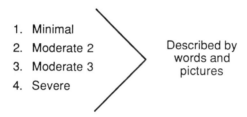

1. Minimal
2. Moderate 2
3. Moderate 3
4. Severe

Described by words and pictures

Fig. 2. Attributes scale.

The area between pass and fail in this situation may seem to involve much subjectivity in characterizing the attribute. A deep scratch is clearly a problem, but what about a scratch that is very slight, visible only in certain lighting conditions? Is it rejectable? The inspection guidelines should provide the key to consistent inspection results. Good guidelines should hold enough words, examples, pictures, etc., for all inspectors to make the proper evaluation. Of course, training and experience are important, but a properly written inspection standard or criterion should enforce as much objectivity into the inevitable "gray" areas as is possible.

Inspection personnel, perhaps Quality Engineers, can also delineate degrees of severity in surface scratches, and assign an arbitrary numerical scale to the varying degrees of scratches, say, from one to four. A "one" rating would indicate a minimal scratch, and a "four" could denote the other end of the range, the severest scratch or condition. With this scale, inspection personnel could sort or isolate scratches according to severity, and perhaps develop criteria for a minimally acceptable scratch by balancing these limits against customer expectations. (Note: this type of system is very difficult to administer among different personnel.)

Why Tolerances are Specified

For these reasons, tolerances are applied to the dimensions of all manufactured parts. A tolerance indicates how much a part may deviate from its ideal or basic dimension and still function properly. A bearing, for example, may require a 1-inch diameter bore to accommodate a 1-inch diameter journal. To make the bearing exactly 1 inch would be impossible, and even if this could be done, the bearing would wear in actual use to a somewhat larger dimension. In the light of these facts, what tolerance should be applied to the 1-inch diameter bearing bore? Suppose that experience shows that the bearing has to be replaced when it has worn 0.005 inch oversize. Certainly, then, the tolerance would not be 0.005 inch since this would permit the manufacture of bearings which were, in effect, worn out before they were ever used. Should 0.004 inch be specified? If so, the bearings produced would range between 1.000 and 1.004 inches in diameter and those bearings which were 1.004 inches in diameter would have only 0.001 inch of wear life left (1.005 minus 1.004 = 0.001). Suppose that 0.001 inch is the tolerance that is finally decided upon, that is, the bearing can be made from 1.000 to 1.001-inch diameter. This tolerance would represent a compromise between two things; the desired life of the bearing and the cost of machining it to a tolerance of 0.001 inch. It is possible to machine the bearing to a closer tolerance, but it may be that it is more economical to hold the 0.001 inch tolerance and replace the bearing sooner when it is worn than to hold the closer tolerance. The tolerance decided upon, therefore, represents a compromise between the accuracy required for proper functioning and the ability to produce economically this accuracy.

In design and preproduction phases of product development and manufacturing cycles, inspection provides critical feedback regarding tolerances and their impact on *product quality, reliability, and performance*. In modern manufacturing environments, the earliest stages of both product and process design and conception are tested and modeled to assist the easy and cost-effective transition to a full manufacturing schedule. Can the product be manufactured efficiently and in a cost-effective manner? The questions regarding the ability of processes to meet the tolerances prescribed by design groups can only be addressed with extensive inspection data.

In this type of approach, tolerances are tested and explored on individual assemblies, the combined prototype unit, and the process in which that unit will be built. Strict tolerances, specifications, and requirements are more expensive and usually more difficult to obtain and hold in manufacturing. This early "inspection" or testing of design limits and specifications ensure a match with the theory of the drawing board and the reality of the production floor.

Unilateral and Bilateral Tolerances

The term *unilateral tolerance* means that the total tolerance, as related to a basic dimension, is in *one* direction only, as shown in Fig. 1. For example, if the basic dimension were 1 inch and a tolerance of 0.002 inch were expressed as $1.00 - 0.002$, or as $1.00 + 0.002$, these would be unilateral tolerances, since the total tolerance in each case is in one direction. On the contrary, if the tolerance were divided, so as to be partly plus and partly minus, it would be classed as *bilateral*, as shown in Fig. 1. Thus, $1.00^{+0.001}_{-0.001}$ is an example of bilateral tolerance, because the total tolerance of 0.002 is given in two directions — plus and minus. (See diagrams, Fig. 3.)

There are different ways of expressing tolerances or of giving allowable dimensions. These different methods should be understood in order to properly "read" or interpret the dimensions on a drawing. When tolerances are unilateral, one of the three following methods should be used to express them:

(1) Specify limiting dimensions only as

   Diameter of hole: 2.250, 2.252
   Diameter of shaft: 2.249, 2.247

(2) One limiting size may be specified with its tolerances as

   Diameter of hole: $2.250 + 0.002, - 0.000$
   Diameter of shaft: $2.249 + 0.000, - 0.002$

(3) The nominal size may be specified for both parts, with a notation showing both allowance and tolerance, as

   Diameter of hole: $2\frac{1}{4} + 0.002, - 0.000$
   Diameter of shaft: $2\frac{1}{4} - 0.001, - 0.003$

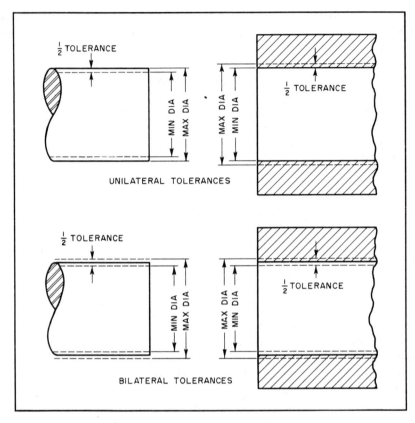

Fig. 3. Diagram illustrating unilateral and bilateral tolerances. Solid lines show basic diameters, dotted lines allowable variations.

Bilateral tolerances usually have plus and minus tolerances of equal amount. An example of the expression of bilateral tolerances follows:

$$2 \pm 0.001 \text{ or } 2^{+0.001}_{-0.001}$$

Bilateral tolerances are not always divided equally as in the preceding example. To illustrate, if the total tolerance is 0.003 inch, it might be given as plus 0.001 and minus 0.002, in which case it would be written as follows:

$$2^{+0.001}_{-0.002}$$

In general, if a greater tolerance is permissible in one direction than in the other, this may indicate that the tolerance should be unilateral instead of bilateral.

Positive and Negative Allowances

No mention has been made of the term "allowance." Let us consider the journal which is to run in the 1-inch diameter bearing, which can vary from 1 inch to 1.001 inch because of the tolerance of 0.001 inch. The journal also will have a tolerance applied to it. Assume for convenience that this tolerance is also 0.001 inch and that the shaft, therefore, can be made from 0.999 to 1.000 inch in diameter. If, in assembling the bearings and journals, it happens that a 1.000 inch diameter bearing and a 1.000 inch diameter shaft are picked at random, it would be found that if they were assembled, the journal will not turn properly in the bearing. A journal could be selected, say, one that is 0.9999 inch in diameter, and this would fit in the bearing. To eliminate the need for selecting mating bearings and journals by trial, it could be specified that no journal is to be made larger than 0.9999 inch. The result of such a specification would be that any journal chosen at random would be loose in the bearing by at least 0.0001 inch (1.000-inch bearing minus 0.9999-inch journal). This intentional looseness between mating parts is called the allowance. What determines the amount of this allowance? In the case of a bearing and journal, space must be provided for an oil film, otherwise proper lubrication cannot be attained. In the case of a shaft which is to be fitted into the hub of a pulley, a snug fit is desirable, therefore no allowance or even negative allowance may be provided. (Negative allowance instead of producing looseness results in interference between the metal of the mating parts and therefore requires the application of force to assemble the parts.) The intended function of the assembly, therefore, determines the allowance or type of fit that should be used. Certain types of fit have been standardized and allowances and tolerances for these may be found in the American Standard for Tolerances. Allowances and Gages for metal fits.

As a further example of how allowance and tolerance are applied to metal parts, consider a shaft dimensioned 0.874 inch and a hole dimensioned 0.875 inch. This represents an allowance of 0.001 inch. (The same hole with a shaft dimensioned 0.876 inch also represents an allowance of 0.001 inch, but, since the shaft is larger than the hole, it is negative allowance.) In manufacturing these parts the dimensions could not be produced exactly. For this reason allowable variations (tolerances) must be provided. If the tolerance required for each of the parts is 0.001 inch, then the shaft would be dimensioned 0.874 plus .000, minus .001 inch and the hole 0.875 plus .001, minus .000 inch. The greatest looseness between these mating parts would therefore be .003 inch and the greatest tightness would give a clearance of 0.001 inch.

Geometric Dimensions and Tolerances

A new system of dimensioning and tolerancing is enjoying a wide use in industry — geometric dimensioning and tolerancing is enhancing the technical descriptions of physical objects, primarily on blueprints. Regular tolerancing and dimensioning methods describe these physical realities in terms of the size limitations for a particular feature. Pointedly absent is an identifiable reference point or datum from which to locate these dimensions in space, and a standardized visual system for describing simple and complex geometric relationships.

Geometric dimensioning and tolerancing uses a system that employs a feature control system — a system for locating in the physical realm of three planes. This system combines numbers, letters, and symbols. These represent the following descriptive features and characteristics (Fig. 4).

(1) The characteristic symbol.
(2) A tolerance value.
(3) Datum reference.

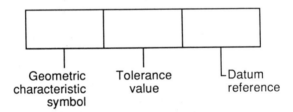

Fig. 4. Feature control symbol.

This system of coded reference characteristics has been in use since 1982, the ANSI Y14.5M-1982, upgraded from the previous and original ANSI specification, which describes this system, established in 1972.

The datum reference indicates a starting or reference point. Datums are shown as letters in the coded system, while the second item in the system, a tolerance value, is of course expressed as the number or dimension that the part may vary from the intended value. Datum points and tolerance values are then linked to the third part of the triangle of information — the (geometric) characteristic symbol, in order to describe the whole relationship.

These (characteristic) symbols are comprised of various descriptions and relationships of form, profile, orientation, location, and runout. These geometric features are further grouped into categories described as individual or related (features), meaning that the feature pertains only a particular indicated surface itself, or to one surface in combination with other surfaces, in order to complete the description of the feature (Fig. 5).

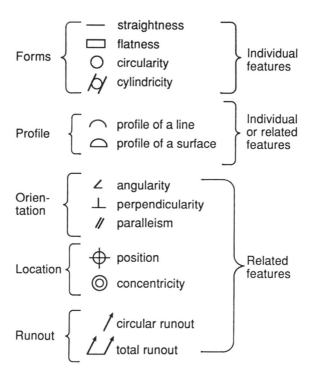

Fig. 5. Geometric characteristic symbols.

Geometric dimensioning and its related tolerancing abilities, like the real physical world, can be simple or very complex depending on the nature of the part or assembly being described. The important thing to be indicated in this text is a basic understanding of how the system works; reference books and specialized instruction can lend familiarity and ease of use to the system through practice and study.

Datum references can likewise be single or multiple, and are arranged in order of importance if there is more than one (datum). Additionally, supplementary geometric symbols add more detail to the basic list of five characteristic symbols (Fig. 6).

Two important supplementary symbols used in this system are the "MMC" and the "LMC" symbols — "Maximum Material Condition," and "Least Material Condition". MMC is a stipulation of sorts that is

Ⓜ   MMC

Ⓛ   LMC

Ⓢ   RFS Regardless of Feature Size

Ⓟ   PTS Projected Tolerance Zone

∅   Diameter

S∅ Spherical diameter

R   Radius

SR   Spherical Radius

( ) Reference Dimension

☐   Basic Dimension

Fig. 6. Supplementary symbols.

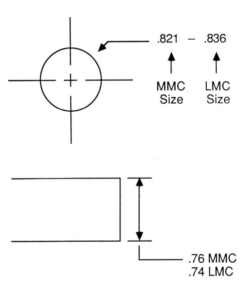

Fig. 7. MMC and LMC sizes of a part.

described by a circled M, and specifies a condition in which the component feature has the maximum or most allowable amount of material (according to the size limits specified). This supplementary or conditional symbol is indicated with the associated tolerance value, or the datum reference, or both. Conversely, the LMC modifier is also used on parts that vary in size, but for the opposite purpose. The LMC stipulates the least material within the specified limits of size (Fig. 7).

A circled S indicates another stipulation — called the "Regardless of Feature Size," and means that the indicated tolerance value applies to whatever the actual size of the feature is. A circled P sets forth a condition specifying the height of a part mated with a toleranced feature. Again, it is another coded version of a physical relationship, designed to indicate much geometric detail in a simple and concise manner. This symbol is called a Projected Tolerance Zone.

Other supplementary symbols consist of: spherical diameter, radius, spherical radius, reference dimension (), and basic dimension [].

CHAPTER 4

How Standards Aid the Inspector

In the industrial sense, standards are practices or limiting conditions, established or authorized as models for comparison. They may have become established as a result of the habits, usages, and traditions of one or more manufacturers or they may have been developed to meet a specific need by a technical committee representing the various interested parties. Standards are representative of acceptable and unacceptable ranges of variation.

Many standards of manufacture are international in scope; others are national or industry wide. The American National Standards Institute, a federation of over 100 national technical and trade organizations, serves as a clearing house for national standards. More than 1000 of these standards representing, in each case, a general agreement on the part of maker, seller, and user groups as to the best current industrial practice have been approved by this association. Manufacturers use these national standards to facilitate production operations, to lower production costs, and to eliminate controversies between buyer and seller. Then there are local standards, under which a single shop, or even a department within it, manufactures. A few standards are individual, a Stradivarius violin being a good example of the latter.

Standards are, in a sense, tried and true versions of what the customer and the marketplace will accept or support in the way of form, fit, and performance. Many (product) conditions become standards because they outstrip the competition in the competitive arena.

Emergency Standards

In addition to local standards, which have been established carefully and deliberately, there are those that are born hastily in emergency. These seldom appear explicitly on blueprints or on specifications. Perhaps parts failed to fit together properly or wore out too rapidly. Or a customer complained, whereupon an executive rushed out into the shop with an order. After the tempest blew over, the particular order was forgotten, at least by those in authority. It was intended only as a temporary decree anyway. But a routine had been started. Routine grows rapidly into habit and habit becomes tradition.

Standards set up under emergency conditions are not always the most satisfactory; nor are standards that have grown up and spread like weeds. The present-day tendency is to recognize that standards are as essential and useful as specifications. They are studied, analyzed, evaluated, revised, and demonstrated, and made official, in writing.

It is essential to formalize, in writing, by picture or example, or by blueprint and measurement specification, criteria for variation, or standards. Through standardization of criteria, those criteria are better understood and applied by inspection forces.

Example of Manufacturing Standard

An example of one of a group of standards for manufacture used by the Jones and Lamson Machine Company of Springfield, Vermont, is given directly below.

Keyway Tolerances

Keyways must be in the center of the shaft within the following tolerances:

Shaft Diameters (inches)	0 to 1	1 to 2	2 to 3	3 to 4
Tolerance (inch)	.005	.010	.015	.020

Woodruff keys must be parallel with the shaft within .002 inch per inch of length.

Square keyways must be parallel with the shaft within .001 inch per inch of length and not out of parallel more than .005 inch in total length of keyway.

Depths of keyways must be according to drawing.

Long keyways in shafts that have sliding keys, such as cross feed screws, etc., must have a smooth finish on both sides. Maximum 15 microinches.

Bottom of these keyways may have feed marks but the marks must be even and free from chatter.

Standards Not Always Precise

Many standards can be precisely defined. The fact that a dimension appearing on a blueprint without tolerances is subject to the general shop tolerance of, say, ± .005 inch, as mentioned in a preceding section, is a precise standard of practice. Other standards are more illusory or more difficult to demonstrate. Personal opinion is involved. Is the gray enamel used on a product the same shade as it was a month ago? What precisely is meant, for example, by the statement that certain work is not up to our usual standard?

The practice in a shop may be to undercut all shoulders and bevel the edges of all bores in order to ensure a close fit of mating parts. The fact that there is to be an undercut is a precise standard, but how much of an undercut or bevel may not be clearly defined.

Does it matter? Is the extent of the bevel or undercut important to the customer? Does the customer mind if the bevel varies within the shipment he receives, or from one shipment to another? These are questions that need to be addressed by groups of functions within the organization that have or can find the answers.

Seeking feedback such as this from customers is a charter of many modern Quality Assurance groups. Customer feedback is critical to a competitive position, and data on product performance and customer satisfaction from the field are as important as feedback from the final test or inspection area. It is more inefficient to correctly measure incorrect criteria than not to measure for a certain standard at all.

Furthermore, the narrowing or isolating of these standards, through accurate description, is essential in making the standard usable to people in manufacturing. The more succinct the standard, or criteria, the more effective it is as a model.

And standards are not born by themselves — rather they are the result of on-going cross-talk and feedback between the producer and the marketplace. More importantly, realize that today's standard is tomorrow's relic, so the concept of change through assessment and improvement is central to the maintenance of all standards, at whatever level they are applied.

Surface Finish as Standard

The subject of surface finish holds a good example of the relationship between standards and industrial practices of evaluation, assessment, and categorization of variation. Very few discussions of shop standards take place without the subject of surface finish coming up. It is one of the most common wrangles in the machining trades. Is the finish fine enough or are the tool marks too deep, too coarse, too apparent, for a surface to be acceptable in appearance or for use? Should the machine be stopped and the tool sharpened or the wheel dressed? Is the operator making the cut too rapidly? Almost as soon as he appears on the factory floor the inspector confronts problems of surface finish, and right here is a good place to discuss a subject that is far from settled in industry. Only in recent years has it been possible for the engineer to write practical specifications for surface finish and for the shop to work to more definite standards.

Recently it has become possible, with sophisticated and sensitive inspection devices (tools that are usually quite expensive), to define properly and characterize such difficult but important judgments as surface finish. The end of the line definition is not easy but essential — it is the customer — what does the customer demand? This has to be clearly understood and balanced against the capability and cost of producing such a finish.

Once this homework is completed, the work can begin — the work of identifying what this means in the production line. The units to be judged will be "eye-balled," and related to that calibrated marketing concept (the customer's requirement) for verification. Once the standards have been set, production and measurement can begin, training for visual identification of the right finish is based upon the standard and balanced against unacceptable versions of a failed attempt at a proper finish.

Enter the modern contribution — electronics and automation, which through optics, electronics, and a host of other sensing, comparing, and reporting features, can actually quantify the criteria or requirement *better* than the human eye. These tools can be employed directly in the inspection process, as training and verification tools for inspectors, and in other applications that make possible consistent and quantifiable calls around the clock, even on the third shift, when the master inspector's perfectly calibrated eyeballs are asleep.

Importance of Surface Finish

Surface finish is important not only as a matter of appearance or expert workmanship but, in the case of mating surfaces, it has a positive and prolonged effect on product wear and usability. If two surfaces bearing against each other, such as a shaft turning in a bearing or a piston rod reciprocating in a gland, are too rough, unnecessary wear will take place. The turning shaft can act like a reamer and the piston rod like a broach. Where the roughness is excessive, the moving parts can heat up, bind, and freeze. Excessive surface roughness on shafts and in bearings on, say, an electrical or motor driven household appliance will require more power — the appliance costs more in kilowatt hours to run. Research has shown, too, that where parts have fractured or ruptured under strain, the fracture itself frequently started at some surface irregularity and that the shaft with the rough surface fractured sooner — under less strain — than the shaft with the "super smooth" surface.

Poor surface finish many times neutralizes the effect of tolerances. Suppose a measurable surface finish of 100 microinches (100 microinches is really .0001 inch) appears on the surface of parts made to .0005-inch tolerances. As far as the close fitting together of mating parts is concerned, the operation has used up .0001 inch — one-fifth of the tolerance — in surface roughness. If at the same time the operator works to the high side of the tolerance (the usual condition), the presence of .0001-inch surface roughness may throw a high percentage of pieces actually into the oversize class, though, basically, the cut is at or just within the top tolerance.

If surface roughness on a lathe-turned piece is considered as, in reality, a very fine screw thread and if the shaft is assembled in a tight fitting hole, some of the sharp peaks of the "screw thread" will be bent or burnished down in the mere act of assembly. As the shaft revolves in the hole, the remaining peaks wear or burnish down very quickly. Soon a tight fitting assembly is loose. Automotive engineers recognize this condition and demand — advertise in fact — "superfinish" on many moving parts.

General Categorization of Surface Finish Detail

A machined surface appears exceedingly complex when viewed under a high powered microscope. What seems to be a smooth surface may contain several hundred thousand irregularities to the square inch. They take the form of rough cavities, pot holes, crevices, ridges, valleys, and peaks. Some apparently smooth surfaces have the same appearance as a level, ploughed, field that is viewed from an airliner.

Surface roughness on a cylindrical piece turned in a lathe, with the tool or wheel traveling transversely, takes the general form of a screw thread or helix. The turning of the surface of a disk-shaped part held in a chuck, with the cutting tool moving steadily in from periphery to center, shows "phonograph record" lines. Surfaces of pieces from planer, shaper, milling machine, and surface grinder compare with the straight ploughed field.

A common way of illustrating the elements of surface conditions is shown in Fig. 1. The drawing shows irregularities in surface texture that are deviations from the geometrically ideal form. The several conditions indicated are defined as follows.

Roughness is defined as finely spaced surface irregularities, usually in some sort of consistent pattern, produced by machining action from cutting edges, abrasives, burnishing, or rolling. Each type of cutting tool and machining action leaves its own individual markings; each type of material — castings, ductile materials, iron, brass, aluminum, etc. — also reacts differently to cutting tools and further develops a unique pattern.

Waviness is an irregular surface condition of greater spacing than roughness. It is not usually caused directly by the cutting edge but by work or machine deflections, vibration, and the like. Irregularities that are geometrically similar may occur from warping or strains in the material. Waviness also usually shows a consistent pattern. Roughness is considered as superposed on waviness as may be seen in Fig. 1.

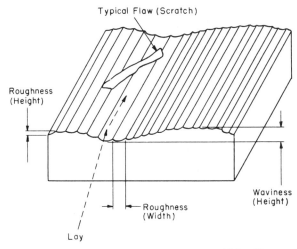

Fig. 1. Diagram showing some of the factors which affect surface finish.

Flaws are scratches, digs, holes, peaks, ridges, cracks, or checking that occur at one place or at relatively infrequent intervals on the surface, usually without consistent pattern. The surface of a casting, for instance, may display waviness and flaws but not surface roughness as defined above. Machine the cast surface in some manner and the waviness and flaws may or may not be retained. The original waviness may be superseded by a new pattern and new flaws may be added. Regardless of these changes, there will have been superposed a degree and pattern of surface roughness.

One other definition should be offered here. *Lay* is the direction of the predominant surface pattern. In other words, the lay of surface roughness and waviness on a lathe-turned cylindrical piece will be in the form of a helix. On a planed or surface ground piece the lay will be parallel to one edge and perpendicular to the other or it may run at an angle, depending on how the piece was located for machining. The lay may be radial; it may spiral like the ridges on a phonograph record.

Measuring Surface Conditions

Surface roughness and waviness can be measured. Roughness may be felt with the finger nail. It may be seen with the naked eye. Then again the surface may seem smooth and polished to the eye or the finger nail but surface roughness still exists, and it can be measured by optical and mechanical means (Fig. 2).

Ordinarily waviness cannot be distinguished by eye or the finger. It must be detected by mechanical measurement — with an indicator or possibly an optical flat. Figure 1 shows that any pointed mechanical instrument for measuring roughness must have a stylus point finer than the width of the finely spaced irregularities while a similar instrument for measuring waviness could bear on the machined surface with a much broader point.

This is rough, or this is not rough — categorization of the attribute of surface finish, and arbitrary, at best without a standard or precise description of what is meant by these two "rough" categories.

When increased detail is needed to clarify an attribute, a continuous or direct measurement is sought. Through the miracles of numbers and divisions of numbers, or fractions, great detail can be assessed against the general attribute or category of "roughness." Once such measurement detail can be assessed and recognized, the path to control and improvement is opened.

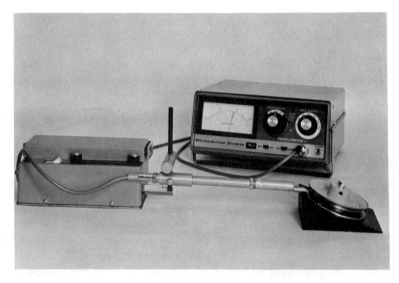

Fig. 2. Measuring the surface condition of a workpiece with a commercial microinch surface apparatus.

Surface roughness is measured (with instruments) in millionths of an inch — microinches. (Where the metric system is used, the micron — one thousandth of a millimeter — is the unit of surface measurement.) The width of surface roughness is measured in thousandths or tenths — as .0022 inch or .0008 inch, for example — and waviness is also measured in thousandths or tenths.

Figure 2 shows a part being checked for surface condition with the microinch finish of the part shown by the amplifier meter. The surface finish reading could also be recorded on a paper chart.

Surface roughness is also measured qualitatively by comparing the specimen surface with a standard surface whose actual roughnesses are known from precise measurements. This is done with the use of replica blocks or surface roughness comparators, so-called. One such commercial set is illustrated in Fig. 3.

Using roughness standard blocks or cylinders, the surface condition of the machined part may be compared to the replica block by eye, but the more accurate way is to scrape the surface of the workpiece with the finger nail and then scrape the finer nail across several of the comparator blocks until one test block is selected whose surface seems to compare closest with the machined part. The surface of the

Fig. 3. Replica blocks for comparison with workpiece surface finish.

workpiece can then be established in microinches, within about 25 per cent accuracy, by reading the measured microinch legend for the comparator block selected.

In Fig. 4 is shown a stereoscopic comparison microscope, which permits three-dimensional comparison of a master roughness specimen with the finish of the work. Such a comparison is shown in the inset at the upper right in Fig. 4. The optical system in this microscope is such that when the standard and the work are viewed alongside of each other, the resulting image produced in the eye pieces is without any demarcation line between the two.

Where a factory is equipped with accurate microinch surface measuring apparatus and where reasonable success is obtained in controlling surface finishes at the machining operations, the engineering department and its draftsmen symbolize the degree and direction of allowable surface roughness with a combination of symbols like that shown in Fig. 5.

These symbols on a drawing not only indicate to the machinist the amount of surface roughness desired but also anticipate the position of

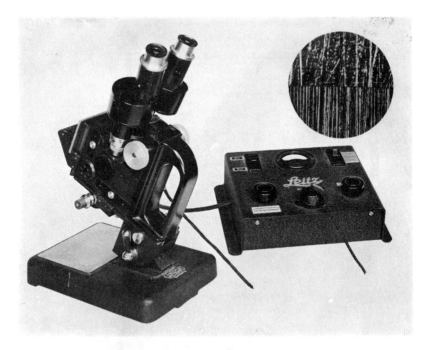

Fig. 4. Steroscopic microscope for comparing a workpiece with a standard of surface finish. Inset shows workpiece and standard surface finishes as viewed.

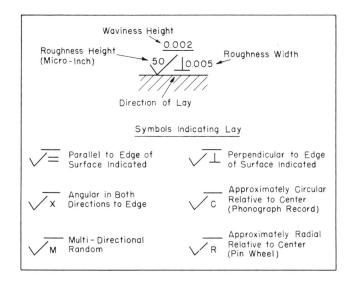

Fig. 5. Symbols used to designate surface finish.

the pieces in his machine or the sort of pattern the machine is liable to produce. Part of the inspector's job in judging conformance of such parts, of course, is to determine how closely the dimensional degrees of roughness and waviness have been obtained and also that the direction of roughness conforms. This he can do with considerable precision if he has the use of mechanical or optical surface measuring apparatus. If he is using comparator or replica blocks, he must reach a decision regarding conformance with his finger nail and eye.

Where only the symbols *f* and *ff* appear on drawings or where the engineer issues no specific information in connection with surface conditions, another type of inspection problem appears. A great many shops have no surface condition measuring apparatus or methods at all. In other factories, apparatus may have been installed and surface specifications issued, but solely for use in an individual department or on a limited class of parts. Either way, inspectors in many departments find themselves faced with decisions over the quality of machined surfaces with no definitely official standards established or apparatus available for measurement. When this happens, time and money are wasted. Standards *must* be derived, or inspection personnel, and others who are attempting to assess a condition "blindly," are wasting their time and their employers' money. It is the responsibility of the Quality Control or Quality Assurance group to clarify these definitions and have them accepted.

Establishing Comparison Specimens for Inspection Reference

A natural and logical solution to this problem is to secure a supply of parts previously machined in the particular area and examine them. The inspector can, for instance, sort and catalog the parts by eye according to surface finish (his opinions of surface finish), placing the poorest or coarsest specimens at his left hand, in manner of speaking, and the best at his right, filling in between with a sort of graduated scale of specimen surface finishes.

By such a procedure, the inspector will find himself able to make a more practical, comprehensive, and accurate decision on a standard of surface finish than by a hurried, random selection of representative components. The eye should perhaps be supplemented by finger nail scratching across the surfaces of the specimens when lined up.

Having built up an orderly demonstration of surfaces, from the worst to best ordinarily available, the inspector will find that others in the area (including perhaps his own supervision and representatives from engineering as well as production people) can then more sensibly express opinions on desired or obtainable surface finishes. Within the time limit or facilities that may be available, the sort of "committee" action implied will be found valuable. Lacking definite surface specifications, it is better to attempt a standard from agreement of all concerned than to establish what might turn out to be only a random, arbitrary standard.

With one or two specimens earmarked as standards for surface finish, the inspector can then proceed to compare and judge surface conditions by eye or finger nail, using the specimens as if they were registered replica blocks.

Economics of Surface Finish

Whether a surface finish standard is determined by an engineer or a committee or from the more or less arbitrary opinion of an individual — whether it is to be measured by accurate instrumentation or by eye and finger nail comparison — the cost of securing and maintaining such a degree of finish throughout continuous production should be kept prominently in mind. It is perfectly natural to set the sights high when it comes to establishing a standard for surfaces. But the extra fine, smooth surface may not be worth it.

Better surfaces can be produced by slowing down the machine, perhaps, or the rate of tool feed or carriage speed. The more frequently the machine is stopped in order to sharpen the tool, the finer the surfaces secured. Changing from a coarser grit grinding wheel to a finer grit wheel will give a better surface, but fine wheels cut metal away much more slowly than coarse wheels. Honing a hole will leave a very smooth surface, but it is a much slower process than broaching, boring, reaming, or drilling. Often, superior finishes can be secured only by supplementary operations such as fine filing, polishing with emery cloth, lapping, honing, or buffing.

All of these may mean slower production, less total quantity at the end of a day, and a consequently higher cost per produced unit. The value of a finer surface must always be compared to the cost of obtaining that surface. Ball bearings are a good example. The inner ball races, and the balls themselves, should be superlatively smooth. But the outer surfaces that are to be pressed into housings or over shafts need no such care or cost. Nor do the housings and shafts themselves require the highest type of finish.

Personal Standards of Surface Finish

The idea just discussed should be reemphasized. The person who is new to industrial inspection will instinctively look for, and wish for, better surface finish and appearance and just as instinctively reject work that is inferior in respect to surface appearance. The production operator naturally reacts to the rejections by trying to produce better surface finish, but in so doing may reduce his production and to that degree increase the cost of the parts and the products. Personal "ideals" of surface finish may not be commercially feasible or commercially necessary.

So-called "personal standards" that may be weighted to any inspection criterion must be weighted against the consideration of the marketplace or customer, and agreement must be obtained on this standard within the manufacturing environment. A personal standard of acceptance or assessment, if it is an appropriate standard, will soon become everyone's standard.

Even the experienced inspector may keep the pressure on production unnecessarily for improved surface conditions. On the other hand, his familiarity with the job may cause him gradually to neglect surface finish and allow a decline in surface standards to the point of shoddiness. Surface finish has much to do with a product's reputation for expert workmanship, or the lack of it. The answer, of course, is to have suitable standards, review them frequently, and use them to compare current work with.

Then, there is the situation where the inspector is faced with the question of surface finish and is at the same time utterly bereft of surface measuring equipment, replica blocks or even homemade standards for comparison. Frequently, too, there is not even the semblance of specifications to guide him. He is suddenly handed a specimen of machine work. The surface — is it passable or not? An immediate decision is demanded.

Experience as a Guide

About the only way to forestall this sort of dilemma is to gain experience as rapidly as possible. Forewarned is forearmed. The new inspector can be observing, absorbing, the general grade and type of surface finish and appearance while he is receiving instructions concerning the routine of his new job; his eyes can be busy from the first minute he steps into the new department. In other words, the unwarned and inexperienced inspector becomes so intent on the instrumentation, mechanics, paper work, and other particulars of the new task, if not overwhelmed by them, that he readily forgets or neglects the constantly important inspection items of appearance, of surface finish, until they are brought up abruptly to him for decision.

These items of appearance must be quantified, detailed, communicated, and applied evenly and fairly. The attribute must be describable from each end of the spectrum — from the "perfect" side and from its opposite side, "rejectable." The description must then move toward the middle in an increasing level of detail — with the so-called "gray" area in the middle bounded by so much description and detail that it is as small as possible. In other words, the attribute, good or bad, will not escape the attention of someone who has been properly trained.

In situations where a hurried and more or less arbitrary decision has actually been made under pressure, the inspector might try what is frequently an amusing experiment. The suggestion about to be offered applies especially where the subject of surface finish, or any other visual standard for that matter, is the cause of frequent or regular disagreements. Suppose a standard of surface finish, based on visual opinion, has been adopted on a certain day, perhaps after a bit of discussion and argument. The record of the decision is a selected specimen of work. Suppose then the inspector stores this standard piece in a drawer or locker for several days or longer. When an exactly similar wrangle occurs again, as it frequently will, he brings out the standard piece.

Almost invariably, it will be found that the standard of workmanship being argued over now varies from the standard so solemnly adopted a few days or a week past. It may be that the production people have urged a softening of the standard. Many times the inspector finds himself unconsciously insisting on a higher standard than he had chosen the last time. Bringing forth the sample workpiece whose appearance everyone had sworn to follow shows quickly and amusingly how readily standards change, even over night, and how readily opinion can be biased by circumstances, especially those potentially unfavorable.

Separating Good Work from Bad

The main difficulty in establishing a standard usually lies in setting the line of demarcation between good and bad work, between acceptable and rejectable units. The transition from good to bad is seldom abrupt.

An experiment made in a psychology laboratory supplies an illustration of the difficulty. The psychologist had rigged a caged-in platform for a rat. By turning a valve he could direct a jet of compressed air at the rat which would make the creature jump. In front of the rat were two white cards as barriers. On one card the psychologist had drawn a circle and on the other an ellipse. If the rat, impelled by the jet of air, jumped against the card with the circle on it, the barrier would fall down and disclose a supply of food. If, however, the rat leaped toward the card with the ellipse on it, he bumped his nose because that barrier was firmly fixed. After very few trials with the compressed air the rodent learned to jump toward the circle rather than the ellipse even though the experimenter interchanged the position of the two cards.

The psychologist then erased and redrew the ellipse so that it was less egg shaped and more nearly a circle. But the rat could tell the difference. Again the ellipse was redrawn more nearly circular. The cards were interchanged.

The experiment was continued to a point where the ellipse so closely resembled the circle the rat was unable to distinguish between them. Yet he was impelled by the merciless air jet to jump. He was forced to make up his mind how to avoid bumping his nose. But the ellipse and the circle looked so much alike he simply could not decide. The creature's solution of the completely frustrating situation was to roll over on his back, curl up his toes, and sink into a coma.

Setting Up a Standard

One systematic solution of the standards quandary is suggested in the following procedure. In looking over groups of work there can be found examples that are definitely acceptable; also unquestionably rejectable units. Samples of the acceptable group can be classified by degree, or graduated in an order from wholly satisfactory to the worst possible degree that would still be accepted. In like manner, rejects can be graded upward by steps to the best appearing units that would still be rejectable. By looking at the work and the matter of standards from these two divergent angles, the boundary between good and bad work, many times, stands out more sharply. If the "worst acceptable – best rejectable" procedure is used, it has been found from experience that the standard for workmanship is better based on the "worst acceptable" classification, on the theory that should the standard weaken or depreciate, consciously or unconsciously, the grade of work still accepted is not so poor as it would be where "best rejectable" marked the line and the quality of work slipped lower.

Where the quality of work is considerably a matter of opinion, the inspector can set up mentally a sort of numerical scale to help him. Suppose it is a matter of stamping a model or serial number on a nameplate or directly on the frame of the equipment. The inspector establishes (mentally) No. 5 as the standard, professional looking job. The letters and digits in such a legend are evenly spaced in a straight line and stamped sharp to a uniform depth. The No. 4 work is not quite so good and a No. 3 label contains almost an excess of errors. And so on down the mental scale. Letters and digits in the No. 2 class would begin to reel up and down hill; they would be unevenly spaced and illegible to the extent the stamps are dull; while a No.1 job would not get by a drunken man. By "zoning" appearances in some such fashion, the inspector accepts No. 5 and No. 4 work, issues a warning when it slips off to No. 3 quality, and sharply rejects No. 2 and No. 1 grades. He has established a system much more consistent than independent spot judgment. In a short while, production personnel get on to the system and take to lettering in the assured No. 4 and No. 5 zones.

Demonstrating Standards of Workmanship

Bernard Baruch once remarked that the ability and facility to express an idea is almost as important as the idea itself. Originality, ingenuity, vision, aggressiveness, and persistence are just as valuable in inspection as in any other field. Perhaps in no other branch of inspection work itself do some of the qualities just mentioned appear to better advantage than in selecting, proposing, and illustrating standards. A decision is always necessary. It is better based on careful, unbiased study and analysis, and the courage of convictions always helps.

Where standards can be reduced to figures or dimensions, demonstrating them becomes easier. Surface finish, for instance, can be expressed in microinches and so measured by an instrument. Allowable taper or out-of-round may be limited to half the tolerance. But there are other conditions, especially of appearance, for which scientific measuring apparatus has yet to be devised. The sample piece has already been discussed, a sample that is "worst acceptable," if that procedure is adopted. One trouble with the single piece or specimen used to demonstrate a certain standard is that one gets used to it — so accustomed to it, that its message is readily ignored. The sample piece becomes dust coated, dirty, perhaps finger marked, and greasy. Or it rusts and tarnishes. Where samples are used to demonstrate standards, they should be kept clean and fresh. It is a good thing to renew the specimen every month or so.

Teaching, transmitting, and otherwise communicating zones or levels of quality requirements is the work of the inspector. Change in these requirements is inevitable, and must be anticipated constantly. Inspectors must strive to quantify and describe standards of assessment in increasingly objective detail. When such standards of quality or excellence become everyone's job in the production arena, a new plateau of measurement and control will have been reached that kicks off the next cycle of improvement — making it one more step better.

Use of Standards Boards

So-called standards boards are excellent means for illustrating standards. An example is shown in Fig. 6. In this particular case, specimens were selected, generally in pairs, illustrating passable and rejectable degrees of each of the several types of defects potential in the operation.

Where a standard is to be illustrated by identifying or mounting specimens, it has been found, too, that a selection of three specimens offers the quickest and most accurate decision. One specimen, while it contains some degree of the defectiveness in question, is selected because it is nevertheless commercially satisfactory. Another specimen marks a completely rejectable degree of the defect. Then a third, the borderline specimen, is very carefully agreed on. This shows a degree of the defectiveness under question such that if it were discarded as rejectable there would be little or no argument, or if it were found in the acceptable work there would be likewise no complaint. Perhaps the diagram of Fig. 7 makes this conception a little more graphic.

The careful arrangement of a three-piece sample board seems to permit a faster and more accurate decision. If a sample of subsequent

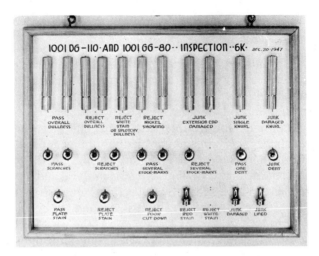

Fig.6. One type of standards board which shows various acceptable and unacceptable conditions for inspector's reference.

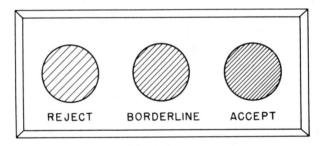

Fig. 7. Schematic arrangement of a three-piece sample board.

work is held up beside the three piece display of degree of defectiveness, the classification of the workpiece as either definitely rejectable or satisfactory seems to show more readily than where the work is compared to, say, a single acceptable sample or even to the paired, rejectable–acceptable samples.

The standards board idea is limited in practice to smaller units. Larger forgings, stampings or castings, rod or sheet, or any sort of workpiece larger than your hand, would ordinarily be too unwieldy for display, although sometimes sample boards are made up of small sections cut from the big pieces, the samples containing the sort of defectiveness occurring in the manufacturing operations.

The principles of selecting and demonstrating standards are not confined of course to surface finish. They may be applied to almost any inspection situation that cannot be solved by instrumentation. Another bone of contention common to manufacturing is the subject of burrs. Wherever metal is cut, ground, formed, punched, or sheared, burrs are likely to form and a burred edge of any sort is troublesome at subsequent operations or on a completed product. Deburring means extra operations and added cost. On many occasions a slight degree of burr may not be entirely objectionable. So, again, the inspector is frequently faced with the necessity of saying how much burr can be allowed, keeping in mind always the added cost of preventing, eliminating, or removing objectionable burrs. Essentially, then, the problem is no different than surface finish.

Standards for Various Finished Surfaces

Standards usually have to be established to prevent objectionable variations in the type of finish on painted, varnished, and enameled surfaces as well as for variations in color from batch to batch. Nickel and chrome plated surfaces present very special problems along with their close relatives, polishing and buffing. The list is long and embraces virtually every type of industry — leather, plastic, and textile surfaces; bubbles and flaws in glass; tool and chuck marks, chatter marks; roll and diemarks, rubs and burnishes, on rod and sheet stock; wrinkles, pits, scratches, and nicks; wear of engraving dies; pits, blow holes, and shrinks on castings. The test or standard may also include such items as softness, brittleness, uniform penetration, flexibility, ring or tone, taste, or odor.

Figure 8 illustrates an inspection test on spotlight types of electric light bulbs. At an early step in manufacture the interiors of the bulbs are "silvered" to form internal reflectors. If the reflecting metal is coated on too thickly, extra and unnecessary manufacturing cost is added at the production of each bulb both from the extra time taken to coat each bulb and also from the extra materials used. On the other hand, if the coating is too skimpy, light can shine through the reflector thus reducing noticeably the effectiveness of the lamp as a spotlight. The inspection standard set is based on the number and size of "pinholes" discernible from strong light shining through the reflecting surface. The lamp A in Fig. 8 is "worst acceptable," while the lamp R illustrates a reject.

Judging Color

Questions of color, shade, sheen, or dullness of paint or enamel are frequent sources of dispute in factories as are the appearance of nickel,

Fig. 8. Inspection samples. Lamp A is "worst acceptable"; lamp R is a reject.

chrome, and other electroplated finishes. In other industries there is often to be settled the question of similarity or difference in color or shade of goods coming from separate batches or dye mixes. Some people are the opposite of color blind; their eyes are quite sensitive to slight differences in shade. From experience and native ability they become quite expert judging colors. For the ordinary person, however, a couple of simple rules may lead to more accurate, consistent, and dispassionate decisions.

In judging color, sheen, or drabness, the matter of light is important. If a sample from a painted, dyed, or plated batch is decided on as a standard, then subsequent test specimens should be viewed under exactly the same amount of light as the standard. A piece first viewed under a 50 watt lamp can look different under 200 watt power or under fluorescent light. Window light may give it another appearance. The sample piece should be held each time at the same angle and at the same distance from the same power light as the standard specimen.

Because of oxidization, chemical action within pigments or plating, or the accumulation of films of dust or moisture, the appearance of a standard specimen will change in time. Usually, at least, the color or sheen becomes duller. Hence, where painted, enameled, dyed, or plated surfaces are to be judged, the standard specimen should be renewed frequently.

Artificial aids help at times. The familiar ghastly beam cast by a mercury arc lamp is especially helpful in analyzing the condition of a polished, buffed, or plated surface, although some inspectors swear by "daylight through a north window at mid-morning of a cloudless day." A high powered glass aids decisions concerning dullness. The system is a little like taking a blood count. By roughly counting, in the field of a microscope, the minute scratches, pits, and irregularities that show on a "lively" surface and then making a similar rough count on a duller surface, it will be found that the count of scratches, etc., per unit of dull surface is much higher than for a lively surface. The microscope count can be established as a standard, as a method of establishing the line between dull and satisfactory finishes.

Standards Applied to Phases of Manufacturing Other Than Appearance

Where — as in America — so many products are mass produced, local, national, and industry-wide standards have had to be adopted for a great many items in order to prevent confusion, duplication, and unnecessary manufacturing and assembly losses. The more standardization of certain types of parts we have, the more readily, for instance, service and repairs can be offered on such products as automobiles, vacuum cleaners, and plumbing fixtures. The nationwide job of standardizing many common products is far from complete but new standards are steadily being added and old standards are being improved.*

Screws, nuts and bolts offer an everyday example of standardization. For instance, you can find in most mechanics' handbooks the American Standard for No. 10 round head wood screws. A table in the handbook will indicate the prescribed diameter of the head, the width of the slot, the diameter of the screw under the head, etc. Certainly, the inspector should have access to one of the standard mechanical handbooks, such as *Machinery's Handbook*, and study it enough to know where to find information pertinent to the products manufactured in his own shop.

As an example of handbook data on a type of established standard appears in the table below. This is presented merely to substantiate the suggestion that for many occasions, where some standard is under dispute in the inspector's shop, the best available answer may be found in a handbook.

---

*A catalog listing all American National standards can be obtained from the American National Standards Institute, 1430 Broadway, New York, NY 10018.

Snug-fit Tolerances and Allowances (inches) —
American Standard†

Closest fit which can be assembled by hand. It should be used where moving parts are not intended to move freely under load.

Diameters	Tolerances				Min. Allowance	Max Allowance
	Hole+	Hole−	Shaft+	Shaft−		
1/4	0.0004	0.0000	0.0000	0.0003	0.0000	0.0007
1/4	0.0005	0.0000	0.0000	0.0003	0.0000	0.0008
3/4	0.0005	0.0000	0.0000	0.0004	0.0000	0.0009
1	0.0006	0.0000	0.0000	0.0004	0.0000	0.0010
1 1/4	0.0006	0.0000	0.0000	0.0004	0.0000	0.0010
1 1/2	0.0007	0.0000	0.0000	0.0005	0.0000	0.0012
2	0.0008	0.0000	0.0000	0.0005	0.0000	0.0013
2 1/2	0.0008	0.0000	0.0000	0.0005	0.0000	0.0013
3	0.0009	0.0000	0.0000	0.0006	0.0000	0.0015
4	0.0010	0.0000	0.0000	0.0006	0.0000	0.0016

†Taken from *Machinery's Handbook*, 14th Edition, The Industrial Press. See later editions for the latest standards information.

Handbooks and standards, including elaborate scales and measuring mechanisms for manufacturing standards, are cropping up everywhere. With the influx of new technology and the ability to turn almost anything into continuous or number-based data, it is now possible to compare across measurable spectrums that were once only a twinkle in the eye of the master inspector.

A good example of this progress to better definition of requirements, and the ability to measure definitively on a large scale or range of variation in a previously uncharted environment, is that of contamination control and measurement. Once not too important, environmental cleanliness is now crucial to the production and operation of many products and assemblies.

Clean or controlled levels of atmospheric or surface level cleanliness can now be measured and specified accurately — all for purposes of control that grew from a new need — a need to specify a realm that was previously unmeasurable and, therefore, uncontrollable. Standards such as these are becoming available wherever there is a need — from computer environments to the newest fields of biological engineering, the need is always there to measure where no measurement has been done before.

Allowable Taper, Out-of-round, or Eccentricity

Mention has already been made of standards required for allowable taper, out-of-round, or eccentricity. It is just about impossible to manufacture parts without some degree of these digressions from the perfect geometrical shape. In fact, it is usually unnecessary and uneconomic to try to eliminate them entirely. Hence, standards for allowable taper, out-of-round, or eccentricity should be established in every shop and, where necessary, for various individual parts. If the type of machined piece can give trouble at assembly or in shortened running life of some product because of one of these digressions, it is better to have the engineering allowance for it shown directly and concisely on the blueprint.

But where the inspector is unable to secure any ruling as to the amount of taper, out-of-round, or eccentricity allowed on acceptable components, he can frequently safely establish the rule that this amount may range up to half the published tolerance for the particular dimension. Some plants have a rule that these digressions may equal but not exceed a half-thousandth on any dimension whose tolerance spread is a thousandth or greater. A rule of this nature covers the evidently ridiculous situation where the tolerance allowed is perhaps − .010 inch and taper, ovality, or eccentricity of .005 inch (following the half-the-tolerance rule) would be evident even to the naked eye.

Interpreting Blueprint Tolerances and Limits

Tolerances on blueprints are, of course, definite, published specifications, but there are occasions in many shops where a little escape from them as rigid boundaries is condoned. Where such a policy prevails, the interpretation of tolerances calls for a shop "standard." Suppose the print calls for .435 inch ± .001 inch. Will this mean that a piece which measures .4361 inch — only .0001 inch oversize — will be rejected? In some shops the answer is firmly yes. There, the operator would do well to manufacture to a .4359 inch upper limit. In other shops some digression even beyond the tolerance limit is permitted. All right then, if so, what is the line? If .4361 inch is permitted, why not .4362 inch or .4363 inch? Or even .4365 inch? If the local practice is to condone some indefinite digression from published tolerances and if the inspector simply cannot get those in authority to establish a "standard" for such digressions, he will be forced to set his own standard. In the case mentioned above he might set, for example, his own rule that any work digressing beyond .4362 inch will be rejected, or, using the half the tolerance rule, he would accept up to .4365 inch but sharply reject anything larger.

If the situation implied above seems illogical, how many human compromises are purely logical?

Background Knowledge Helpful to Inspector

To a very considerable extent, standards cannot be intelligently established without some knowledge of how and where the part, subassembly, or material is going to be used. One of the best lessons for a floor inspector is a few hours work as an assembly hand, using the parts he customarily inspects. Usually, where the workers and inspectors in a factory department have only an obscure idea of what will happen to the parts after they leave their hands, the standards of workmanship are established at one extreme or the other — either too severe or too slack. This is apt to be true where the manufacturer is "subcontracting" work, which is shipped to another plant, parts that disappear into some product the other fellow half way across the continent is making and marketing.

In this regard the inspector should be objective and practical. An assembly operator transferred to a parts department as an inspector will at first, consciously or unconsciously, try to tighten up the standards, inadvertently interrupting or slowing down production and increasing costs. When an inspector's mind fastens on standards, his primary reaction is to demand higher, better, tighter standards. From the thoroughly practical point of view, however, he should perhaps deliberately reverse himself and consider the possibility of existing standards being unnecessarily severe. If he will counter a natural tendency for perfection with an open-minded analysis of what the situation really demands, he will usually reach not only a balanced conclusion but probably a more correct and practical opinion.

In one department of a factory making small, portable, pumplike machines for compressing and straining liquids, short lengths of pipe were cut off and threaded. The work was done in lots of several hundred at a time as the stock inventories required replenishing. One length of pipe was used inside the machine where it was concealed from view. Another of the lengths of pipe appeared on the outside of the aparatus to be used purely as a handle to carry the portable pump from place to place.

The floor inspector rejected lots of the internal, concealed pipe, objecting to chuck and tool marks on the surface, but overlooked the burr set up in the mouth of each piece of pipe when it was cut off. Inconsistently, when the same machine made a lot of the very nearly similar pipe lengths that would be assembled for handles, he paid no attention to objectionable external tool marks but insisted that bore burrs be carefully reamed out — in this case by hand. In other words he had the essential inspection standards completely reversed. The surfaces of the pipe lengths for handles should have been smooth and of good appearance; the bores of the handle pieces could have been plugged solid with burrs and chips since no liquid would flow through them. On the other hand, all sorts of surface tool marks on the internal pipes did no harm (provided the pipe lengths were not actually weakened) while the bore burrs gave a great deal of trouble in the field.

Standards Should be Consistent

If the situation just described seems somewhat improbable, only a little experience at inspection on almost any shop floor will bring to light many examples that are comparable, if not more ridiculous. One lesson can be learned here, however. Standards should be consistent. It might have been wiser in the case of the compressor pipe nipples to have required the same general standards for external appearance and bore burrs whether the pieces were to be used to conduct sludge or as handles. The matter of habit should be considered. To let word out on a shop floor that type A pieces can be run off without regard to quality but that great care must be used on the considerably similar pieces of type B, because the latter are to be used for certain special purposes, creates a sort of confusion. The operators get used to slashing out type A pieces and the slack habits established carry over inevitably into the manufacture of type B pieces. It is like "company" table manners. If we allow slurping coffee from the saucer at home, we shall be awkward and unnatural about holding the coffee cup by its handle and properly curling the little finger when we go to dinner in polite society. And probably unconsciously make a slurping noise!

The more uniform and consistent, but practical, manufacturing standards can be across the board, the better the overall grade of workmanship and, frankly, the steadier and higher the production.

An example of the good effect of an inspector's knowing something about the use of a product occurred in a factory manufacturing textile machinery. Certain long rails or beds on spinning frames were being machined. Machinists carefully trained through apprenticeship and years of work felt, rightfully, that the rails displayed expert workmanship. From the point of view of a machinist they did — square corners, long surfaces smartly machined free from waves, chatter marks, and blemishes. The inspector, however, nosed around the test floor where the machines were tried out under textile mill conditions. He noticed that the fine threads being spun and spooled snagged every so often, snarled, knotted, and broke, as they happened to dip and touch the expertly machined rails. Every time this happened the spinning machine operator had to clear out the snarls, lint, and slubs formed and tie a knot to start the spooling again at that station. The trouble was cured by breaking the sharp edges of the rails and by literally buffing them with a portable polishing wheel.

The inspector should continually keep looking at the products going by him with the eyes of the customer. Considering sensibly the price, competitive standing, and use of the products, would he buy them without grumbling over them? Is the workmanship as good as it was three months ago? From time to time you buy a certain brand of canned corn, hack saw blades, or socks basically because their quality is uniform and consistent. Part of an inspector's responsibility is to secure, through his influence on the situation, a desirable uniformity. He must not sanctimoniously tighten up on standards one day and carelessly ignore sloppy work another time.

Catching the Unusual Defect

He is faced, too, with the occasional, the unusual, and the intermittent digression from a product standard of appearance or workmanship. Some undesirable deviations may not appear more than twice a year and then only on a few units of the product, but the inspector must be alert. Painters were spraying the ceiling and walls of a factory passageway. A trucker passed through on his way to the assembly department with several skids of newly enameled black frames. At the end of the assembly line, a little later, an inspector caught six of the household gadgets, dispersed among a day's output of several hundred, dotted with tiny blobs and spatters of white paint. A plastics manufacturer suffered a rash of warped covers which was traced to a single hour's moulding, a trouble that had never occurred before and that never appeared again. Aerospace manufacturers have adopted the term "gremlins" for inexplicable, erratic troubles. When the heat is turned off in the spring and summer's humidity appears, rust will suddenly show up, seemingly in less than an hour's time, on all iron and steel parts unless they are constantly treated for rust prevention — a seasonable difficulty. The list of "sabotaging" gremlins is legion: blisters, flaking, ripples, blow holes, soft spots, broken braid, corrosion, smudge marks, tiny cracks, foreign substances, grit — things that occur only once or so seldom that the causes of them cannot be readily traced or in any manner assigned to the regular process. Discovering the unusual and abnormal is part of the inspector's job.

Discovering the problem, or the unusual defect is the easy part, and something that goes on every day at manufacturing facilities. The difficult or unusual aspect, and a prime responsibility of inspection personnel, is to understand, report, and deal with the new situation so that the *system* will benefit, not just the current inspection operation.

After the identification, the impact of the change should be assessed, using data and measurements (or at least detailed reports from the inspector), and enlisting the aid of support engineering groups to evaluate the situation. Perhaps a special test needs to be run to assess the change, to understand the impact. Following this data analysis and problem solving session, the solution must be implemented. If the solution means changing the inspection criteria, and communicating those changes, the follow up to this detail is as important as uncovering the variation.

The Inspector Should Not Establish His Own Standards

In discussing tolerances, in an earlier chapter, the implication was that tolerances are established by design, engineering, or similar groups in a plant. In other words, an inspector would not be expected to say what the tolerances should be for a certain type or piece of work. Exactly the same practice should be followed with regard to general standards, although unfortunately such is not usually the case. No inspector should write or establish specifications, tolerances, or standards, on the theory that a man cannot fairly judge his own legislation. We all have read of the situation in some villages where one man as, first, selectman, voted a speeding ordinance and then as constable arrested a speeder and finally, assuming his role of local magistrate, decided on the guilt of the hapless tourist.

From a practical viewpoint, the inspector who sets tolerances and standards will almost inevitably either judge work too severely or too readily hedge from a previous decision. It is easier to stop smoking when the doctor orders it than it is to live up to your own New Year's resolution. The inspector's observation of work and his judgment, even in borderline situations, is much more objective, dispassionate, and fair if standards have been established by someone else. In common industrial practice he is not often required to establish tolerances or similar definite specifications, but more times than not he is looked to for setting and demonstrating standards. Therefore, he should, as far as possible, arrange for having each and every standard he needs to use established by a supervisor, by some sort of committee action, or by the equivalent engineering action. Even though the inspector should be in a good position to recommend a standard, the opinion of someone separate from routine inspection or production should also be obtained as to its practicability and value.

When the Inspector is Overruled

Inspectors, even the best of them, are subject to being overruled. At any time one of his decisions may be reversed by someone of greater authority. The fact that the work is okayed after an inspector has rejected it may not stem from any lack of skill or judgment on his part. He may be overruled because of conditions beyond his control — the fact, for instance, that sheer production requirements or economics demand the acceptance of definitely below standard work for a temporary period. Or the work may be borderline, anyway, and one man's opinion is thought to be about as good as another's. Changes in tolerances, specifications, and standards may have been authorized without the inspector knowing it. Human inconsistency, whimsy, snap decisions, or some form of what might be called "local politics" too often play parts in the overriding of inspectors' decisions. Again, the decision may be to try to salvage the rejected work.

The reason, however, this discussion is brought up here is to make two suggestions to an inspector. One is that where the reversals are occurring too frequently and too consistently on any particular item, the inspector should make every effort to get the standards officially changed — usually broadened — to accommodate the line of thinking in vogue. The other recommendation is that the inspector must make clear-cut decisions. If he frets over potential reversals of his own findings, he soon reaches a state of mind where he is utterly unable to make close decisions. The act of judging conformance should be fenced off from the function of disposing of the goods afterward.

In judging conformance the successful inspector attempts first to secure the facts. Secondly, he tries to be consistent. What was rejectable yesterday is rejectable today. Finally, he makes decisions based on the situation at hand as independently as possible of contingent or extraneous conditions.

CHAPTER 5

Basic Procedures

In the previous chapters we have discussed the role of inspection in industry, as well as basic inspection criteria and the fundamentals of assessment or inspection methodology. The subject of inspection methodology, or inspection *procedures*, is an extensive and rapidly changing field of application. In this chapter, different types of inspection procedures or formats will be briefly reviewed, contrasted, and brought to light.

This chapter deals with procedures in a fashion similar to topics from other chapters — standards of measurements in industry are also held by the quality and inspection industry itself. The tradition of inspection involves many tried and true methods, such as 100% screens of some variables, as well as modern standards of inspection, such as sophisticated sample plans and statistical evaluation models, which are designed to make decisions based on a minimal input.

The science of inspection and quality is rapidly changing — industrial competition is driving inspection to make better and more accurate decisions with less direct inspection, at decreased costs. This rule of the marketplace drives an evolving science, which ensures less variation through more information from less inspection.

Inspections cover all types of testing — electrical tests, hardness tests, spray tests, trial runs, tests for noise and vibration, accuracy tests, and many others. Such inspections may be only visual. They may involve only gaging. Many times both operations are combined.

The type of test or inspection may be purely functional. In such a test the product is actually used, run, worked, or stressed at inspection, in about the same manner the customer will use it, or it will be tested in the way it will enter an assembly. Running a ring thread gage over a screw is essentially a functional test, although screwing on the actual nut that will eventually assemble to the screw would be the ultimate and purely functional inspection of the screw.

Many inspections are semifunctional. A number of gaging operations fall into this class, the gage taking the place of the mating part. The final factory test on a vacuum cleaner includes wattmeter readings and a dialectric test to ground, plus coupling a vacuum gage to it; these tests do not simulate actual working conditions but indicate, from previous engineering laboratory experiments, that the cleaner will do the work expected of it without breakdown or shock hazard.

The keys to such inspections are relevant performance criteria that match or exceed customer requirements applied in a uniform manner so that two things can happen: early failures in the life of the product can occur within the manufacturing plant (not in the customer's house), and problems of early failure can be fixed guided by the data that indicate their failure.

A hardness test will indicate whether or not a heat treated shaft, for example, will withstand wear from friction or abrasion because there is a definite relation between the hardness of steel and its resistance to wear.

Whatever the test used or the method of assessment that is employed, it must reflect the criteria that are being evaluated, and these criteria must be what the customer wants to buy, at the right cost. In other words, the hardness test must not only evaluate the expected level of hardness, but that level of hardness must be the proper level for the market niche of the product.

Inspectors, of course, will not make that assessment on their own, but it is clear that the demands of the inspection environment, or procedure, are essential to many other people doing their job well, such as marketing. The key is communication, and inspection personnel must be as equipped to translate and transmit the results of their assessment as they are in performing the measurement work. Just as importantly, as the marketplace demands change, the system must exist to support inspection by setting revised requirements to the shop floor as quickly as possible.

Many inspections and gagings are made to prevent waste, scrap and lost time at subsequent operations or assemblies. If a shaft is rough turned to too large a diameter, the following grinding operation will

- Define problem
  define critical measurements

- Operationalize measurements (inspection)

- Inspect — take measurements

- Analyze data

- Identify causes of variation

- Find solution to variation
  fix problem

Fig. 1. Variation improvement cycle.

require too much time for taking off unnecessary metal. If the shaft is rough turned too small, it is scrap; the grinding wheel does not put on metal.

Historically, inspections probably started with the complete observation, test, and functioning of each unit of the product just before it was packed for shipping, a shop routine usually named final inspection. Then inspections moved back into and through the shop to the several locations where detailing or screening seemed necessary. The final step was to subject purchased components and materials to extensive inspection.

This type of inspection relies on a thorough screening of the product, or components or assemblies of the product. As information is gathered on the goodness or unacceptability of the unit, good units continue, and the bad ones are sorted or put aside. Then the bad units are evaluated, and manufacturing personnel try to figure out why a certain number of parts need to be scrapped, and perhaps this cause may be fixed, after the fact.

A more modern approach to inspection involves perusal and inspection of the manufacturing process itself — to measure the *source* or product variation, as it occurs, for purposes of immediate feedback

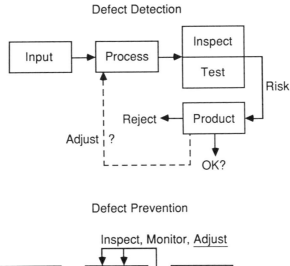

Fig. 2. Defect detection and prevention.

and resolution, within the process itself. The result of a well-tuned inspection system that is keyed to primary variables on a real-time basis is less scrap from fewer bad parts, and a process that is understood and controlled because *it* is the object of the inspection.

Today, the *prevention* of problems or nonconformances is deemed more important in industry than the *detection* of defects. Consequently, the inspection of the key *process* points, or evaluation of the actual performance of the process, is critical to a controlled and efficient manufacturing operation. All inspection data, whether in-process or any other type, must be aimed toward this purpose; the control and reduction of critical variation.

Feedback from processes regarding the performance of the process can be derived from product inspection data — both variables and attributes data — it can come from measurement of the process directly. Either way, inspectors or manufacturing personnel are inspecting work *as it occurs*, and can often identify problems or offer causes or solutions before the situation becomes critical or costly.

New types of detail, then, are important to setting up and maintaining systems of defect prevention. These details are used by many manufacturing groups and disciplines to ensure that process changes or corrections are instituted, and that they continue to ensure adequate quality levels as measured by inspection techniques.

Manual Inspections

Manual inspection is a term derived from comprehensive, product-oriented inspections, in contrast to automated inspections of the product. But principles of manual inspection also apply to assessments and measurements that occur on either the process or the part, close to the point of operation in the process, using a real-time or on-going measurement and recording system. In this sense, general considerations to industrial engineering techniques apply in such a manual set-up and operation, whether it is product or process-centered.

Whatever type of inspection is used, whether it be a complete manual *or* automatic gaging setup, or whether the inspection is a simple visual assessment of a sample of cabinet finishes on an appliance, the rules are the same — have an organized proven plan for the inspection, and follow it with discipline as long as it serves the needs of the appropriate inspection.

From a process point of view, the location of the inspection itself, whether it occurs in an isolated receiving inspection area or out on the line, affects the physical setup of the continuing inspection, and in many cases can be a prohibiting factor in an efficient setup. Therefore, the area of the process where the inspection occurs is critical in proper inspection planning and consideration.

The first step in manual inspection is practical preparation. The setup for an inspection is just as important, from the point of view of efficiency, as the preparation for a machining operation. The principles of work simplification* find an immediate application in most inspection work.

Any inspection department that lacks expertise in workplace layout will do well to consult with the specialist in the Industrial Engineering or Methods Engineering departments for help in setting up an inspection work area. The object of the inspection operation is, perhaps, to check a certain diameter on a series of workpieces. The actual final adjustment of the piece on the gage anvil plus the flick of the indicator hand may mean only a matter of a second, but getting the piece *to* the gage and *away* from the gage is frequently another matter. In work simplification terms, the "make ready" and the "put away" might consume from 10 to 20 times the amount of time and energy used for the "do."

Work Arranged Properly Versus Improperly

A simple diagram, as in Fig. 3, illustrates a common error. Suppose the work is to lay on or in the gage as shown at *a*. Usually the rodlike workpieces would be piled on the bench neatly enough perhaps and accessible, but in the direction shown at *b*. Too many times they are piled over in back of the gage as at *c*. The inspector not only reaches for each piece but he must turn it, juggle it, to get it around into the *a* gaging position. Then just as frequently he deliberately turns it 90 degrees again so that it will pile neatly at *d*.

Suppose the work were piled as diagrammed at *e*, and repiled after gaging as at *f*. Several arm and wrist motions have been eliminated because the work can be slid straight across the gage. The work motions have been streamlined. Perhaps it would be easier to have the "put away" at position *g* so that the instant the indicator registered the measurement, the workpiece might literally roll off on to pile *g*. The arrangement shown in Fig. 4 is an example of good planning. Attention to such details saves a lot of human energy.

Many times inspectors seem to "cross hand" themselves. They persist in removing each piece from the gage with the right hand and reach over across the left hand to dispose of it. Or they reach to the

---

*Any inspector who can arrange to attend a course in modern work simplification will find he has added a worthwhile industrial asset.

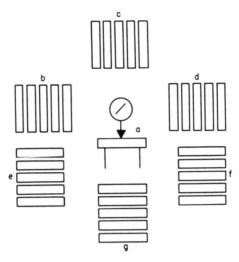

Fig. 3. Possible workpiece arrangements relative to the gaging device which may be used in an inspection procedure. Positions *b*, *c*, and *d* require extraneous hand motions. Positions *e*, *f*, and *g* reduce the work motions.

Fig. 4. An example of how a good handling set-up reduces the expenditure of energy in moving a workpiece to and from the gage.

floor with the left hand for the work, gage it, and again reach to the floor on the right to dispose of it. If inspection work were a matter of gymnastics or slimming exercises, many operations would be complete successes.

Concentrate on the Act of Measurement

The trick in manual inspection is to concentrate all the attention and energy possible on the measuring act itself and as little as possible on all else. Unnecessary hand, arm, and body motions are distracting they take energy, time, and attention that are better devoted to the work place. The more automatic the necessary gaging motions become, the better inspection job will be done. The inspector should never reach an inch farther or higher than necessary in order to pick up the parts he is gaging nor reach over or down in order to dispose of them. Attention to such details can convert a wearisome, even exhausting, day into an easier one and, in addition, save many minutes of time.

The proper position of the gage is essential, of course. If it is too near or too far back on the bench, too high or too low, aching shoulders may be the penalty after an hour or so of inspection. The indicator should be at natural eye level. If possible, perform manual inspections while seated in a chair of the proper height.

It is also desirable to locate manual inspections so that the inspector is not directly facing the glare of windows or unshaded lights. The same general sort of psychology applies in attempting to avoid a work area beset with the continuous vibration and thump of machinery or where the ventilation is poor. To perform manual inspections near employees who talk and chatter all day is extremely and unnecessarily tiring. Remember that, at best, repetitive manual inspections are monotonous; it takes enough of a particular kind of stamina to withstand the boredom anyway without adding unnecessary nerve strain.

In a timely, process-centered inspection sequence, an important point to plan for in manual inspections is the measurement of parts *as* they are produced. In this fashion, variation in parts can be detected as they occur, and related to the aspect of the "process" (machines, procedure, worker) that produced the variation, and the problem or cause can be more easily fixed. To sequentially inspect a pile of 200 parts for specification adherence will only result in a pile of good ones and a pile of bad ones.

Use Two Hands for the Workpieces

Where hand gages, such as micrometers, snap gages, and depth gages, are used the tendency is to hold the gage in one hand and pick up the pieces to be measured with the other. One hand, then, is acting simply as a vise. It is almost always possible to secure or devise simple clamp stands to hold the gages. Then both hands are free. Some shops equip gages with foot pedal devices for opening and closing the jaws.

The inspector should train himself to pick up the pieces with one hand and apply them to the gage. In the meantime the other hand is either picking up the next piece or disposing of the last one. Good manual inspection displays rhythmic, alternate come-and-go motions of the hands.

In addition to efficiency gained, there is another good reason for having a gage clamped in some sort of stand. Where a gage is held continually, the heat from the hand is transmitted to it and as a result of internal expansion the measurements may become inaccurate.

Most inspection and gaging apparatus is designed to readily accept the piece, part, or component in the correct manner. The inspector should avoid twisting, cramping, or misaligning the work in the gage. This means he first must have been well instructed in the use of the gage. Second, he should relax; let the gage do its own work.

Also, inspectors should be cognizant of contributing wear, damage, or critical contamination to parts and assemblies that are inspected. A major potential source of problems with components or assemblies themselves lie with the inspector's involvement with those parts — if the inspector follows a safe and methodical routine with the work-pieces, problems will be avoided and costs will be cut. And, of course, this routine must reflect special precautions that may be used in order to ensure the integrity of the part or component after it has been inspected.

In fact, the advantage of assemblers or operators as inspectors here should be obvious — no one knows the process, and the problems of the process, better than the people who have to perform the process. The advantage of inspecting work for proper conformance to specification as it is produced is derived from the availability of the operator in respect to the cause of the measured variation.

Avoid Three Types of Errors

At manual inspections three common types of errors should be avoided. The first, *parallax*, is due to reading a scale or dial at an angle so that the reading seems to be higher or lower than it actually is. Position the gage directly in front of the eyes.

The second common mistake is known as *rounding off*. For instance, the reading for a diameter might land between .758 inch and .759 inch. The inspector would read, say, .758 inch. In individual cases, the error might be trivial or harmless, but in repetitive inspections, especially where precision is required, rounding off can lead to unjust rejections or to the acceptance of substandard work. The answer to rounding off is to secure gaging equipment with the required discrimination.

Another error is caused by what is called *flinching*. It is the deliberate practice of giving the product the benefit of a doubtful reading. If, in the example above, the .759 inch reading would bring the part within specifications, some inspectors would decide on the .759 inch. Or vice versa. Both rounding off and flinching are slack traits, which unfortunately tend to grow worse until the habit becomes so fixed that practically all of the inspector's readings are undependable.

Use of Multiple Gages

In modern factories, attempts are made to reduce much of the effort and time used in manual inspection operations by the adoption of multiple gages, which are illustrated in Fig. 5. Multiple gages are made

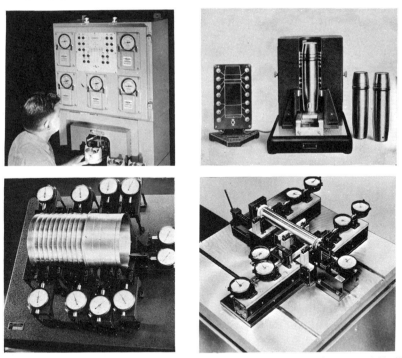

Courtesy of Sheffield
Courtesy of Colt Industries, Pratt & Whitney Machine Tool Div.
Courtesy of Federal Products Corp.

Fig. 5. The use of multiple gages that measure several dimensions simultaneously reduces the number of handling operations required for complete inspection of a workpiece.

that register simultaneously anywhere from two to a dozen dimensions at once. At first thought, the inspector would seem to be thrown into a quandary as to how to read correctly several dials staring at him from all angles.

For the first few pieces gaged, his eyes *do* travel around the circuit of indicators as they did when he mastered the gage originally. Usually when the gage is mastered, gage adjustments can be and are made at each station so that each indicator registers the correct basic dimension and each indicator hand is pointing in the same direction. As the workpieces are placed in the gage, the indicator hands then take different positions, since few workpieces ever duplicate the master dimensionally. However, the inspector soon gets used to the indicator dial sectors within which the indicator pointers would ordinarily come to rest when the workpiece dimensions are within tolerance. In fact his eyes become so accustomed to the appearance of the gage's indicators with in-tolerance pieces that they immediately and readily register any indicator reading falling outside the normal dial sectors. Frequently, the indicators are equipped with so-called tolerance hands or masks. As long as the indicator hands flick inside the tolerance hands, no impression reaches the attention of the experienced inspector. But where a reading shows outside the tolerance sector, his eye and consciousness immediately register the discrepancy in the overall appearance of the gage.

In recent gage designs, electric lights, digital readouts, or bull's-eyes have been added. Figure 5 shows two examples of this sort of equipment. When the gage itself registers any one, or several, dimensions as out-of-tolerance, a bull's-eye lights up, usually a red light. The human eye, of course, detects a lighted bull's-eye more quickly and easier than it can the position of an indicator or meter needle. On some designs, the refinement is added of having an amber light signal undersize, a red light oversize, and either no light or a green light for correct size. (It is equipment of this type to which bells or buzzer tones can be added to direct a blind person employed at manual inspection.)

There are many occasions where manual inspection seems better performed where the inspector is standing. The weight and sizes of the workpieces may preclude sitting down. In general, if the workpiece is considerably heavier than the gage, it is easier of course to bring the gage to the work than the reverse. The work may be traveling along on a conveyor or for production reasons it may be lined upon a bench. But, unless the manual inspection job is temporary or intermittent, it is far better to rearrange the mechanics of the situation, somehow, so that the inspector can be sitting. Anything that tires an inspector unnecessarily, like continuous standing in one place, only drains away from the energy, concentration, dexterity, and accuracy he requires for performing the consecutive measurements he is expected to make. Chutes, belt conveyors, high stools, or similar aids surely can be planned for the inspector who is to do detail work all day long.

Visual Mechanical Inspection

One of the more common inspection operations is the examination of finished products, parts, components, and materials for visual defects. Appearance is important as well as function and various degrees of mechanical perfection. The rule applies to automobiles or wrist watches, to carving knives or rubber boots. Wire and cable in reels, paper on rolls, cloth in bolts, rubber, plastics, table salt, or packages of cigarettes, all must have a certain eye appeal.

Visual inspections are also used to detect mechanical deficiencies as well as lack of merchandising appearance. A burr on a press part may prevent ready assembly; a blow hole in a casting is not only unsightly but very often weakens the piece structurally. Neither are readily measurable in the sense, say, of gaging a length, outside diameter, or thickness. The color of brass forgings can indicate whether the brass is too hard and brittle. Bubbles and waviness in blown glass distort the television picture. And how can the strength, security, or completeness of soldered electrical connections be quickly measured better than by an experienced eye?

Somehow, the allowable extent of such a defect — like porosity in castings — must be quantified. The detail of this defect, and when such a defect is indeed considered to be rejectable, must be described and outlined in objective terms by Quality or Support Engineering groups. Of course, to eliminate the problem would be preferable, but, if that is not possible, then an accurate description of that "gray" area, where acceptable meets rejectable, must be defined so that the producer and the consumer (and the customer) all get what they pay for as they work their way toward cost-effective improvements in the process that created the porosity.

It is impossible in a book of this size to name or describe more than a very few examples of visual inspections. However, the fundamentals and principles connected with the mechanics of routine, repetitive visual inspection apply practically throughout.

Knowing What to Look For

The first thought of anyone about to start a visual inspection should be about what to look for. The conditions, the appearances, and the items or elements governing rejection or acceptance of materials, parts, or products — the standards in other words — are established in industry by the engineering department, by the laboratory, by the sales or service departments, frequently by the chief inspector or quality manager, occasionally by the manufacturing departments, and sometimes out of sheer necessity by the inspector himself.

As a prerequisite for effective visual inspection, an inspector does well to secure, somehow, enough time to become acquainted thoroughly with the manufacturing process through which the goods he is inspecting are routed. To know more or less exactly what causes the sort of visual defectiveness he is supposed to look for is a big help. His study should probably go back to the operations immediately preceding the inspection for underlying causes and conditions. The final trouble may start from the condition of materials and parts coming in the receiving door. Scratches, pits, scale, and tarnish on silverware pieces can be traced many times to the original sheet stock; slubs, snarls, knots, and raveled yarn in high price worsted suitings originate mostly at the cards and spinning frames.

In a similar manner, the inspector should know where the parts or materials he inspects are going. If he is inspecting the interiors of tire valve bodies for cracked and split fiber washers, he will appreciate the reason for his inspection work more after he has suffered from a slow leak and a flat tire on his car. A period of work as an assembly operator helps. Salesmen make meticulous inspectors because they have so many times faced the complaints of irate customers over substandard and shoddy merchandise. The inspector in the woodworking shop soon learns what to look for after he has worked in the paint shop.

To establish some of the basic principles and techniques of visual inspection, let us use as an example some small, simple metal ferrules shaped something like, say, fountain pen caps. We could consider any similar sized machine or plastic parts for the purpose. Or gaskets, glass marbles, paring knives, playing cards, bronze bearings, ad infinitum.

Preparation for Visual Inspection

The first step in preparing for a visual inspection task has already been suggested. Be sure that the extraneous physical circumstances surrounding the job have been made as comfortable as possible. Chair, bench, or table should be of the right height. Light should be adequate with no direct glare. The inspector should not be facing extraneous light as, for instance, a row of windows over beyond his work place. The remarks about vibration, noise, neighboring human chatter, and poor ventilation already brought up in the discussion of manual inspection apply here also, of course. Freedom from drafts and other forms of personal discomfort is essential. Visual inspections ought to be located away from through aisles if possible. The intermittent rumbling of hand trucks and shop tractors going by is distracting, as are interruptions of passersby.

Remember that basically only the eyes are needed to do the work of visual inspection. Try to eliminate everything and anything that delays or distracts them. Remember too that, basically, there are as many decisions to be made as there are pieces in the lot being inspected. If the inspection is proceeding at the rate of 1000 pieces per hour, 1000 decisions per hour are made. As each piece passes before the eyes, a decision is made that it is acceptable or rejectable.

Assuming that adequate forethought has been given to reducing general physical discomfort and strain to a minimum, several practical steps can be taken to lower eye strain and nervous tension. The trick in continuous, repetitive visual inspection, strange as it may sound, is never to see, almost literally, the good work — never to be conscious of it.

Reporting and Follow-up to Visual Inspection

Once again, it must be stressed that the reporting of the results of a visual inspection, and the subsequent follow-up to these results, is just as important, in fact more important in the long run, as performing the inspection properly. Inspection at a high rate of speed means that the results of the inspection need to be available almost immediately — preferably on a "real-time" or "as-it-is-done" basis. The availability of these results will allow the manufacturing operation to adjust or make changes as discrepancies are produced, before a large pile of scrap material accumulates.

From this standpoint, adequate preparation for timely recording of the inspection results and some painless method to convey these results to support engineers and managers are needed to make the inspection worthwhile. As much time and planning as that devoted to the actual inspection has to be afforded to the data collection and reporting system. Many times inspectors will breeze through a number of production units, properly sorting good from bad, and then go on to the next task, only to realize that no data — no information — are available. Or the unaware inspector may record results well after the fact, relying on memory to commit the details of the inspection to the record. When this is the case, valuable information is lost, and corrective action becomes difficult or impossible — the inspection produced only a pile of good ones, and some bad ones. This scenario will be repeated.

Data collection requires much time and forethought — whether done with paper and pencil or with computers and automatic input as recording tools. Whatever the means, the concerns are the same as those used in inspection routines — the method must be understood and flow easily without a large, additional investment in time, and the results must be verifiably accurate and readily available.

Basic information must comprise the inspection data, such as the name of the operator/inspector, the station and the time. If attributes data are being sought and recorded — good or bad — then an accurate and timely count is needed of the calls, as well as a description of the defect and any notes that might be helpful in understanding the problem. Variables or numbers data — the real stuff of inspections — must of course be accurate and readable. The analysis of these numbers will guide important and costly business decisions, and the numbers also must be tied with useful information that will lead to an understanding of the particular variation at hand — from the inspector's point of view.

Therefore, the inspection routine today takes on a different twist — inspectors will spend more time working with support groups and engineers — the real-time flow of their data, and the unraveling of that data from a manufacturing systems point of view are very important parts of the inspector's work. Mere classifying and sorting is a thing of the past in manufacturing inspection, and will only bring benefits if the causes of discrepancies are understood and fixed by the system.

A very important factor in the success of rapid visual inspection is the use of effective inspection tools and fixtures to assist in the finer discriminations of the inspection. The "gray" of the gray area can become more visible if an inspection tool can narrow the area between the black and the white — for instance, many go/no go gages define the limits of an inspection measurement by homing in on the desired measurement from each side — the black and the white; the area bounded by both extremes becomes an easily identifiable gray area, and the inspector is relieved of having to agonize over the boundaries without the proper tools.

Sorting Slows Down Inspection

One of the monkey wrenches inspection management occasionally throws into smooth, skillful, visual inspection operations is the demand that the rejections be classified. Put the splits in this box (we want to get a report back to the press room) and the off color pieces here (they can be stripped and replated; besides, the plating room ought to know about it), and all the dents and bent lips over there (that new conveyor I griped about does all that damage). And so forth.

Attribute — Qualitative data	Variable — Quantitative data
Pass / Fail Good / Bad Go/ No Go	6.3 inches 0.0073 inch 0.64 mm 8.1 kg
Inspected 100 units 80 passed 20 defective units 25 defects	Inspected (Measured): 100 units Data: 100 numbers
Percentage of defective Number of defects	Average Range Standard deviation

Fig. 6. Inspection data types.

So long as an inspector merely needs to extract the bad, the defective, the shoddy, or the substandard from the good work, high efficiency in both speed and clean inspection is obtained. Add a classifying job, set the inspector to worrying about this particular kind of a defect and just that sort of scratch, and the production rate drops. Furthermore, the quantities of actual defectives that "get by" an inspector will increase.

It is far better for the inspector to work 10, 20, 30 minutes or 1 hour steadily separating the goats from the sheep, simply following the screen system. Throw *all* the defectives in one till. Then stop inspecting for a few minutes and make a special task of sorting the defectives into the required classifications. Incidentally, a few "good" pieces that accidentally slipped into the defect box will be saved but, by making a special task of sorting defectives, many fewer defectives will have previously slipped through into the good work. Some shops are wise enough to use "salvage" inspectors whose main duty is to classify defectives from reject boxes. In those shops the inspector assigned to routine visual screening (frequently on piece rate or incentive) has only a "good" box and a "reject" box — no set of classifying tills.

Sorting should be interesting to Manufacturing and Quality Engineering — sorting means classifying, and through classification of defects around good/bad/borderline conditions, dispositions of rework, scrap, or return to vendor can be established, and feedback to the appropriate processes can be instituted. The purpose of constructive sorting, then, is to provide more detailed information to the interested parties — that process responsible for the defective material.

Today, many industrial inspectors are involved in problem solving that is based on a regular review of their discrepent material — in as timely a manner as possible. Representatives from Manufacturing, Manufacturing Engineering, the Quality group, and maybe Materials and other representatives, typically gather to review the results of a certain time period's production. At this meeting, various members of the groups, including inspectors or line workers involved in inspecting parts and material, verify rejects and attempt to describe a valid cause and effect relationship between the defect and its source. Following this procedure, corrective action is usually ascribed that would prevent the problem from recurring. In this sense, the benefit of sorting and classifying defects is brought full circle to prevent the same problem from repeating.

If, however, the sorted discrepent material is *not* used for analysis of the source of the variation — to establish cause and effect relationships in the process that created the bad parts — the manufacturing unit will fail. Only those factories aggressively pursuing resolution of their problems, and more definitive, on-going process control, will succeed in today's marketplace.

Lighting Should be Specified in Appearance Standard

The use of light is important many times in establishing or displaying standards. One of the touchy subjects in visual inspection is the matter of finish, such as nickel and chrome plating, paint, and enamel. Color is another troublemaker, the exact shade of dye, stain, or lacquer, for instance, often being the subject of almost bitter dispute. Plated pieces may be examined under tungsten lights in the plating room and a decision made there that their "color" or appearance sets the standard. The same two pieces can look entirely different upstairs held up by the window in the boss's office. In fact the two may not even look alike in daylight. Take them over under the fluorescent fixtures and the high light intensity at an inspection bench and a third opinion might be formed. Incidentally, under the same light they may not seem the same a few hours or a few days later; perhaps they sparkle less. Factory dust, vapor, or oil film has already settled on the samples. It is good practice to wipe a visual inspection sample clean before using it as a standard.

To be completely technical about setting standards of abstract appearance or color, the same type and intensity of light source should shine on the work being inspected as shone on the example "master," or standard originally offered the inspector as a guide. Even the matter of angle of incidence is important. An attempt to delineate what is meant appears in the diagram of Fig. 7. If the master standard has been studied at position *a*, the work should be inspected at the same location. To examine the standard at position *b*, for example, and the work at position *c*, might produce different appearances or apparently different shades of color (because of the variable angles of incidence and reflection) and work might be either rejected or accepted in error. To maintain the exaggerated technical perfection that Fig. 7 suggests, the inspector's forehead could rest against a "stop" as at *d* in order to maintain the optical angles. Furthermore, the wattage of the lamp should neither be increased nor decreased after the master has been observed.

The above is of course cutting things pretty fine if the suggestions are followed literally, but if practice is sufficiently close to theory, many petty shop wrangles and errors over rejections may be avoided. But

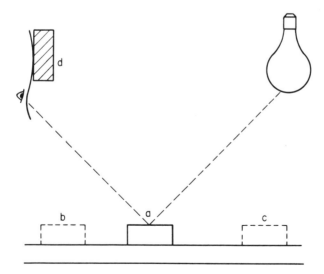

Fig. 7. When inspecting a workpiece for color, the same type and intensity of light should shine on the work being inspected as shone on the matter. The angle of incidence should also be the same.

these details are important to inspection consistency — what distance from the object to be inspected and what lighting should be present are basics to setting the inspection criteria and definition.

At best, any human, over a period of time, will unconsciously modify the standards theoretically set for visual inspection decisions. On strictly borderline work, what is acceptable today may be rejected tomorrow, or even an hour from now. Another time, the inspector is more self-righteous than slack and his standards are temporarily tightened up. All inspectors' opinions (and that is all that visual inspection of marginal work is: opinion) should be refreshed regularly with reviews of the concrete standards.

Improving Mechanical and Physical Circumstances

Recognition of the part that monotony, fatigue, ineptitude, and rhythm play in inspection work should lead to better training and instruction, improved mechanical and physical circumstances surrounding an inspection job.

Speed and physical handling, particularly of small pieces, in visual inspections can be aided by the use of inspection trays. Ordinarily these are metal trays shaped like a household dustpan, without a handle, but somewhat smaller in length and width. Using both hands the inspector scoops up a layer of the parts from the pile on a bench or out of a box and, by shaking and rolling the parts on the tray, detects the defectives and picks them out.

Inspectors may sit on each side of a belt conveyor and arrange for a steady flow of the parts, subassemblies, or assemblies onto one end of the belt. As the pieces travel by, the inspectors sample the work. Inspectors are stationed at the regular materials handling conveyors with the same purpose in mind. One example is the inspector working at the end of a bottle-washing machine picking off chipped, cracked, and dirty empties. In a somewhat similar manner, inspectors watch the output of continuous processes like wire drawing and insulating, the rolling out of plastic, rubber, or metal sheets, or they watch the fabric rolling off a loom. Visual defects are "rejected" by chalking the defective stretches of product or they are marked with a spray gun.

Generally, it is not cost-effective to use a single inspector to perform in-process product inspection unless he also has a role in producing the parts or assemblies in question. When this is the case, the operator's familiarity with the process will make that person a better inspector, and the inspection role of the operator will aid in raising the quality levels of the parts that are produced. The operator has always been the best inspector — legitimizing that role and providing feedback from the operator to the entire manufacturing system, which will not only identify problems sooner, but will ensure that a solution to those problems will be more easily implemented.

Inspection of Large Single Units

Many visual inspections cover larger single units like refrigerator doors, automobile fenders, and the like. Often the inspector is stationed directly following a continuous enameling or varnishing operation, checking off with crayon marks or paint blobs the unsatisfactory units. In textile mills the fabric is made to travel over special inspection perches in front of the inspector.

The sort of inspections just described involve those situations where, because of the area, size, bulk, or weight of the workpiece or material, it is easier to send the inspector to the work than it is to bring the work to the inspector. Many times, too, as in the case of conveyor belt inspections on smaller parts, there is the profitable matter of saving production and handling time in a manufacturing cycle by not bottle-necking the product for extensive inspections. But the inspector must not lose sight of the principles of visual inspection that apply.

Just because the inspection is performed somewhere out on the line is no reason why the lighting should not be correct and uniform for the inspection purposes. Following the reasoning illustrated by Fig. 7, the inspector should maintain a definite location in front of the work passing by (perhaps he should paint a target or "box" on the floor). Everything about him should be as relaxed as possible except that part of his body above the nose. He should take whatever definite steps he can to avoid or eliminate tiring physical distractions — drafts, noise, interruption — as much as he would at bench inspection.

The inspection of large finished units and aggregate assemblies is still quite common in industry today and these inspections are often carried out after the fact, by someone other than the person who made the unit. Such inspections are termed "product audits," because they are often performed on sample basis, with a highly structured formula for what is inspected against what is produced, since it may be difficult or too expensive to inspect everything.

In such an inspection, several things must be taken into consideration. The first is that these inspection routines must always be based on a clear and consistent routine — there may be 100 – 200 items to check visually on such a large inspection. When this is so, the inspection must always be systematic, and defined so that any one inspector will actually follow the same *order* of inspection as any other. In this way, time is regulated, and engineers who design the inspection routine or route will be assured that all relevant detail is given adequate weight or attention, precluding the possibility of an inspector dwelling on his favorite defect.

Also, these types of inspections need to be reported on and dealt with properly and tactfully, with an understanding that such inspection routines may well point to individual operators or assemblers as the cause of some defects. Of course, the cause for a problem may also lie in a tool or a process, but whatever the source of the problem, the inspection system must be set up to deal with it. In the case of an operator-caused problem, a system must be set up to verify that the operation was indeed the cause, and then a formal and acceptable feedback vehicle, involving the supervisor, must be invoked in order to correct the problem. Once again, without consideration to these important details, the impetus and advantages to such large-scale product inspection must be called into question.

As an example, consider switches or relay mechanisms. The visual specifications on one such mechanism called for the plastic container or shell to be free of burrs, chips, checks, or cracks. An inside wall or partition should not come from the mold broken through. The switch lever must be in the off position and an adjusting screw is to be tight and securely soldered. Certain lock washers are to be surely compressed to the point of flatness.

Here are a half dozen specific elements to be looked for on each switch, a situation where a single embracing glance will not suffice. The eyes must travel from item to item — take an inventory of each piece.

This operation, then, calls for the study of about a dozen types of defects, any one or all of which may happen to show up on any unit inspected. Some of the defects are fairly self-evident if present, such as the location of the adjusting screw; other defects involve close and critical examinations (is the adjusting screw really tight — securely soldered?).

It is significant to point out once again the difference between defects and defectives — for purposes of understanding a manufacturing process's true capability to produce a certain quality level of parts and for purposes of charting the progression of these defects or defective rates on control charts, which have found so much use in modern manufacturing as a tool to contribute to understanding process capability, and effecting process control.

Defects are indeed the problem that besets the product — the porosity on the shaft or the hole in the can or the crack in the bottle. Defectives are counted as the can with the hole or the bottle with the crack, and the difference is significant. Suppose in a report to my supervisor my shift allowed five defects to be produced Monday, and only three on Tuesday. Suppose again that my supervisor was so enamored with my group's progress that he told me to take Wednesday off, and on Wednesday the Production Control person told him that we produced one hundred units on Monday and only ten units on Tuesday. Obviously, Tuesday's defect rate far exceeded Monday's rate, but the boss did not realize that was the case until he understood *both* defects and defective (units). So the decision to follow or plot one or the other (or both) is important. Generally speaking, defects should be plotted or charted if there is likely to be more that one per unit of production, whereas following defective units is better if they are likely to contain only one (or no) defects per unit.

Establish an Order of Checking

The inspection of this part illustrates another set of worthwhile habits the inspector should acquire. As the eye and mind range more or less automatically over each piece, the items or elements making up the inspection should be taken each time in the same order. Look at the bakelite case, for instance, always first and then at the switch; next examine the set screw, the soldering, and finally the washers. Follow the same order, mentally and physically, with each succeeding piece. Don't look at the soldering first, on one piece, and then at the lock washers first on the next piece; in other words don't let the inspection procedure become a hit or miss jumble of glances like a grasshopper jumping around a field.

In some factories, a sort of operation sheet or typed set of specifications is supplied the inspector. Where it is not, an inspector would be wise to write up his own order of events, his own operation sheet, and review it, say, every day to be sure he is not overlooking an important item. But the world of inspection does not take place in a vacuum. Such a list can be devised, and it is essential that if an inspector instigate such a sheet, he should review it with appropriate support personnel, to ensure that the criteria are correct and desirable at all levels, and to ensure that other groups can have the benefit of increased inspection detail.

The inspection of a switch is, perhaps, simple compared to the inspection of an auto truck cab, for instance. Usually in an automobile plant the vehicle comes along the assembly line equipped with an inspection card, a check list in other words, and the inspector is required to put a penciled okay after each item as he completes each detail of the inspection. Sometimes there are over a hundred items listed — windshield, door latch, horn button, dash fuse, lock washers under sill screws, and soon through the innumerable list of things a truck driver expects to find, operate, and give no trouble.

Final Inspection

If the inspection procedure is not used or condoned anywhere else in a factory, it is liable to be adopted for the inspection and screening of the products made and sold, before they are packed for shipping. The inspector assigned to final inspection lives, in many ways, in a different climate and faces a different set of problems than he would likely find in inspections tucked in among various steps of the manufacturing process.

The primary purpose in the final inspection of a shop product is, of course, to make sure it is "right" before it is packed and shipped. The products passing final inspection are what the customer pays for. The examinations of components at process inspections are not under the same burden, for they may be reworked, salvaged, scrapped, or otherwise disposed of and some of them never reach the customer. People know of a company from magazine, newspaper, or radio advertising. While these sources may influence them favorably, their real judgment, in the last analysis, is based on the product itself, and the company's prosperity depends on the favorable reaction its products make on its customers. Hence, an atmosphere of tension, to some degree, usually pervades, the responsibility of the final inspection area.

Mechanically, final inspection normally can be divided into three classes of observations: visual, manual, and testing. In another sense, there are three other classes of attention. Foremost of course, the final inspector wants to be sure no blemish or defect or missing part will gain customer disapproval. Second, assuming all the components and subassemblies were satisfactory as each left its process or operation, the inspector looks in particular for errors or damage that may have been caused solely in the final assembly operations. Final inspection is usually the only check on assembly department quality. Third, he searches out hidden defects or possibilities of future breakdown of the product. For this latter purpose, the inspector may rely on special tests or on experience that develops an instinct for trouble or hidden substandard work.

The final inspection of many products consists totally or solely of visual inspection. Other products demand final gaging, testing, or manual inspection in some form. In the latter event, it is of course a matter of efficiency that the detail of visual inspection be done as the manual inspection is performed.

Final inspections are often 100 per cent inspections. Each and every unit of a product may be examined or tested. Some products, however, are so uniformly produced that sampling inspection can be used. Where a shop makes machine screws, for example, the inspection of a random handful of screws can designate the conformance of the entire order.

The final survey of other products involves what is called destructive tests. Certainly it would not be feasible for a confectionery manufacturer to bite into each candy bar to be sure of its goodness before it is wrapped and packed. Hence, sampling must be resorted to and depended on in many circumstances. Substitute or equivalent tests can be devised many times. One way of testing ball bearings might be to give them a "life" test — running each bearing for many, many days under nominal load. Commercially, however, they are run for a few seconds in a dynamometer vibration tester. If the needle of the tester swings over too far, indicating excess vibration, it is known from previous research that the bearing will not last as it should in ordinary service and it is discarded. This link between functional testing and product quality and reliability must be formally established and understood by all parties in the manufacturing environment.

More than anything else, perhaps, the visual inspection of finished products has to do with appearance, with the impression the product will make on the customer's eye. For this reason, the final inspector must know thoroughly his shop's standards for finish. In detail, he should become expert at distinguishing, for instance, between a lively and a lifeless painted or enameled surface and know what happened back in the shop to cause an insipid appearance. Similarly, polished and plated and buffed surfaces are either cloudy or bright. If the product has, for example, two or three different nickel or chrome-plated plates or strips on it (frequently only as a matter of decoration to lend merchandising appeal), one of the plates may be dull or cloudy and another brilliant. The contrast will give the product a shoddy appearance, and the reason for the trouble may be that the pieces were plated and buffed at different times in the shop, one of them in substandard fashion.

Rust, corrosion, dirt, and grease are taboo on finished products, at least to any appreciable degree; likewise dents, scratches, pits, digs, nicks, rub and score marks, and blistered and peeled spots. The inspector is usually instructed to look for sharp corners, knife edges, splinters, or any similar condition on the product that might damage the customer's person or property. The inspector is expected also to look for missing screws or missing parts. The assembly and shipping specifications of a motor, for example, may call for a Woodruff shaft key to be supplied and a set of toe bolts. Such parts, frequently contained in a tagged cloth bag wired on the motor frame, are easily forgotten by assemblers or get torn off. Other items include illegible lettering on name plates or the undesired presence of tape, metal strip, wire, or other material used in the manufacturing or handling of a product that have crept through on the components.

The object of an effective final inspection is to prevent defective units from leaving the manufacturing plant. One hundred percent inspection is not always an effective tool against producing defects, but many factories use this approach to ensure that "all the bad ones are caught." Usually the inspectors continue to "catch (most of) the bad ones," and Manufacturing continues to produce "bad ones." Nevertheless, this approach is still used, depending on the customer demands and the producer's performance to those demands, to measure and control outgoing quality.

Generally, the worse the outgoing quality is, the higher the inspection rate that is employed. Proven high levels of product quality going out the door (and coming in the door, for that matter) will be treated to the opportunity of decreased inspection (rates) or sampling. In some real sense, product sampling is earned, based on past performance, and the confidence that the producer and consumer have in the continued high level of product performance or quality. The same is true of testing product going out the door. The average percentage defective going out the door represents the risk the consumer takes in accepting that product.

The product inspection sampling rate is based on an assessment of that risk, called the AQL (acceptable quality level), which balances the risk to the producer and the consumer in using a certain product inspection plan. This relationship is described by rules of statistical probability, and helps determine an adequate inspection rate on a sliding scale that defines the risk of buying or selling defective units according to the rate of sample in the inspection. The AQL considers the rate of defects that are encountered in inspection, set against the probability that a given sample rate and lot size (or product group) will display a given number of defects.

Sample size and an acceptance number help steer the various sample plans described in the Military Standard or Specification Sample and Inspection plans that succinctly lay out the grid work of inspection rate, referenced to the results. The overall approach is simple, though the mathematics that form the basis for these statistical sampling programs is very complex. The better or fewer defects displayed by the inspection, the less the sample. As defects are encountered, and their rates of occurrence increase, the sample plans adjust for increased inspection rates, which continue to ebb and flow as more and fewer defects are encountered. (See Figs. 8 and 9.)

Process Inspection

The term "process inspection" is deliberately used here as a generally inclusive expression. In manufacturing plants and industries

Single Sampling Plan (from Mil-Std-105D)

Acceptable Quality Levels (normal inspection)

↓ = Use first sampling plan below arrow. ↑ = Use first sampling plan above arrow. (Ac = Acceptance number, Re = Rejection number)

Code	n	0.010 Ac	Re	0.015 Ac	Re	0.025 Ac	Re	0.040 Ac	Re	0.065 Ac	Re	0.10 Ac	Re	0.15 Ac	Re	0.25 Ac	Re	0.40 Ac	Re	0.65 Ac	Re	1.0 Ac	Re	1.5 Ac	Re	2.5 Ac	Re	4.0 Ac	Re	6.5 Ac	Re	10 Ac	Re	15 Ac	Re	25 Ac	Re	40 Ac	Re	65 Ac	Re	100 Ac	Re	150 Ac	Re	250 Ac	Re	400 Ac	Re	650 Ac	Re	1000 Ac	Re
A	2	↓		↓		↓		↓		↓		↓		↓		↓		↓		↓		↓		↓		↓		↓		↓		↓		0	1	1	2	2	3	3	4	5	6	7	8	10	11	14	15	21	22	30	31
B	3	↓		↓		↓		↓		↓		↓		↓		↓		↓		↓		↓		↓		↓		↓		↓		0	1	1	2	2	3	3	4	5	6	7	8	10	11	14	15	21	22	30	31	44	45
C	5	↓		↓		↓		↓		↓		↓		↓		↓		↓		↓		↓		↓		↓		↓		0	1	1	2	2	3	3	4	5	6	7	8	10	11	14	15	21	22	30	31	44	45	↑	
D	8	↓		↓		↓		↓		↓		↓		↓		↓		↓		↓		↓		↓		↓		0	1	1	2	2	3	3	4	5	6	7	8	10	11	14	15	21	22	30	31	44	45	↑		↑	
E	13	↓		↓		↓		↓		↓		↓		↓		↓		↓		↓		↓		↓		0	1	1	2	2	3	3	4	5	6	7	8	10	11	14	15	21	22	30	31	44	45	↑		↑		↑	
F	20	↓		↓		↓		↓		↓		↓		↓		↓		↓		↓		↓		0	1	1	2	2	3	3	4	5	6	7	8	10	11	14	15	21	22	30	31	44	45	↑		↑		↑		↑	
G	32	↓		↓		↓		↓		↓		↓		↓		↓		↓		↓		0	1	1	2	2	3	3	4	5	6	7	8	10	11	14	15	21	22	30	31	44	45	↑		↑		↑		↑		↑	
H	50	↓		↓		↓		↓		↓		↓		↓		↓		↓		0	1	1	2	2	3	3	4	5	6	7	8	10	11	14	15	21	22	30	31	44	45	↑		↑		↑		↑		↑		↑	
J	80	↓		↓		↓		↓		↓		↓		↓		↓		0	1	1	2	2	3	3	4	5	6	7	8	10	11	14	15	21	22	30	31	44	45	↑		↑		↑		↑		↑		↑		↑	
K	125	↓		↓		↓		↓		↓		↓		↓		0	1	1	2	2	3	3	4	5	6	7	8	10	11	14	15	21	22	30	31	44	45	↑		↑		↑		↑		↑		↑		↑		↑	
L	200	↓		↓		↓		↓		↓		↓		0	1	1	2	2	3	3	4	5	6	7	8	10	11	14	15	21	22	30	31	44	45	↑		↑		↑		↑		↑		↑		↑		↑		↑	
M	315	↓		↓		↓		↓		↓		0	1	1	2	2	3	3	4	5	6	7	8	10	11	14	15	21	22	30	31	44	45	↑		↑		↑		↑		↑		↑		↑		↑		↑		↑	
N	500	↓		↓		↓		↓		0	1	1	2	2	3	3	4	5	6	7	8	10	11	14	15	21	22	30	31	44	45	↑		↑		↑		↑		↑		↑		↑		↑		↑		↑		↑	
P	800	↓		↓		↓		0	1	1	2	2	3	3	4	5	6	7	8	10	11	14	15	21	22	30	31	44	45	↑		↑		↑		↑		↑		↑		↑		↑		↑		↑		↑		↑	
Q	1250	↓		↓		0	1	1	2	2	3	3	4	5	6	7	8	10	11	14	15	21	22	30	31	44	45	↑		↑		↑		↑		↑		↑		↑		↑		↑		↑		↑		↑		↑	
R	2000	↓		0	1	1	2	2	3	3	4	5	6	7	8	10	11	14	15	21	22	30	31	44	45	↑		↑		↑		↑		↑		↑		↑		↑		↑		↑		↑		↑		↑		↑	

Fig. 8. Single sampling plans for normal inspection.

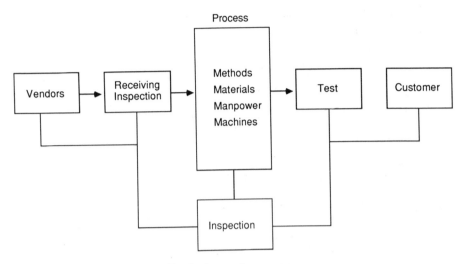

Fig. 9. Inspection process.

across the country, the sort of inspection to be described or implied may be called by a variety of names and cover a number of specific types of inspection functions. Local terminology includes such names as batch or lot inspection; departmental inspection; terminal inspection; patrol, roving, machine, or floor inspection. The name of the process inspection may come from the type of manufacture where it is performed, such as foundry, welding, assembly, lehr, strip, or card inspection. The location often affects the name of the inspection, as in subassembly inspection, conveyor or hopper inspection, stores or stock inspection. All of these are varieties or classes of process inspections.

Perhaps one orderly way of presenting and developing process inspection is to follow the general historical trend of events in inspection, as a whole. Almost universally, inspection started in industry as a 100 per cent final inspection. There came the time and the need for inspecting, examining, testing, and culling the final product just before it was shipped out. Then, naturally, the economy of introducing similar inspections "down the line" on essential parts, subassemblies, and operations became apparent, and from such experiences came what will be termed here "batch inspection."

The "shift-left" theory describes a simple relationship in manufacturing, quality and inspection, and cost. Basically, it provides the insight into the relationship between problem detection, problem prevention, and cost: the longer a defect survives in the manufacturing process, the more expensive it becomes. The best place to find a problem is where it began because it is easier to recognize and less expensive to fix.

Batch or Lot Inspection

Even in the most continuous manufacturing processes it has always been expedient to break the output up into more or less natural units of manufacture or production. The process in a wire mill for economy reasons is about as continuous as any, yet the quantities of wire are eventually cut and subdivided into reels or coils for convenience in shipping and customer handling. So the ordinary mileage of wire in a reel becomes a sort of natural unit that is used back through the whole process for pricing, costing, and inventory. Where castings are made, the contents or "heat" of a furnace may make up the natural unit for the lot size. The number of pieces a machine can make in an hour is sometimes a batch unit. The sales quota of a machinery manufacturer may be, for example, 240 machines a year. Hence, he plans to make up 20 machines every month and his lot sizes of parts, castings, and subassemblies have a natural tendency to number 20 pieces of each kind.

The scientific study of lot sizes is a very important economic factor nowadays since it affects the purchase and use of equipment, the setting of incentive rates, machine speeds, tool and die life, set-up costs, quantity purchases of material, inventory requirements, and a large number of other factors. Lot size means that continuous flow of the material to be inspected is interrupted; instead of looking at individual units to make decisions on passing or holding material with defects, the lot is considered. Some sample plans relate to the continuous flow of materials. Others relate to lot-by-lot inspection. The manufacturing setup has to facilitate the concept of moving by lots, so time is involved — within the lot, the units are usually identified as coming from a certain production run or period of time, so that problems can be traced back to a lot. Conversely, decisions of control based on lot movement can come somewhat after the fact, depending on the process flow characteristics. It is possible to subvert the advantages of in-process, real-time inspection by queueing units for inspection somewhere in the process.

Other Factors Determining Lot Size

The preceding has been mentioned only to indicate that usually the lot or batch size has been established for the inspector by factors out of his control. Many times the inspector's batch size is determined simply by the number of pieces that happen to fill a tote box, truck, or pallet, or by the number of yards in a roll. In general, then, the inspector adapts his procedure, sample sizes, timing, and other factors to what the shop is naturally using in one form or another as batch units. Occasionally, he is forced to be arbitrary and establish lot sizes of his own. He might decide, for instance, to inspect each half-day's production of a machine, whether it filled one or a dozen boxes or whether the half day's work represented one-tenth or ten times the amount specified in a production control order. Next to actual box, truck or pallet counts — physical quantity units — some natural time subdivision is perhaps the favorite measure of lot quantities or units used in inspection.

The concept of "rational" or "homogeneous" subgroups of work or process delineation must be understood here. Inspection data are designed to isolate variables within a sequence of operations separated by time. In this sense, time is the only thing that should separate groups of data or inspection samples — everything else should remain the same, or homogeneous. When such a natural subdivision of activity or product has been established, then data from that process subdivision can be examined for change or associated variation relative to previous or subsequent inspection efforts. When change is found, that variation or change can be based in an undesirable cause that has found its way into a process point or group of products that was not previously experiencing the variation — an opportunity to fix a problem.

As far as possible, the inspection department will render great assistance if it will insist on an attempt to establish as small batches as is possible or reasonable. Furthermore, each batch inspection should be made as soon as possible after the manufacturing operations are completed.

Basic principles of inspection are at work here — it is necessary that inspections of batches or lots render information that is accurate and usable — to the purpose of the inspection. Therefore the inspections must be both speedy and timely, in order to ensure an economical contribution to the system. The cost of variation is high, and the cost of inspection can be high also. In order to properly balance a company's "cost of quality," the data derived from inspection must be highly reliable and address the relevant points of the criteria or standards with irreproachable integrity.

Where batch inspection is an established factory inspection routine, the work is ordinarily automatically routed to an inspection station. Such a station may be a certain bench, a crib, or a special set-up area. It may be located adjacent to the manufacturing department or off on another floor; again, batch inspection occurs, in some instances, at the end of a conveyor line and, occasionally, right beside the machines. The more common procedure is to have the manufactured pieces taken from the machines in some type of pan, tray, box, truck, or conveyor to the established cleaning or degreasing process and thence routed to the inspection station, crib, or room. (See Fig. 10.)

Fig. 10. Typical large batch inspection area.

Information Needed for Batch Inspection

To competently perform batch inspections the inspector should also be equipped with three general classes of information. He should be sufficiently conversant with all the preceding manufacturing operations to quickly recognize substandard work and to know pretty well what caused it. He should have, of course, up-to-date blue prints and specifications and be thoroughly instructed in the technical details connected with the immediate inspection. And he should have learned in reasonable detail where and how the parts will be used subsequently.

The inspection system has at its heart the measurement or assessment itself. And on the front end of the system is a fully integrated operation sheet or inspection procedure that details how the inspections should be performed. Equally important is the report or data that are generated to summarize the results of the inspection and that are fed back to the groups responsible for the work that was inspected.

Data reports, from batch inspections should cover the basics immediately — how many units were inspected, by whom, when, and according to what procedure or process sheet. Then comes the payoff — how many units passed? How many units were defective, and what were the defects contained in those defective units? Simple and usable statistics can cover this information easily — what was the per cent defective (units) and defects? Finally, what were the most numerous defects present, for these could almost certainly be the problems that need to be investigated further in a cooperative effort with other support groups.

Inspectors performing batch inspections usually handle a variety of parts and components and the requirements mentioned above should be constantly reemphasized in his mind. In addition, he should be conversant with the shop's general standards of manufacture and appearance. Where batch inspections call for the use of gages or measuring and testing apparatus, the inspector will, almost without exception, use inspection department equipment. Care must be taken, then, that inspection gages, say, compare as exactly as possible with production gages.

Calibration specifications and standards, as well as a system to guarantee that calibration cycles are being met, are crucial to the effectiveness of inspection, and the ability to relate inspection results to the actual production effort.

Three Functions of Batch Inspection

Sampling and statistical quality control techniques are being introduced increasingly into batch inspection procedures. As a result the viewpoint has developed that it is seldom economical to try to make 100 per cent of the parts or products up to specifications. Some practical percentage deviation is allowed away from the absolute perfection of making every single component within tolerances. Hence, batch inspection has two main objectives. One is to detect substandard and out-of-tolerance work and the other is to determine whether or not the quantity of defectives in a lot has reached an unprofitable level. The third duty is to, once again, report data from the inspection sequence — become involved in analyzing the data, and eventually to help fix problems of excessive variation. The identification of causes of variation must be a concrete part of the inspection contribution. Solutions are easy to come by, if concentration is put on the proper causes of variation. The identification of the *right* causes of variation in a system is assurance that the variation can be prevented from recurring.

Generally, in today's factory these objectives are attained by first sampling the batch, using a suitable sampling plan or table at some prescribed quality level, and secondly, *detailing,* that is, 100 per cent inspecting those lots that the sampling plan does not accept.

The surviving principles of "batch inspection" are more adequately described by the term "process inspection." Essentially, the product flow is divided into various important process stages or operations. Within these operations, several different types of inspections can take place.

It is now less common to see the archetypal batch inspection where a load of parts is carried off to a separate area for inspection; this separate inspection could be total 100 per cent inspection, or a sample of the "lot" that has been identified and pulled aside. More likely is an on-going inspection of work in process, usually in two stages, centered first amount (piece) parts coming *to* the process operation, and second around parts *leaving* the operation for the next manufacturing routine. In both cases, the assembler or work-person from the station generally performs the inspection at or very near the work station, to save time and increase efficiency.

The inspection can involve the measurement and control of attributes data — where pass/fail goodness or badness is qualified in an inspection per clear defect criteria standardized for the inspection. In this case, the operator will keep a count of the good/bad ratio for purposes of statistical control charting. When the rate of defectives hits a specified point and becomes abnormal, the operator recognizes the general abnormality and begins looking for causes with his support groups. In such a process inspection, the inspection usually deals with *all* of the parts or assemblies, in a continuous-flow basis.

If the operator is measuring directly — taking numbers as the point or purpose of the inspection — the rate of inspection will not be 100 per cent, rather some rate that is much less, but sufficient to indicate a statistical change or shift in the dimensions and the average and range of the dimensions that are being taken. Likewise, at the point at which normal statistical limits of the (average and range) indicated dimensions are exceeded, the operator would take action to understand the excessive variation.

Dangers of Monotony in Batch Inspection

In some instances, batch inspections are simply inspections of successive lots of the same part. In such a case, the bench layout, the facilities for handling the boxes or containers, the gages and testing devices, will remain the same, day in and day out. The inspector must be alert to the danger of overconfidence where batch inspection is of this repetitive and frequently monotonous nature. His own personal conceptions of the standards are likely to become dulled and blunted. If the standards at production are gradually relaxing, the inspector is liable unconsciously to start sliding down the same slope. If the print calls for a diameter to be, say, .265 inch — .268 inch, he would more naturally notice occasional pieces of .2679 inch to .2680 inch if most of the batch came along .265 inch to .2655 inch, but where practically all the pieces were .2678 inch, .2679 inch, or .2680 inch, he could more easily fail to observe .2681 inch or even .2682 inch work. White has a tendency to become gray, fine to become coarse, and round to become oval, and if the transition is gradual enough, the inspector will fail to notice it. So, the batch inspector should institute a definite routine under which hourly, daily, or weekly he pointedly refreshes himself by reviewing the standards; where he deliberately brings himself up short and checks his own inspection methods.

The same condition applies to gages, testing equipment, and inspection apparatus. Unless there is an established routine for having the accuracy of micrometers and indicators periodically checked, they too will soon wear and become as sloppy as the inspector using them. Where an inspector uses a microscope regularly, as an example, how often does he deliberately stop and think to wipe the dust, sweat, oil, and fog off the lens?

Dangers of Variety in Batch Inspection

Happily, batch inspection more often covers the examination of a variety of parts and materials and the shortcomings of monotony are lessened, only, however, to have one or two other potential dangers substituted. One is the decided tendency to be lazy about looking up specifications and standards, for the inspector to believe he remembers the details of the inspection from the last time he performed it. Is he sure, for example, that the tolerances weren't changed since the last batch came through? Better get an up-to-date print every time! The same type of slackness applies to gages and equipment. The particular gage is used, for instance, only on this particular component. It was accurate a month ago and it hasn't been touched since; why bother to have it checked!

The seasoned inspector watches one other element in connection with batch inspections of a variety of materials and components. It takes time, effort, sometimes a great many footsteps, to collect together the gages and apparatus necessary for a single type of part and to set it up ready for use. Sometimes it takes longer to make ready for an operation than to perform it and frequently the time loss is unavoidable. Figure 11, as an example, illustrates the collection of apparatus necessary for inspecting a batch of a certain type of coiled spring. To minimize this time and energy loss, the inspector and the department can make sure that intermittently used apparatus is conveniently stored in readily accessible locations. But more to the point, the individual inspector can materially help the situation by the manner in which he puts the apparatus away each time *after* he is through using it. A minute or two spent cleaning and readjusting apparatus just at the time he is finished with it can save an hour, perhaps, the next time he needs it. In fact, the apparatus usually can be stored away carefully set and immediately ready for use the next time. The worst thing an inspector can do is toss his apparatus into a cabinet the way a child discards his toys and runs off to other interests.

Fig. 11. Some batch inspection operations, such as those involving a certain type of coil spring, require many pieces of inspection apparatus.

The inspector today is bound by rules and systems like everyone else in the factory. To obtain and preserve consistency and reliability in inspection, the rules, tools, and methods of quality inspection belong to the system — not to the individual inspector. More employees are needed to define and maintain the modern quality system than are used in the inspection act itself. Procedures, reference documents, and training tools, the control of process documentation itself, calibration schedules, and data collection and reporting schemes all need to be standardized and fully detailed. The inspector who has to choose whether or not to follow an acceptable pattern of behavior on the line is working for a company that will soon be less than adequately competitive.

Process Audits

Depending on local conditions, this type of inspection may be known as roving inspection, floor inspection, machine inspection, line inspection, departmental inspection, and other similar names. In process auditing, the inspector goes to the work, in contrast to batch inspection, where, for the most part, the goods are routed to him.

Process audits include not only a decision as to the conformance of products to specifications, but also a preview or forecast as to how the operation will probably proceed. Being right at the source, the inspector has an opportunity to look for surrounding or impending

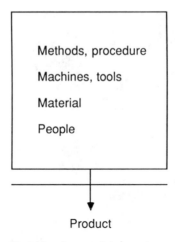

Fig. 12. "Process" inspection.

problems for substandard work. At batch inspection the work is completed — sometimes long since done. There is more of an air of finality at batch inspection. The inspection operation is physically separated from the manufacturing operation. So batch inspection decides only that the goods do or do not conform and that they can or cannot proceed in their present condition to subsequent production or assemblies. The evidence is all in at batch inspection and is represented by the goods themselves; in a process audit an inspector can be guided in his decision to some degree by circumstantial evidence.

Process and sometimes roving product audits are today mostly performed by the professional Quality Inspector — sometimes referred to as the auditor. His primary job is to check out and report on the status of the process — the process as a whole — the material, the manpower, the machines, and the methods of production. Following a systematic format of auditing, these employees check the operation and mode of everything at the work station, and generally publish a report that receives wide distribution in the factory. For this is the information that engineers and support groups depend on — what *really* happens on the shop floor. This is true and pure process inspection, and relates directly to the quality and performance levels of the assemblies produced by that process.

Inspection Should be Timely

Inspection, in general, to be most effective should be timely. In the sense of uncovering and remedying substandard design, manufacturing, and production conditions, the inspection of a run of product occurring hours or days after the work is done has lost, perhaps, its most valuable element. True, a post mortem inspection will cull out defectives, scrap, and junk, and present a cleaner product to the next operation, to assembly, or to the customer. This is something gained, of course, but it is probably the most inefficient manner of gaining the end result of all manufacturing.

Inspection Should Give Warning of Trouble

Management and industrial people outside the fold of inspection generally hold one of two more or less correct views concerning inspection. One is that it is the inspection department's job to sort, to cull out defective work. Another is that this department has the duty of constantly checking the flow of work to make sure that the work is not deviating from prescribed standards and tolerances and to give warning if it does. Without doubt the latter is the most valuable service inspection can perform in industry.

The question is frequently asked: If extensive inspection is not 100 per cent effective, if a screening inspection is bound to miss some substandard pieces, how then can we be sure of getting a defect-free product through to our customers?

There is only one obvious but impractical solution: Do not manufacture any substandard, defective, or out-of-tolerance product at all!

And right here is where process audits fit. Properly performed, they can be a powerful agency toward preventing deviations.

The most effective service inspection can offer industry, however, is some form of preventive therapy. It should expertly examine products and diagnose conditions at times and places where it can give warning that standards, specifications, and tolerances are not being met, or are about to be exceeded, so that costly and wasteful amounts of substandard and defective products, parts, and materials will not accumulate.

Process Audits Should Be Orderly

Efficient and effective process audits are always orderly. Like a watchman, the process auditor makes his rounds and works on a schedule frequently as exacting as a railroad timetable. On his route he traverses a prescribed series of operations or machines or a mixture of both. Ordinarily his "territory" is so mapped out that he completes a round once every half-hour or once an hour.

Sometimes such a process audit may include an element of a planned "surprise" inspection, to assess a process that employees may not be expecting an assessment. In any case, communication that is usable and correct is the goal of such an inspection format.

Almost without exception, he makes his observations and reaches his decision by means of samples. Today, the process auditor's sampling is being based on scientific statistical techniques. The matter of sample size and the decision to be reached from a sample will be given little specific attention here. More emphasis will be placed on the actual mechanics of process audits. Most plants have their own systems, and the study of statistical quality control methods can give the inspector a more detailed description of sample sizes than space here permits.

It is essential to remember that the inspector is referencing the expressed variation from his inspection results against a sample of an overall population. As such, it is important to know the size of the sample in relation to the size of the group or population from which it is drawn. The so-called "confidence" to the producer and consumer from this relationship can be statistically derived and manipulated. However, it is the responsibility of the inspector to provide the pure numbers, and those numbers must be correct in order to derive decisions from the final formulas.

Two Objectives in Process Audits

In general, there are two objectives that process auditing may accomplish. Either or both may be aimed at, and this fact governs the size of sample and the way it is taken.

One objective is simply and directly to determine the condition of the work that has been completed during some arbitrary period which has elapsed since the last process audit. This is usually a half hour, an hour or so. Does the work conform to specifications or is there too great a percentage of defective units to permit passing it on to the next operation? This objective applies particularly where it is suspected that the operation may be erratic in performance. Process audit sampling of this sort resembles batch inspection, and a larger and more random sample is needed than when the second objective is, alone, aimed at.

The second objective is to determine the trend of the quality of the work being produced. Thus, from the standpoint of this objective, at any one visit by the process auditor, the work from a machine or operation is considered not only representative of the conformity of output (or lack of it) at that time, but is also descriptive of the output for some time past and indicative of how the work is expected to come off the machine for some period in the future.

With his eye on the second objective, the inspector is more interested in what the machine or operation is doing and will do, and is less concerned with the condition of the batch of work itself that has been produced since his last visit. He can reach an accurate enough conclusion with respect to this second objective by taking samples of the work as it comes off the machine, the number of pieces in this sample being less than was required for the first objective.

Restating these objectives it may be said that the first concerns itself with the acceptability of a lot prior to its transfer to another station for further processing. The second concerns itself primarily with an evaluation of the trend of the operation, that is, whether or not it is producing and will continue to produce satisfactory pieces.

Finally, it must be remembered that most operators are also gathering numerical data on the product as they are producing it. In this sense, the roving auditor performs a check on the check — a verification not only of the product by sampling the dimensions, but also on the *process* that checks the data, and regularly ensures control.

This is one more extension of the process audit — where the methods, material, machines, and people who work the process are regularly evaluated and reported on.

Interpreting Process Audit Data

Suppose a process auditor's sample of work, where the blue print calls for 1.555 inch — 1.556 inch, shows the following five-piece reading:

> 1.5555
> 1.5553
> 1.5555
> 1.5554
> 1.5555

He could reasonably assume that for the past few minutes the machine had put out similar size work and that it would continue to do so. Also, considering that the size differences between the pieces he examined did not exceed .0002 inch, he could also reasonably assume the machine was inclined to be steady, to produce reasonably uniform work.

If at a second later visit he found the machine producing pieces that measured:

1.5556
1.5558
1.5557
1.5558
1.5559

he should then realize that the work done for some interval in the past was probably within tolerance but that very soon the machine could turn out some oversize work.

However, if his readings showed:

1.5552
1.5558
1.5554
1.5558
1.5553

he should suspect that the machine was not too capable of holding within the 1.555-inch — 1.556-inch tolerance and that there was a chance, with a .0006-inch size difference between specimens in his sample, the machine could have already made oversize work.

When process auditing is functioning as it should, the inspector is not only making an estimate of the condition of the work done since his last round, and passing judgment on it — making the decision, per-haps, that it is sufficiently contaminated to warrant segregation or withdrawal from the production flow for screening or salvaging, but he should be also carefully observing the trend or tendency of the work. It is just as important to warn of impending trouble as to report the presence of defective product.

This is the contribution of statistical process control — the ability to plot and identify trends before they cause a direct violation of specifications is what makes real-time charting of defect trends so attractive to inspection groups. See the shark before it grabs the product and causes revenue loss — predict and see ahead through simple statistical applications, and then *act*; that is the key.

The sort of judgment a process auditor must use frequently can be simply illustrated by considering the milling of a steel plate to blue print thickness requirements of .750 inch ± .005 inch. Such a specification technically permits the machinist to make plates .745 inch up to .755 inch thick and still not be "breaking the law." However, it can be found out by both the inspector and the operator that the rough milled pieces are to be hardened and, afterward, surface ground to meet a final blue print specification of .745 inch ± .0005 inch.

The knowledge that the plates are to be hardened and surface ground to fairly fine tolerances may modify the inspector's judgment or decision at the milling machine. Suppose the milling machine operator is taking complete advantage of the .750 inch ± .005 inch tolerance and milling the parts .745 inch thick. This, then, would leave the surface grinder operator only .0005 inch in which to clean the surfaces flat and free from hardening scale and to overcome any taper or warping. Perhaps the inspector should consider such work unsatisfactory until official action had been taken to improve the dimensions and tolerances in a direction to avoid the dilemma implied above.

The judgment used at process auditing includes a little wider field than the literal, exact, and technical comparison of products to the particular blue print or specification. The patrol inspector should have a thorough and accurate knowledge of just how and where a part is to be used and of the succeeding operations to be performed on it.

This latitude is not to be taken to excess. Any change from the expressed process operations sheets is unacceptable. And any suggestions of change or reinterpretation of the *process sheet specs or instructions* is unacceptable. Only after input to and appraisal from those responsible for maintaining the process inspection sheets should such changes be made. The value of data, numerical data, and their statistical validity comes from their accuracy. It is of paramount importance that the inspector reflect this numerical accuracy in his readings, so proper interpretation can ensue.

Visual Inspection on a Process Audit

Visual inspections are an important item in the process auditor's agenda. He usually must make decisions as to surface finish, tool marks, burrs, and fillets. He must know, for instance, the effect that excess surface roughness in the parts he has sampled might have on the efficiency of subsequent machining operations on the same pieces. Or if they are not to be subjected to further finishing, what affect surface defects would have on the appearance of the finished, assembled products.

While the examples of a warning trend in dimensional changes offered above relate to machined parts, the principle of inspection illustrated can be applied to almost any type of work. The detection of rubs, burnishes, tiny scratches, burrs, or minute edge splits may well serve as warnings that a punch press die should be dressed, sharpened, or adjusted. An examination of several units of soldered connections, showing, for example, tendencies toward carelessness or a cold iron, could well indicate to the experienced eye certain timely corrections in the operator's technique that might prevent defective connections. Mispix just beginning to appear in the width of fabric being woven at a loom should be sufficient warning, even though the general appearance of the fabric is still commercially satisfactory.

The Sample Should be Small

At a process audit the inspector is on his feet. The gaging and testing are normally performed at the machine, although there are occasions where the sample is taken to an inspection bench for special work or observation. The inspector uses as far as possible the same gages, testing apparatus, and equipment the operator is using rather than special, duplicate, or more precise inspection equipment. Hence his sample should be small and his work done as quickly as possible so as not to interfere more than necessary with the machine operator.

If the inspector feels that three or five pieces gives him information enough, he need sample no more; if he is in doubt, he should continue to examine enough pieces to make up his mind. Where the machine is multistationed, like a six spindle screw machine, for example, the sample taken should equal the number of stations or be a multiple of it — in the case of the six spindle machine a 6 or 12-piece sample, for instance is chosen.

In fact, the inspector does not need to "feel" how many units are proper for a valid inspection and statistical statement. A sample size of three to five units is fully sufficient when plotting the range and average of measurements on an $\overline{X}$ and R chart. The only question that arises is that of sample frequency — how *often* should a sample be taken (a sample of three to five units). The statistical validity of this sample size (n) will hold for true variables measurements and interpretation. Sample frequency will depend on the output/flow of the line itself, and the separation of one "subgroup" of samples from another.

A subgroup of data, or samples from one subgroup, is based on the notion that there is a normal or "rational" grouping of parts in the assembly line — when some parts are produced, when are we likely to see a change in the detail of these parts — at shift change? from one hour to the next, or as operators trade work stations? These things are factors that help engineers design inspections in terms of sample frequency based on rational subgroups.

The work done since the inspector's last visit may be found piled upon or near the machine or bench. It may be in tote boxes or pans, thrown into barrels, stacked on skids, or, as in the case frequently of assembly operations, pushed along to the end of the bench. A great

— Inspection Record —
VMI

Name _____

Date _____

Station _____

ID Number	Defect	Time	Comments

Number units inspected ____   Paretto analysis

Number units passed   ____   ① _____

Number units defective ____   ② _____

Total defects ____   ③ _____

Fig. 13. Inspection record.

deal depends on the shop's material handling methods. The more modern shops have conveyor systems for taking away an operation's products. But of course, the modern successful factory environment will have reduced inventory flows, and perhaps operational "pull" systems on the line that prevent work and assemblies from piling up. "Make one, move one" — the advantages of such material control are clear in modern manufacturing and inspection environments — if there are few parts to evaluate, problems will be more conspicuous and easier to remedy through cause-and-effect investigations.

Taking a few pieces as they fall from the machine will give the inspector an idea how the operation is going at the minute and also how the operation is likely to proceed for a while to come. But to make more certain of what the machine has been doing for some time past, the inspector will need to dig deeper into the pile. ("Digging deeper" in the case of conveyorized parts may be an impossibility of course, in which event the inspector must form his opinion from the production immediately at hand.) A good sample will represent as nearly as possible an honest cross section of the work already done between inspection visits, and it should foretell to some degree the type of work that will be done until the next visit.

Errors in Process Auditing

Three errors are common in process auditing sampling. The inspector secures an inadequate sample — inadequate because the number of pieces is too few. Another mistake is due to a degree of over-conscientiousness perhaps, or inexperience. The inspector spends unnecessary time on too large a sample. If the work were perfect, one piece would tell the story as accurately as a hundred. He should stop inspecting the minute he knows an operation is satisfactory, and probably will be, or that the work is rejectable and the operation should be stopped. The third type error appears when the sample includes pieces from work that has been previously sampled. Suppose an inspector approves an operation at nine o'clock, but by ten o'clock it is running just out of specification. If the inspector happens to scoop up already approved nine o'clock pieces among his ten o'clock sample, his judgment — his decision — can well be confused.

Use of Operator's Gages by a Process Auditor

The inspector using an operator's own gages and apparatus, checking the work in the same manner the operator does, usually begets greater confidence in the inspection. For the auditor to appear with his own micrometer, gages, and apparatus (which are usually cleaner and shinier than those on the machine) sometimes rouses resentment or suspicion, especially when, by unfortunate mischance, the inspector assumes some slight air of superiority.

The human relations situation just implied places at least one of two responsibilities on the auditor. Where he is using the operator's gages or equipment to check conformance, he must be sure of their accuracy and reliability. After all, the inspector's basic responsibility is to compare, accurately, the product's conformance to specifications. The mere fact that an operator's micrometers say a piece is to size may not be sufficient. If the local circumstance is such that the inspector uses his own equipment, he must be sure above all else that it is accurate. It needs only one situation where the inspector's and operator's gages disagree and where the inspector's gage is the one in error to make the inspector's observations and decisions to some degree impotent from that time forward.

Performing a Process Audit Away From the Machine

Of course, a process audit right at the machine can be a nuisance at times. There may not be adequate room for both the inspector and the operator at the machine itself or the production rate may be so rapid that the inspector's presence brings about unwarranted interference. In such a case, of course, the auditor secures his sample and retires to one side to make the necessary checks with his own gages. For this purpose, the inspection department is usually assigned a convenient location in the manufacturing area with room enough for a bench, or benches, chairs or stools, a stand for a surface plate perhaps, and cabinet space in which to store inspection apparatus. Where the use of an optical projector, for instance, is prescribed, the instrument is usually located along with the other inspection equipment.

One reason for performing the inspection on sample pieces away from the machine is the fact that the particular test, observation, or gaging may be considered a little too intricate or time consuming for the operator to do it. There is a point in manufacturing economics where it is less costly in the end to free the operator from certain inspections and have them completed by another man — by an inspector. A simple example appears in the sketch of the screw machine part in Fig. 14.

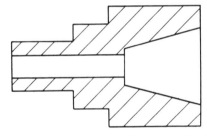

Fig. 14. It may be more economical to have an inspector check a piece-part, such as that shown, than to have it checked by the machine operator.

In this case the internal taper section was fairly critical, the tolerances so exacting that the use of a simple taper plug gage had been unsuccessful. The tapered cutting tool plus the machine setup could be depended on for a period. So an auditor took periodic samples and checked the taper and depth to shoulder on a surface plate. Incidentally, he made timely checks on new and resharpened taper tools, keeping a supply ready for the operator when the tool in use became dulled, as indicated by the period process audits on the tapered parts being produced.

There are many occasions among manufacturing operations where the particular inspection may not be made necessarily on the components — some occasions, for instance, where a drill jig might be checked rather than the pieces being drilled. The examination of a gear hob is another good example, or checking broaches or milling cutters.

Tracing Causes of Defective Work

Another function of process auditing is that of ferreting out causes of bad work. This is frequently the case on gang or sequence operations. If the pieces coming off a reaming operation do not conform, the basic trouble may actually lie in the previous drilling operation. Go back to Fig. 3 in Chapter 1 for a brief review of this possibility.

The trail back, many times, is not so obvious or simple. In one carpet mill, unreasonable variations in spun yarn thickness were traced back through miles of spinning and roving operations to unequal loading of raw wool hoppers on the far side of the first row of carding machines in the mill!

Another trial the auditor faces is the disposition of parts that are defective because of some operation preceding the one he is inspecting, an earlier operation over which he has, perhaps, no jurisdiction. Suppose, as a simple example, an inspector is examining parts for plating, polishing, and buffing defects and discovers a tray of them on which two countersunk holes are missing. Obviously the batch of parts slipped by the drilling operation way back down the line. (They may have been omitted by the vendor supplying the original parts!)

The parts must be rejected, of course, and a scrap or salvage ticket filled out by the inspector with the charges usually being shared by the department responsible for the bad work and also the Inspection Department for allowing the defectives to pass through.

Auditing a Continuous Process

Process auditing includes checks on continuous processes. Figure 15 shows part of an extrusion insulation machine with a continuous gage measuring the overall diameter of the insulated wire. Readings at stated intervals are made by the patrol inspector at this gage, supplementing the more or less continuous observation of the insulating process by the operator.

In a somewhat similar manner, auditors will frequently check temperature readings, make grain-size tests, chemical titrations, hardness tests, and other inspections of a more or less strictly technical or laboratory nature.

First-Piece inspection

One common division of process inspection in machine shops is known as first-piece inspection or setup inspection. The names used

Fig. 15. Process audit includes checks on continuous processes such as is taking place on the extrusion-insulation machine shown.

practically describe it. It is not necessarily an inspection of absolutely the first part made, for the operator may spoil several pieces before his machine, tools, or apparatus are properly adjusted, but it is the formal examination of the official first piece that the operator considers representative of what he intends to do. The inspector may check a succession of "first pieces," working along with the operator until both are assured the setup is satisfactory for continuous manufacture. Many times the inspection must cover checking the jig, tools, or fixture.

The piece may be the result of several operations performed on it more or less simultaneously as in a screw machine, on a boring mill, or at an assembly bench. Again, first-piece inspection might cover only a single operation.

Where the first-piece routine is practiced, the inspection is expected at the start of each new production order. In the case of continuous production, the formal first-piece check may be performed at the start of the day or shift and again perhaps at any time immediately after tools, equipment, or the operator have been changed.

Requirements of First-Piece Inspection

In some shops the operation is held up after the presentation of the official first piece until the inspection has been completed and the go ahead signal given. Usually, however, the operator proceeds, and stops his work for further adjustments if the inspection report is adverse.

Either way, first-piece inspection should be prompt as well as thorough. The inspector needs to be available, timely and quickly. First-piece inspection must be especially accurate and reliable in order to avoid any tendency on the part of the manufacturing department to place the responsibility for any subsequent bad work on the inspector.

The inspections are liable to vary in intricacy from a simple visual examination or the check of a dimension with micrometers to complex surface plate setups or the use of special gages, comparators, and testing apparatus. The work can be better done, in many instances, right at the production operation. Time and effort are saved, many times, by checking a piece when, for example, it is clamped against the lathe face plate. Obviously, also, a machine bed weighing a ton cannot be readily shifted to an inspection station; the check up must be made on the floor. The operator and the inspector working together can develop mutual confidence in the outcome of the inspection. On other occasions, the parts are carried to an inspection station. The latter alternative has the merit of getting the inspector and the operator out of each other's way, especially if space around the machine is at a premium. The inspector is then freer to do more precise work.

Objectives of First-Piece Inspection

Most inspections are made simply to determine the bald fact that components do or do not conform to specifications. At first-piece inspection, however, the inspector should keep firmly in mind the important object of making sure the setup will very surely produce what is wanted.

As an example, the dimensions of a workpiece may correspond properly to blue print requirements but the inspector observes chatter marks, which are a symptom of something loose as well as a sign of impending out-of-tolerance work. The operator may hand him the first piece from a cylindrical grinder where the piece itself may be okay. But is there out-of-round? Is there taper? Have the countersunk center holes in the ends of the piece been properly lapped smooth? What provision is the operator taking for dressing the wheel and for properly sparking off? Is the coolant flowing rapidly enough? Is it the right coolant at the right temperature? Is the machine in the line of direct sunlight or drafts? Some of these questions may seem foolish but not too zany for .0001-inch tolerances. If first-piece inspection is to be worth the time spent at it, more than just the bare conformance to specifications must be observed.

One object of first-piece inspection is for the inspector to observe finish, appearance, and workmanship. If the first piece does not conform to the general shop standards for surface finish, tool marks, and chatter, correction should be made. Perhaps the stock is not "cleaning up." Where the first piece involves a threading operation, for instance, the thread form should be examined. In the case of a tapped hole, perhaps the chips are not clearing out properly.

In many operations, especially at boring mills and similar machines, the primary function of a first-piece inspection is to check layout. The inspection may be a first-class long surface-plate problem on hole location. Angles and tapers may require a sine-bar setup. Along with location, on some types of components, it is up to the first-piece inspector to remember to check concentricity or squareness or parallelism.

First-piece inspection is especially useful in the departments where stamping, punch press, and draw press parts are made. Duplication of parts is very nearly achieved on punch and die work. In addition, press work proceeds usually at a rapid pace. Hence, if first-piece inspection confirms the setup, an undue amount of unintended scrap may be avoided as well as a series of subsequent checking inspections. Some shops, in fact, rely on the first-piece inspection and a *last-piece* inspection for press work, thus eliminating intermediate process audits, on the theory that if the first and last pieces conform, all the work in the batch between should be satisfactory.

Visual Examinations in First-Piece Inspection

First-piece inspection of press parts should include examinations for certain unsatisfactory conditions other than purely dimensional conformance. This can be illustrated by reference to the part shown in Fig. 16. The size, alignment, setting, and adjustment of the punch, dies, and tools may spoil the pieces for some of the following reasons.

If the corner of the die or punch has been machined too sharply or if the punch shoulder travels too close to the die, the flange, *a*, is formed too square. Wrinkles or corrugations are sometimes indicators. A fatigue crack or split may be started at the corner *c*. One test is to clamp the flange in a vise and attempt to tear the body away from it with pliers or a bar.

Similarly, the tear test should be used to discover fatigue cracks at the corners of the lugs at *d*. Splits are ordinarily discernible to the eye, but the tear test is perhaps more reliable.

If the punch and die are not aligned, if either is egg shaped, a thin wall, as at *e*, can result. The visual evidence is a series of parallel streaks of rubbed or burnished metal, called die marks, and frequently, a sag or elongated lip formed on the thin side by the squeezed metal as at *f*.

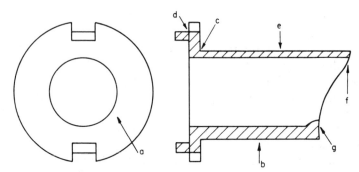

Fig. 16. First-piece inspection of press parts should include an examination for unsatisfactory conditions other than purely dimensional conformity.

The end of a punch may be nicked or chipped a trifle. The trouble can be diagnosed from the appearance of a "fill" or "plug" as at *g*.

Deep scratches along the inside or the outside of the body, *b*, indicate sharp burrs on the punch or die, respectively, or the presence of dirt, metal, or grit particles.

To discuss particular receiving inspection procedures in a book of this sort would be misleading because in one plant an "incoming" specialty may be lumber, in another, rod and sheet, in a third, tank cars of a chemical. In one shop attention may be paid to quantities of screw machine or press parts and its next door neighbor may be concerned with timing mechanisms or condensers. The checking of newly bought cutters and tools is frequently an inspection department responsibility.

Receiving inspection, of course, deals with purchased parts, subassemblies, materials, and goods coming in from outside the shop. Orders and contracts are placed with other manufacturers, subcontractors, suppliers, jobbers, and merchants for the various things the shop needs that it does not make itself but that either enter the manufactured product, are supplied with it, or supplement the manufacturing process. If the factory is one of a series owned or controlled by the same company, receiving inspection may entail the acceptance of parts, subassemblies, materials, and work from other branches.

As a general rule, receiving inspection is performed in an area close by the factory receiving department, if not actually in it. In the larger plants incoming inspection is a separate division and function of the inspection department and the factory's materials handling routine definitely includes the receiving inspection step. Usually, too, the inspectors assigned to incoming inspection carry on no other inspection work. In the smaller shops, the division is not so sharp. There may not be enough receiving inspection but what properly trained inspectors can be shifted over for a part of each day or week to clean up the accrued receiving work and then be returned to other routine inspections.

The conclusions arrived at by receiving inspection are usually based on samples and seldom on 100 per cent inspections. Nowadays the sampling itself is usually based on one of the available scientific, quality control sampling tables.

There are three reasons for performing receiving inspection operations, and in any good manufacturing system, these reasons often overlap. The first is to accept (goods), the second is to reject, and the third is to assist the vendor in gaining control of his process. The last reason is of course the most important — to help the vendor get control of his process, and cut everyone's associated costs.

All of which requires the receiving inspector to obtain truly random samples. Since the goods to be inspected may be baled or boxed, packed in barrels or bags, or arrive on pallets, in rolls or reels, or in bundles, the inspector frequently faces a mechanical task of getting a satisfactory sample of each item. In fact, securing the required sample may take more in time and energy than the actual inspection. Because it is burdensome, however, it should not be neglected.

The next question, of course, is whether or not the goods conform to specifications.

The first-piece, or "first-article" inspection as it is sometimes called, is usually the most detailed and intensive of all possible industrial inspections. The blue print, purchase order, and other documented and approved vendor/customer agreements, are all called out and issued for consideration, particularly in the first-article inspections, and often in the typical receiving inspection routines.

Many of the operations and setups described in this book are typically carried out only in the receiving inspection lab. Although these inspections contain both dimensional and attribute evaluations, the focus is on numbers and dimensional conformance. Often the lab will verify attribute characterizations such as "too dirty" with a quantifiable method that can be correlated with the vendor and the vendor's requirements. This detail is rarely reflected on the line, although some select measurements are of course conducted in routine or in-process inspections. The wherewithal and the time needed for such in-depth inspection are called for only at the "front door," in order to make important decisions that are then supported in a different fashion by process inspection.

The Purchase Order as a Specification

Almost without exception, the purchase order is chapter and verse as far as specifications are concerned. The term "purchase order" is used here in the broader sense, for the goods may come in as the result of a telegram, verbal order, or contract. Parts received from a branch office arrive in response to some interfactory memorandum like a material requisition or production control order.

The purchase order may contain a complete description — the specifications — on the face of the order itself. Customarily, however, it is accompanied by blue prints or special written specifications. Again, reference will be made to a supplier's catalogue number or to some universal commecial standard like, say, S.A.E. specification number 1010 for a particular composition of carbon steel.

Check the Purchase Order Carefully

The point to be emphasized is that the inspector, as the first step in preparation for the receiving inspection of any article, should refer to the purchase order (or dig out its equivalent if the order was verbal or wired, or was in the form of a contract) and carefully study all blue prints, sketches, written specifications, or catalogue references in connection with it. The usual danger at receiving inspection is over-confidence — the inspector believes he knows more about the specifications than is actually the fact. Particularly at receiving inspection, he should watch out for changes in the purchase orders or specifications. Usually the receipt of the same kind of parts or materials is not a daily routine. Several days, weeks, or even months may elapse before another shipment of a particular sort of goods shows up. In the meantime, the inspector could well have forgotten details of the specification or the engineering department may have made a slight but important change. Look up the purchase order each time.

The expression "look up the purchase order each time" is also used in the broader sense. The act of looking up may involve securing specific information from the laboratory or engineering department or a discussion with the manufacturing department regarding their problems in connection with the particular material or components at hand. Sometimes the exact information desired in order to reach a decision at receiving inspection must come from the sales department itself. At any rate, the receiving inspector should be equipped with the latest blue print or set of specifications if they are at all available. Some care has been used to emphasize the preparedness implied in the preceding paragraph or two because, on the whole, the receiving inspector examines parts, materials, or goods that are to various degrees unlike anything made in his own shop.

Make a Full Record of Each Receiving Inspection

One other step, in addition to making technically accurate inspections, that the inspector should surely take is to make an accurate and comprehensive record in connection with each receiving transaction. His written-down data or report should include his own plant's purchase order number, receiving slip number, and any other identifying information; the vendor's name and any order numbers the vendor uses; the date and time, of course; the count, weight, or physical inventory of the lot, batch, or order sampled; the sample sizes used, the count of defectives, or a record of the basis for acceptance or rejection; and at least a brief note as to the technical reasons for rejection. Sometimes a sample piece typical of the rejection is retained for a while.

Two kinds of disputes may arise from the results of receiving inspection. The arguments may come up three days later or three months later. (In Government renegotiation cases, some individual differences have been aired more than three years after the original transaction.) The inspector's own shop may sharply disagree with his acceptance of certain materials or parts. Or, many times, the vendor may object to his rejection. Either way, the inspector needs all the facts, records, and evidence he can command to illustrate and justify his decision. Receiving inspection records have proven valuable many times in legal suits.

As a detail in making receiving inspections and records, the inspector should carefully distinguish between original vendor manufacturing defectiveness and damage or defectiveness caused by packing, shipping, receiving, and careless handling. He may rightfully report critically on the way the goods were packed or handled, he may express an opinion as to whether the damage was done by the vendor's handlers and packers, by transportation agencies, or in his own receiving department.

Examples of Receiving Inspection

As an example of some of the details in receiving inspection, Fig. 17 is offered to illustrate first and primarily the taking of a sample. The item under inspection is a shipment of studs for aircraft equipment. Note that each of the cartons has been opened and a small but random sample from each box has been set out on the inspector's bench — all the groups adding up to a random sample of the shipment. Note, too, the punchboard style tray in which any defective pieces found are deposited. At the moment, in Fig. 17, the inspector is making a

Fig. 17. In receiving-inspection procedure, a small random sample lot is taken from each of the packages received, all of the lots adding up to a random sample of the shipment.

permanent record of his sampling in the modern form of a quality control "lot plot" of the characteristic of pitch diameter.

Good receiving inspectors are better chosen from the ranks of shop inspectors who have had at least some months' experience as process, batch, and total lot inspectors. They should have had considerable practice with gages and instruments. For them to have served stints in the plant's assembly lines, the repair department, in the laboratory, as well as along the machine lines, represents worthwhile experience for receiving inspection work. Even then, the receiving inspector must study, learn, and observe. If the plant is at all progressive, something new and unusual to be inspected and checked is continually coming in the receiving door.

Accurate, economical, timely, inspection records are a most important part of an inspector's job — the records are the results, and the results must be *used*, or the inspection effort was wasted. Inspection results are the second, third, or fourth translation of the direct reading or assessment from the inspection. If the results or records do not accurately reflect the true assessment, the *data* have no integrity — the data are wasted and useless, in fact, more than this, they will lead to incorrect conclusions and assumptions.

The records must be relevant and to the point of the inspection. Not only must the records reflect the *results* of the inspection, these records must contain additional information that identifies time and other circumstances of the production environment that would allow corrective action to follow from the results, should that be necessary.

Knowledge of the "process" is handy for the inspector, and whoever is planning the data collection/inspection effort with the inspector. When the process is understood, whether that process is a one-step operation across the table, or a complicated manufacturing system 2000 miles away, the question is: Who needs to know about the results of this inspection, and what do they need to know to control the quality, or the assessed criteria, of the product under investigation?

Inspection Records

The particular factory or production control order number should be recorded. This might be termed a lot or batch number. The part number or numbers and the material or assembly number should also be noted. Such items as the building and manufacturing department symbol, the machine or apparatus number, the operator's name or number are valuable information on any inspection record. A memorandum of the particular operation should be made, like annealing, cut-off, drill two saddle holes, or operation number 46-A if the process follows a numbered operation sheet.

Of course, never forget the data on any tag, form, report, or memorandum. It is also good practice to note the time of day of the inspection. Additionally, *any* applicable detail such as shift or name (of inspector and/or assembler) should be included on the record or inspection tally sheet. Records can be properly drawn directly from the inspection sheet itself, providing that all necessary detail of information is available from that sheet, or records can be delivered in the form of reports summarizing inspection results. In any case, as few translations as possible will aid the accuracy and integrity of the data. Whenever possible, inspectors should know what information is required upfront, and provide it as directly and economically as possible.

Sample size, in terms of the number of units inspected and the number contained in the lot or batch itself, is critical in maintaining usable inspection records. From this information, process performance and product quality levels can be deduced. Numbers are critical to the proper assessment of systems and processes, and if the "raw" numbers are accessible in inspection records — number inspected (out of how many units?), number defective, number of defects, etc., then more advanced statistical interpretation of the results, centering around such mathematical applications as average, range, and standard deviation, can be accomplished.

An inspector should always make some sort of memorandum concerning the disposition of the goods, wherever he enters the production flow. If he accepts or "passes" the work, it is assumed that the goods proceeded along established channels, but many situations arise where the record of such an event proves useful. The time, for instance, at which a certain lot passed a certain point in the process is frequently valuable. And exactly what happened to the goods where a rejection was made is of course an essential record.

There should be few or no occasions where the inspector would not validate his report, record, memorandum, or recommendation by signing his name or initials. "Whodunit," especially in the case of an inspector's rejecting goods, is a question frequently asked in a factory.

In most plants, the inspection department has its own tags or forms, usually in color, which call the attention of everyone concerned to the inspection activity.

Whatever routine factory forms are furnished the inspector and whatever details of information they demand, the inspector should see that the preceding information somehow appears on any reports he pencils, even though the shop tag does not specifically call for one or the other items.

Extensive records made by a receiving inspector have been found essential in disputes with vendors over goods returned. Sometimes credits are not issued or bills are left unpaid. Original receiving inspection records, in fact, have been accepted as substantial legal evidence in court cases.

Keeping track of time and of lot numbers, machine numbers and other descriptive data is important because discussions and disputes arise many times over the wrong group of work. It might be valuable to know, for instance, and to say, that the work coming from a certain operation (all having the same lot or order number) was satisfactory until about 10 A.M. of a certain day, that subsequent work was defective and rejected, and that the trouble was not cleared up until about noon.

To be able, in addition to accepting and rejecting work accurately and assaying the cause of trouble intelligently, to report comprehensively at some future time, concerning the whole situation enhances the inspector's position and reputation in the shop very materially.

Budgeting Time Helps Inspector Keep Ahead

The amount of work generally delegated to an inspector makes time his enemy. If he doesn't watch out, he will find himself never caught up. Aside from many common distractions, there are frequent legitimate interruptions. He may conscientiously plan to oversee the work at ten machines in the succeeding hour, or to finish the sampling and screening of the batches delivered to his bench, only to be held up by some unforeseen but necessary reinspection, by the need for calibrating a gage, for instance, to hunt up a missing blue print, perhaps, or even to be forced to stop and wrangle over a rejection with a supervisor or engineer.

It is useless for the average inspector to delude himself that his schedule will go along as he has planned it. Far better that he deliberately allow time for potential interruptions, delays, and traffic jams.

Hence the importance of odd moments. As an example, suppose he completes a round of operations or the inspection of a batch sooner than he anticipated. The ten minutes or so gained, say, he might well use to check his micrometers, clean and oil gage blocks, master a comparator, or fill out a report. Previous mention has been made of carefully putting gages and apparatus away in such a manner they are completely ready for the next use. Work of this miscellaneous character should not be left to that hoped-for free hour.

There are frequent extenuating circumstances in industry where overruling an inspector cannot be avoided. The absolute need for the goods in order to maintain subsequent orderly production or to meet stringent shipping schedules can require the movement of the goods even though they contain a high proportion of substandard units. Sometimes engineers and manufacturing people can see ways unknown to the inspector of salvaging a rejected lot or modifying some subsequent production operation to accommodate the substandard work and the rejection order is reversed. But the manufacturing house that does not restore full legitimacy to the quality and inspection need is cheating itself. Time, in terms of operational standards and formal proceduralized expectations, must be relegated to the inspection effort, as it is optimally designed, or it will not take place. Training, calibration, maintenance, data collection, and reporting all take time, as does on-line multidisciplinary problem solving. Unless these time and money needs are acknowledged and dealt with formally, inspection cannot contribute to expectations.

Constructively, there are several things an inspector can do to mitigate such circumstances. Part of the trouble can well be caused by not quite enough personal manufacturing knowledge, or a suitable conception of the commercial standards for the product his company sells, or sufficient detailed information concerning subsequent operations. Ignorance, of course, can always be overcome. Again, if the reversals are based mostly on disagreements over standards, specifications, or tolerances, the inspector can start the ball rolling toward getting official alterations investigated and effected. He might be the one to instigate a formal meeting of minds over disputed standards.

A Little Economics Provides a Better Perspective

People and companies are in business to make a profit. No business can exist long unless the income exceeds the outlay. Losses and expense in manufacturing can be built up in hundreds of different ways. If an inspector's interpretation of surface standards, for instance, unwittingly causes extra and unnecessary dressings of a grinding wheel, an unneeded loss has been incurred. For a company to buy even one wheel per year more than it needs is just one more drop of loss leaking from the bucket. Where an inspector fails to complete an inspection on time and causes a trucker to retrace his steps in order to pick up the delayed batch, a few more pennies' loss have disappeared down the drain.

In a sense, an inspector is constantly in a position to sabotage his company's profits. If he too constantly approves substandard work, then extra effort, motions, time, and cost can build up at assembly or the inspector's carelessness may boomerang in the form of lost trade in the company's market from customer dissatisfaction over inferior merchandise.

On the other hand, he can build up unnecessary losses from being too strict and self-righteous at his inspections. Always, somewhere in between, is the correct balance. No inspector probably ever achieves it, but he can assuredly try constantly for perfect decisions.

More than many others in the manufacturing areas, an inspector should acquire the broader thinking of top management and learn to realize the long-range effect of his succession of immediate decisions on the company's profit and loss statement. Because inspection effort — the actual mechanics and techniques of inspections and gaging — are all too frequently necessarily picayunish and confining, it is all the more difficult for an inspector to "see the forest for the trees." When he is engrossed over distinguishing size differences of a tenth of a thousandth of an inch in parts, the climate is not conducive to remembering the effect of his measurements for good or bad on some product being shipped half way around the world. Dealing daily with only a dozen diverse humans running as many machines, plus perhaps wrangling with local supervision and engineering, presents a very limited industrial horizon. The inspector lives usually in a very circumscribed valley, and he needs regularly to climb the mountains, deliberately to get up on mental heights, and refresh his viewpoint by looking over (if only in imagination) the broader landscape of the business he is in.

Inspection calls for knowledge, experience, judgment, vision, stability, consistency, and horse sense, as well as an ability to expertly manipulate a set of gages. An inspector's conscientious attempt to develop such traits, in addition to manual dexterity and technical manufacturing knowledge, plus his growth in such factors as ingenuity, self-fueling energy, drive, friendliness and ability to influence people will soon work him out of a job — into a better one.

Statistical Quality Control — An Inspection Tool

There are few manufacturing plants or shops in the country into which a new inspector could not step and not hear something said about "quality control" or "statistical quality control." In the more progressive factories, he would soon be introduced to sampling tables and control charts. Statistical quality control uses a mathematical and statistical approach and makes use of that branch of applied mathematics based on the theories of probability. Quality control methods make use of a few samples to estimate accurately the condition of the lot or batch. There is no need to feel alarmed over the term quality control by statistical methods. The procedure is to go ahead and try statistical quality control techniques.

Statistical process control is now an essential ingredient of any human operation producing variation of any sort, manufacturing or otherwise. These techniques are time-proven and industry-successful techniques of getting more value from the inspection dollar. The most powerful statistical tools of data analysis and tracking are the most simple tools, and everyone can and should learn to use them, especially in the world of inspection.

Like all the other techniques of improvement or record-keeping or change outlined in the previous chapters, statistical charting and analysis techniques are only as good as the inspection itself. If the data are not accurate, without bias, and speedy, economical, and timely, then all the charting and manipulation techniques in the world will not assist the recipient of the data.

The rules of the game are simple: Go about inspection or measurement as you would normally: Define the critical criteria or parameters, and apply these through inspection and an operational definition of the inspection. What is to be measured, and how is the measurement to be obtained? These are the essential predecessors of any good data analysis system — good data.

Attack the problems — out of these problems will come opportunities to collect data. After the problem has been defined and the inspection process has been determined, a data form must be designed that is simple, relevant, and timely. Have a plan, ensure that the plan can be instituted through training, and define the proper time to take the inspection.

Define the sample — this is the step that takes ordinary data collection into the realm of statistics. Follow easy and time-proven rules to ensure that the data are obtained from a random sample, and that the data relate properly to the population from which they arise.

Both attributes and variables data can be plotted and analyzed statistically. However, much more accurate and decisive information can be gleaned from variables data, and the statistical study of the behavior of variables data as a description of variation.

Both attributes and variables data can be plotted and analyzed statistically. However, much more accurate and decisive information can be gleaned from variables data, and the statistical study of the behavior of variables data as a description of variation.

In using variables data, the critical parameters or measurements of the product or process detail is simply assessed through inspection and plotted over time against simple comparators such as the average range, and standard deviation of the numbers or data gathered. With such a comparison, assessments or conclusions can be drawn relative to the behavior of the data in consideration of the attending specifications — do the measurements fall within spec? All or part, and how do the run of measurements relate to the overall average of process center, and what is the range of measurements accrued?

Such an on-going measurement and analysis tool soon becomes second-nature to the inspection–manufacturing link. The trends and histories that are detailed through these presentations of inspection data show whether or not the processes are producing satisfactory product.

If variables data are unavailable and attributes or pass/fail characteristics are being plotted or tracked, the principles remain the same. What is the percentage defective rate? What are the average number of defects per 50 units? These and other questions are not so far afield from our everyday analysis of information that is constantly being absorbed.

What is the average time it takes to get to work? And what is the range, from longest to shortest? Of course, if it suddenly takes three hours to drive to work when a typical time is 10 minutes, you do not need a control chart to tell you that something unusual has happened. But in a manufacturing environment, the discipline and overall measurement capabilities derived from simple statistical tools are invaluable.

Once the inspection has taken place, the results are in, the data are reliable, economical, correct, and speedy, then the fun begins. Analyze the data: What are the "top hitters," the most numerous problems — this is a Pareto analysis, a method of ranking and analyzing defects that simply puts the most numerous failures at the top of the list.

Measurement, comparison, and graphic portrayal of defect and defective unit information is necessary to assess the level of variation present, and derive causes for excessive variation, which may lead to solutions. Bar charts, histograms (Fig. 18), and frequency distributions are examples of such statistical devices.

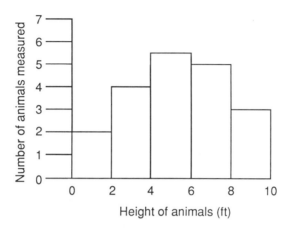

Fig. 18. Example of a histogram.

Control charts that plot the trends of variables data, as well as attribute defect rates, are simple to use and helpful *if* the data are acted upon. In most cases, statistical applications are only helpful if people use them to fix problems. The links among problems of variation, causes, and solutions is often a long and difficult road. But again, simple methods of cause and effect analysis and other problem-solving routines can make quick work of variation *if* the ability to measure accurately describe, and assess that variation are combined in an atmosphere that pursues constant improvement of the system.

Statistical process (SPC) or quality control, then, is simple, more simple than most people believe. Taken by itself it consists of a group of tools that are used to assist in analyzing (inspection) measurements. These tools start out simple and become more mathematically oriented. But they are still just tools — tools for analyzing measurements.

The various SPC control charts used are as follows.

Xbar and R — Variables data, average and range
P, NP — Defectives (attributes data), percentage or raw number
C, u — Defects (attributes), percentage or raw number

SPC becomes difficult when the larger picture is considered — the picture of utilizing these various tools. SPC is often difficult to implement, on a large industrial scale, because of its ultimate focus on problem-solving, on *fixing* things. This tough part usually involves different people's or groups' combined efforts — groups that have to work together with the goal of reducing critical variation in the process and product.

Attributes-based control charts take "normal" inspection record-keeping a step further — based on a period of time, or a volume of parts or assemblies, the "defect" rate, or the rate of "defectives" is identified. This rate can exist as a percentage ratio, if the number inspected varies over time, or can be expressed as a raw number — five defectives — *if* the number or sample inspected is constant over time. Plotting this on a chart simply means showing the ratio over time. The chart will display the average number or percentage defective or of defects, or the number or percentage of defects. Continuous monitoring of these results, and comparison with the overall average and (standard deviation) control limits, will identify acceptable or unacceptable (out of control) levels of defects or defectives.

With attributes data, this is all you need to know — "in control" means: keep working and measuring and recording. An "out-of-control" statement means: stop working, the "normal" rate of defects or defectives (normal is *not* perfect) has been exceeded.

At this point, problem solving begins — the problem is described, in terms of the defect from the Pareto chart (Fig. 19) of most significant contributor to the high rate (most important defect). Then cause-and-effect or other problem-solving methods are used to identify the cause. A solution is proposed, accepted, and modeled — if the problem was solved, subsequent inspection results will be "in control" when plotted on the chart. If the problem was not linked to the particular cause and solution, the rate of defects or defectives will still be up and "out of control." You will have to restart the investigation.

Week One Production Defect Rate

Trim Down

(1) Too large (7)
(2) Too small (5)
(3) Oddly shaped (4)
(4) Others

Fig. 19. Pareto chart.

Another tool that may be used in problem solving is the process flow analysis — simply, a sequential diagramatic picture of the ordered process, whatever that process might be (Fig. 20). Such a representation of the actual (process) situation can assist in understanding and visualizing process relationships that may have contributed to the problem or out-of-control situation.

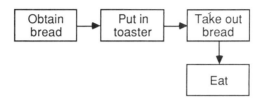

Fig. 20. Process flow analysis of making toast.

Numbers, or variables data, afford more opportunities for analysis than the words that qualify attributes ratings, so they can be more complex as the analysis becomes more sophisticated. But the principles are the same — samples, small samples typically of from three to five measurements are taken of a critical variable or parameter. Remember that this parameter is probably quantified in a specification somewhere — from nominal to the plus and minus ranges of acceptability.

The measurements are taken and averaged (to show the *level*) of the readings from an overall point of view. They are subtracted (lowest from highest) to find the *range* of the sample. These two numerical derivations are the basis for Xbar and R charts — and these charts reference the trends or runs of the separate samples in terms of Xbar (the average of all the sample levels) and Rbar (the average of all the ranges) (Fig. 21). The movement of the sample averages and ranges in relation to the standard deviations of the overall averages and ranges show a sensitivity to change in the parts' measurements that is very important in controlling variation in the parts — before they exceed the specs. In other words, the (standard deviation) control limits show when the measurements exceed their "normal" variation. The comparison of this "normal" variation to spec is the comparison of dimensional reality to what is desired (spec).

Likewise, as parts exceed their normal levels of control (which should be within spec) in the process, problem-solving and experimentation must be initiated to bring the variation back into normal or acceptable limits.

This problem solving can involve the use of other statistical tools — histograms, or picture graphs, which show graphically the

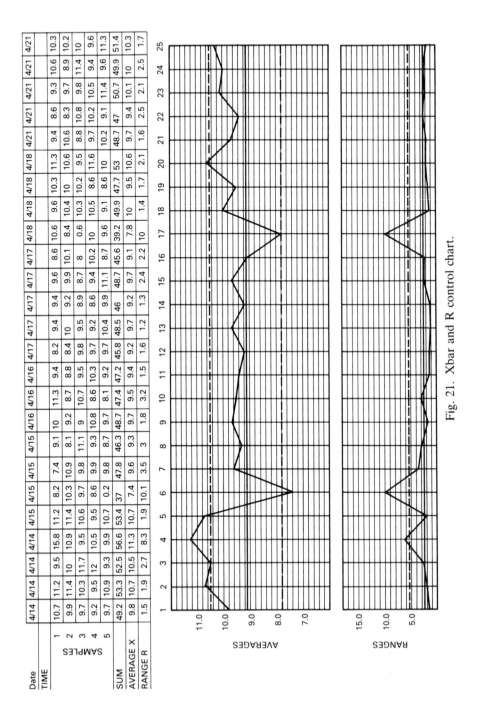

Fig. 21. Xbar and R control chart.

spread of the range of measurements, and the frequency of each measurement as it occurs from a group, and other mathematical interpolations of the relationship of the data numbers.

The scenario is at once simple and complex:

1. Define the (product and process) requirements
2. Define the inspection routine
3. Inspect and record data
4. Analyze data and identify problems
5. Determine causes and solutions to problems
6. Fix problems (restore acceptable requirements)

CHAPTER 6

Basic Principles and Techniques of Measurement

Measurement is the most important function of industrial inspection. The word inspection generally brings to mind precision measurement. While linear measurement is usually thought of, an inspection may mean reading watts, amperes, oscillations or other electrical units. Products may have to be weighed in pounds, ounces, or grams; a process may be timed with a stop watch. Hardness, elasticity, viscosity, torsion, and temperature are common inspection terms. Counting what can be seen in the field of a microscope is a form of measurement. In many plants the inspector delves in the fields of physics, chemistry, and metallurgy and measures density, humidity, alkalinity, or grain size. The whir of ball bearings is registered on an audiometer.

Whenever possible, measurements or inspections should yield a result that is understood quantitatively — through the use of numbers. Numbers are the "stuff" of measurement, and whatever the measurement or assessment, it is described more accurately if that description includes a number that is referenced against a standard unit of measure. To arbitrarily declare a block of stone to be too big or too small will not result in a successful pyramid.

Quantitative measurements are described as variables data — numbers — and offer more tangible information in an inspection than a qualitative assessment of a component or workpiece. These qualitative assessments are called attributes data–pass or fail. Attribute characteristics can only be qualified by words, not numbers, and as such do not contribute as much concrete information as a dimensional measurement — numbers.

The eye and the sense of touch are our basic tools of measure. But successful measurements, that is, accurate and repeatable measurements, are possible only after the addition of a dependable numerical system and a common reference standard to our basic senses.

History of Measurement

Civilization's earliest measurements were merely representations and interpreations of our five senses. Units of measurement or comparison were literally at our fingertips — hands, arms, feet, legs. These became arbitrary standards of reference; often the appendages of famous or powerful individuals, such as kings or other rulers, were established as the "standard" measurement of the land.

Measurements generally combine the standard or unit of measure (foot, fathom) with a count of *how many* standard units are tallied or "measured." Five lengths of the king's arm may even have come to be the standard unit for a "common" or standard building stone. The oversight, of course, rests in the fact that the king often had difficulty getting to all the constructions sites.

The missing link in the creation of reliable measurement systems became fairly obvious over time. The use of the king's forearm was not a bad idea, as long as many standardized *representations* of the king's forearm were available. The manufacture of similarly dimensioned units, based on the standard, was the answer.

Common units of counting and measure have been delivered to the western world down through the annals of European recorded history. The use and distillation of these measurements have resulted in the English system of western Europe, which is manifested in the U.S. Customary (measurement) system today. These familiar units of measure are the results of a long evolutionary measurement system.

Divisions or standards of inches, feet, and yards are summarily divided by what we call "common fractions" (1/2, 1/4, 1/8, 1/16, etc.), or decimal fractions (1 in., 0.1 in., 0.01 in., etc.). These units of measure form the basis of the standard system of measure that most people use in the U.S. today.

Recently, the United States has pledged to change and keep pace with the rest of the world, which uses the "SI Metric System" or the "International System." This system has indeed gained prominence as the foremost measuring system in the world. A common system of measurement, to be used throughout the world, without an unwieldy set of formulas for translation, has an obvious appeal to efficiency and standardization.

Perhaps one of the most useful and attractive aspects of the metric system is the use of the number 10 as the basis or base for its counting *and* measurement operations throughout the entire system.

Many of the standards dealt with in this book are measurements of *linear* units — measures of units of length, width, height, and "straight-line" progression. These units of linear projection or scale are commonly delineated in either the English or the metric scales, and, of course, they are easily translatable.

Other basic units of measure deal with some other mode of description of the physical universe — volume measures three-dimensional displacements of space, which uses linear measurements as part of the measuring formula. Angles and arcs are also identified by their own unique systems of measurement. These measurements and more — area, circumference, and other geometric arrangements — also make up special measurement systems and combinational relationships that are described in our physical world. These unique measurement systems and geometric relationships provide special challenges to the tasks of measurement and gaging.

How to Use The Steel Rule Correctly

The primary factory measuring tool or aid in the inspection of machined products is ordinarily the steel rule. A picture of a standard 6-inch machinists' steel rule (sometimes erroneously called a scale) is shown in Fig. 1, along with two illustrations of common uses.

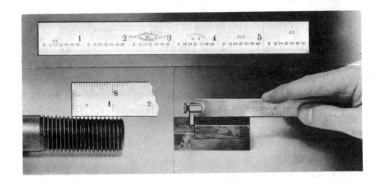

Fig. 1. Typical uses for a standard machinists' rule.

Steel rules (Fig. 2) are simple instruments of linear assessment, and discriminate in the following categories of increments: (1) Fractional inches, usually to 1/64th of an inch. (2) A decimal scale, with a typical determination or discrimination of 0.01. (3) Millimeters and half millimeters.

Steel rules are made of three types or varieties in the western world. These are (1) rigid rules, (2) flexible rules, (3) shorty rule set. Additionally, steel rules are manufactured for a variety of special applications, such as rules with tapered ends for easy insertion into areas that may be difficult to assess with a regular sized and shaped rule.

If you want accuracy, the steel rule should never be used in the manner shown at the left in Fig. 3. It is much better to butt the piece against a knee, or some similar flat surface, and measure as shown at the right in Fig. 3. The point is that the end of the steel rule is liable to

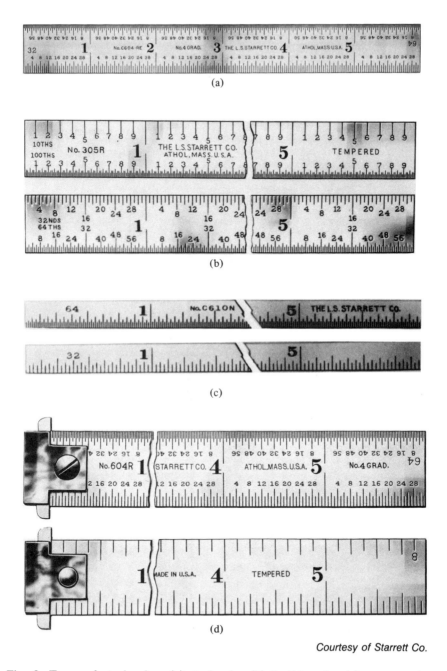

Courtesy of Starrett Co.

Fig. 2. Types of steel rules: (a) steel rule; (b) flexible rule; (c) narrow rule; (d) hook rule; (e) narrow hook rule; (f) rule set; (g) tapered end rule; (h) drill point gage.

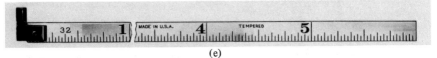

(e)

Courtesy of Starrett Co.

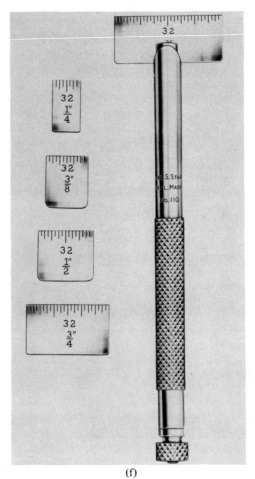

(f)

Courtesy of Starrett Co.

Fig. 2 (*Continued*)

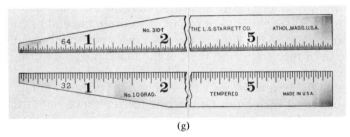

(g)

Courtesy of Starrett Co.

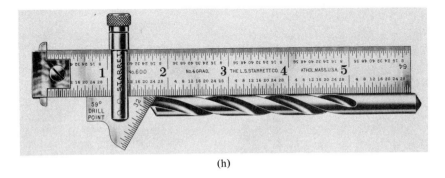

(h)

Courtesy of Starrett Co.

Fig. 2 (*Continued*)

be worn or rounded, or its corners may be crushed in. The condition may not be apparent to the casual glance, but it can exist in sufficient degree to produce errors in reading measurements. (Magnify the zero or working end of a steel rule in an optical comparator. The warning above may then seem worth heeding.) Another good reason for measuring in the fashion just described is the fact that the edge or

Fig. 3. (Left) Incorrect method of using a steel rule.
(Right) Correct method.

corner of the workpiece being measured may not be sharp and square. And still more to the point — it is harder to bring the end or corner of the rule coincident to the edge of the workpiece accurately than it is to coincide one of the graduation lines in the manner of measuring about to be described.

Probably the best way to use the steel rule is pictured in Fig. 4. It is better to use the 1-inch mark on the rule for reference rather than the 0-inch end (not forgetting of course to subtract 1 inch from the reading actually taken). The rule should be stood up on the work more or less perpendicularly, rather than laid flat, as in the left view of Fig. 3.

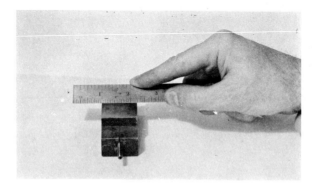

Fig. 4. Best method for obtaining accurate measurements with a steel rule.

Steel rules called hook rules can be purchased, the little bars or hooks on the end of which serve in the place of butt plates (as shown in Fig. 5). Where, however, the hook rule is to be relied on for fairly precise measurements, the "hook" must be checked regularly to be sure it has not loosened, that it is still square with the length of the rule, and that it has not worn back nor rounded off. A general difficulty in connection with various helpful attachments offered with measuring instruments to make their use easier or quicker is the fact that the

Fig. 5. Steel rule equipped with hook for accurate referencing.

accuracy of the attachment itself is too seldom questioned. The attachment, like a hook or a caliper jaw on a steel rule, becomes relied on and yet by its very nature, design, and use proves to be the item that becomes bent, worn, or loose and makes the measurements obtained inaccurate.

Rule clamps (Fig. 6) can be used to hold two steel rules together for straight-line measurement that is not possible with just one rule. Likewise, key seat clamps (Fig. 7) are available for a linear translation

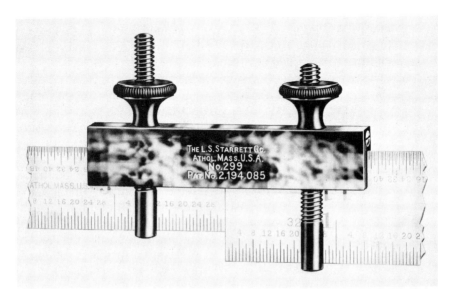

Fig. 6. Rule clamps.

Fig. 7. Key seat clamps.

of the steel rule that can be used on round stock — for laying out keyways and parallel lines on round stock.

Right angle clamps that provide a firm 90° angle in the flat sides of two rules are also available for rough layout work and other shop applications. And, finally, there are rule holders that align a steel rule at 90° for translation work on a surface plate.

From Reference Point to Measured Point

Right here, while the practice of measurement with the steel rule is being discussed, might be a good place to describe one or two elementary and basic principles of measurement.

Any linear measurement can be broken down into two "points": the reference point and the measured point. Actually the "points" may be true points, in the geometric sense, or they may be a pair of lines or edges, two planes or sides, or two circles. But though we measure between two walls or two edges, if the point to point conception of measuring is kept in mind, the results will be more accurate.

We measure *from* the reference point *to* the measured point. In using any measuring instrument we set the instrument first on the reference point and move the instrument — or read along it — until we find the measured point. (In the case of large, heavy gages and measuring instruments, where the work is brought to the instrument rather than the instrument to the work, the same principle is followed. The work is "referenced" first on the gage's reference surface.)

There is always an uncertainty in measurement, derived from an attempt to strictly and succinctly define the geometric characteristics of a physical object. Pure perfection in measurement is a goal to be pursued. The act of measuring ensures the possibility of error in that assessment or description.

The dimension, or the measured description of a physical object, is accomplished through a quantified description of its variables. Those variables or properties that are to be measured or assessed are described through a numbered description of the *form* of the object by referencing the *position* of that object in relation to the measuring system itself. The measuring system uses fixed coordinates that are physically apart from the object being measured, such as the 1-in. mark on a steel rule. The description of the form of the workpiece is accomplished relative to the measuring system that is directly outlined on the increments of the rule.

In the act of measuring, therefore, the use of the steel rule relates the object to be measured to a position point on the scale of the rule. The 1-in. mark referenced to the workpiece sets the stage to describe

the form or dimension (in this case linear dimension) of the object being measured on the standardized dimensional scale. Understanding the relationship between form and position ensures the proper use of a measurement system to assess and describe geometric relationships.

Where, in the right view of Fig. 3, we butt the steel rule against a knee or parallel, we are establishing a firm reference point. If instead of using a butt plate, we index the rule on the edge of the work at its 1-inch mark, as in Fig. 8, the 1-in. mark becomes the reference point. Setting the steel rule on a reference point and holding it firmly there is just as essential for accuracy of measurement as reading correctly at the measured point, if not more so.

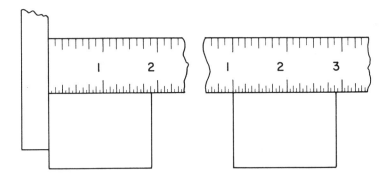

Fig. 8. (Left) Use of butt plate as a reference point. (Right) Use of 1-in. mark as a reference point.

The purpose of the hook rule or a caliper jaw or most any similar accessory is to help establish, conveniently, a firm reference point.

Measuring the diameter of a hole with a steel rule, as illustrated in Fig. 9, forms an excellent exercise for establishing the principle of the reference point and the measured point in your mind. Here, the 1-in. mark is the reference point. The rule is carefully set on one point of the circumference of the hole so that it can be pivoted about the 1-in. mark. To get the measurement or the measured point, the rule is swung back and forth slightly along the arc of the hole opposite the stationary 1-in. mark — the 1-in. mark acting as the pivot or hinge — until the opposite point on the longest chord, which is the diameter, is located, and this diameter is correctly measured and read.

When a succession of similar pieces is to be measured, the accuracy of measurement and the comparison of the dimensions of the piece are facilitated when the reference point for each piece is approximately in the same location and the measurements are all taken in the same direction. This idea is illustrated in Fig. 10 where, because a counter-

Fig. 9. Method used to measure diameter of a hole with steel rule.

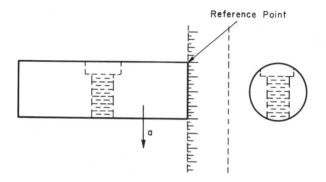

Fig. 10. Use of a similarly located reference point when succession of similar pieces is to be measured.

bored tapped hole marks a direction on a shaft, the diameter measurement is perhaps more consistently taken in the *a* direction rather than haphazardly.

Direct and Indirect Measurement

There are two primary methods of measurements that are used in the shop, the inspection locale, or, for that matter, around the home:

1. *Direct* comparison with an established standard.
2. *Indirect* comparison with a standard, by deriving a *difference* between the standard and the workpiece, which is then measured directly on a calibrated standard.

Direct measurement (Fig. 11a) systems are most familiar in our world. This method uses a direct assessment of the workpiece with a system of standardized increments (such as the steel rule or micrometer) to derive a measurement.

An indirect measurement (Fig. 11b) system identifies the *difference* between the object being measured and a known dimension. This type of measurement is the basis for the use of comparators — those "transfer" instruments that contact and assess the workpiece and then compare that contact to an external scale which determines the amount or dimension of the expressed difference. Other examples of indirect measurement tools include the caliper or the sine bar (for angle measurement).

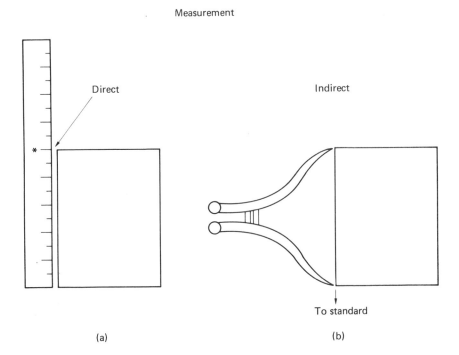

Fig. 11. (a) Direct measurement; (b) indirect measurement.

Repetition is the Test of an Accurate Measurement

The test of a measurement is *repetition,* that is, repeated readings that are the same. If you measure the diameter of a shaft or the width of a bar several times in the same location and get different readings, you are probably not taking the measurement accurately. Thus, improperly using the worn end of a steel rule as a reference point, as in Fig. 3 (left), will probably fail to give repeat readings. The lack of repetition can also indicate that something is wrong with the measuring instrument itself. While such a condition would not ordinarily show up with the proper use of the standard steel rule, lack of repetition will show up loose and worn hooks or caliper jaws.

The graduations or marks on a steel rule have intrinsic width, especially under a glass, no matter how carefully they have been etched or engraved on the rule. Manufacturers of this type of equipment attempt to keep the width of the lines as near .003 inch as possible. Since the steel rule is not supposed to read closer than 1/64 inch (.015 inch), any error produced by the width of the graduations would not exceed one-fifth the natural ability for reading the finest division on the rule. Even so, especially when a magnifying glass is used, the width of the graduation mark might cause an error in a reading, or seem to. Where the lines on a steel rule are engine engraved, the graduations are really sharp V's, the apparent width of the marks being the open or upper ends of the V's. Hence, in using a rule on very fine work, measure from the center of one graduation — the reference point — to the center of the graduation marking the measured point.

Accuracy and Precision in Measuring

Accuracy is represented in the amount of error possessed by an instrument — the instrument error. In assessing the performance or measurement of performance of a measuring instrument, the performance expectations of the instrument must first be considered in terms of *accuracy*. Obviously, the choice of the proper measuring tool for an inspection is based on a prior expectation of that tool's accuracy.

Accuracy equals error, which is very important in inspection results and most acts of measurement. Error, then, is the difference between the *true* or absolute dimension of the variable being measured and the *reading* of an instrument, the perceived dimensional result of the measurement.

Error, as an expression of the accuracy of a measuring instrument, is known as the *uncertainty* of an instrument. No one should use a measuring device that produces an error, unless, of course, the extent of that error is understood. Knowledge of the amount of error inherent in a measuring tool in a particular application is essential and is expressed as a percentage or *number* of units (deviation) from the actual or direct measurement that is being sought through the use of a particular measuring instrument.

Precision is also related to the accuracy or measuring ability of an instrument — in this sense the term is not new to the lay person. The exact meaning of precision, however, rests on the familiar concept of *repeatability*, or how many measurements are the same, from a particular series of readings or measurements. The closeness of the results of those measurements, taken in the same manner under the same conditions and criteria of measurement, represents precision. As measuring instruments "wear out," and need calibration, they lose precision.

Discrimination Means Size of Scale Unit

Steel rules ordinarily have four scales — 1/8 inch, 1/16 inch, 1/32 inch, and 1/64 inch. If a certain scale has been chosen and if the reading falls between graduations, then measure with the next finer scale. Figure 12 portrays what is meant. The number of graduations on an instrument or the degree to which it subdivides an inch, for instance, is known as its *discrimination*. Except where it is absolutely necessary because of lack of equipment, an estimate of the reading between graduations should never be made. In the first place it is a lazy habit; better look around for apparatus with finer discrimination. In the second place none of us are as expert as we think we are at estimating. When the actual reading falls between graduations, make up your mind whether you are going to choose the lower or higher graduation value for your reading, if a finer scale is not available.

The inch on the finest scale on most steel rules is divided into sixty-fourths. It has been found through the years that few humans can determine measurements accurately closer than 1/64th of an inch, at

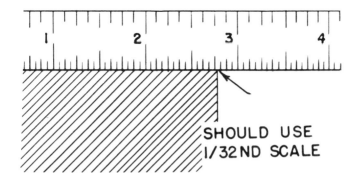

Fig. 12. When a reading falls between graduations, a scale with finer divisions should be used.

least without the aid of a magnifying glass. Then, too, without mechanical aids such as clamping and fine adjustment devices, few of us can hold a steel rule steady enough to get an accurate reading below a 64th. Some mechanics claim an ability to read and measure down to .005 inch, but they have difficulty proving consistent hour-in and hour-out accuracy.

Related Elements of Measurement

Other related elements of measurement are also important to the concepts and understanding of basic measurement techniques and inspection accuracy (Fig. 13). What is the *range* of the measuring instrument; do not try to assess the distance from New York to Chicago with a yardstick. The scale of the measurement system must match the level of variation being assessed — contact points on a gage that open and close to minimum and maximum conditions must be able to encompass the allowable variation of the part being gaged. Measurement devices and instruments possessed of a large range may offer a trade off in terms of discrimination.

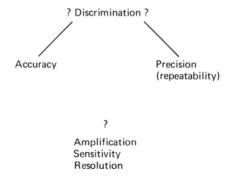

Fig. 13. Elements of measurement.

Amplification is a topic that will be brought up again later — a concept important to measuring and measurement that is closely related to the notion of range in many gaging operations. Amplification describes the movement of a measuring device's contact points in relation to the needle or readout or indication of movement on the measurement scale. How is the actual gage contact to the workpiece variation translated to the movement of the dial or scale? Again, mechanical and electronic instruments having great amplification (and thus more discrete scales of discrimination), have less range.

Amplification of workpiece variation is a prime factor in the high performance characteristics of mechanical, air, optical, and electronic gaging and measurement systems.

The principles of measurement would not be complete without a discussion of the bain of repeatability — hysteresis. Hysteresis is a representation of the differences derived from taking readings in different physical manners or fashions, such as from the top down, and then from the bottom up. Fully responsive to the physical laws of nature, hysteresis dictates that a repeatable and precise reading is hard to come by, and is a factor in most applications of mechanical measurement.

Likewise, the principles of *sensitivity* and *resolution* are key to understanding the measuring ability of a measuring instrument — how close is close enough? Is it "good" or "bad," and how much variation is present? Are the inspection results a matter of luck, of finesse, and how is the measuring ability of the instrument contributing to the results (and the inspector or measuring person)? These questions can only be answered if clear distinctions are understood among the concepts of accuracy, repeatability, precision, and other factors that can result in measurement error or uncertainty.

For instance, the ebb and flow of the tides would not be assessed by a galvanometer, to establish a case in point, because the galvanometer would not be able to measure accurately or even discern the movement of the tides — the important variable to be measured. The correct instrument, with the correct scale of discrimination, must be applied properly in order to obtain accurate, repeatable, and precise measurements.

Sensitivity and resolution take the analogy one step further. What is the sensitivity/resolution of the chosen inspection instrument? The sensitivity of a measuring device is related to its ability to translate mechanical contact with the variation of the workpiece into a movement of the needle on the mechanical dial. A blending of the concepts of accuracy, discrimination, and precision determines whether or not an instrument is sensitive enough mechanically to assess the variation that is present. Resolution refers to sensitivity of a digital or incremental readout. What is the increment of variation or change that is needed to see a change from a "0" to a "1" on the display?

What Measurement Tool to Use?

Discrimination, then, is a major factor in the selection of which measurment tool to use — how small, or how fine or discrete, is the question. Discrimination is the ability of the measuring tool to take an exact reading to the level of dimensional variation that is observed on the measurement tool itself.

The following questions should be asked in setting up an inspection routine or sequence:

1. What is the variable to be measured? What is the discrimination of the tool or device? What level of sensitivity or resolution can be expected? What accuracy and precision is required of the measuring tool, and what conditions are required to use the tool properly?
2. How are these factors to be dealt with successfully — what tools, techniques, and sequences would be most successful in optimizing the inspection or measurement? How should the results be reported or displayed?
3. What inspection system guarantees that this same measurement technique will yield comparable or repeatable results next time?

Parallax Introduces Error in Scale Reading

In boasting of their ability to detect a difference in measurement of .005 inch or .010 inch (1/64 inch is .0156 inch), inspectors forget the error of parallax even though they may be able to hold a steel rule steady enough. Parallax is the apparent displacement or shifting of an object caused by the change in position of the observer (Fig. 14). Look squarely at the face of a clock and read the time. Then walk several feet to one side and honestly read the time there as you actually see it obliquely across the hand. The clock will read several minutes slower or

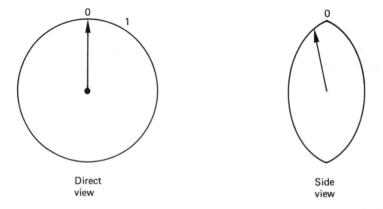

Direct
view

Side
view

Fig. 14. Parallax error.

faster, depending on which angle you view it from. All of us have unwittingly made this error.

If your nose and eyes are squarely over the reference reading on a steel rule, you will make somewhat of an error if you simply turn your eyes and try to read a graduation several inches obliquely along the rule (Fig. 15). Parallax will result in an error of several thousandths of an inch if you are trying to estimate some measurement finer than 1/64 inch. Even though you move your head, nose, and eyes over the several inches from the reference point to the measured point, you may find your hand, holding the rule, has unconsciously traveled with the body motion and moved the rule across the work a few thousandths of an inch, thus making an error in the measurement.

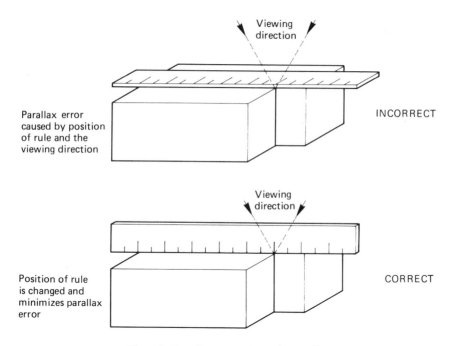

Fig. 15. Parallax error in scale reading.

Look Out for Manual Errors

With whatever instrument you measure, be it a steel rule or a supersensitive gage, let the instrument do the work. Do not cramp its style. Relax, loosen your grip. The hand trembles more readily when you grip a steel rule as if it were a crowbar. The most you want to do in measuring is to set the reference point in position and read the measured point. Hold the rule lightly, but firmly.

Then, there is another common manual error made in measuring, an unconscious one. Suppose it is difficult to decide whether or not the width of a slot is 15/64 inch or 16/64 inch (1/4 inch), and suppose the specifications call for 1/4 inch or 16/64 inch. And more particularly, suppose you really want the slot to measure 16/64 inch. Unless you realize the possibility, unless you make some degree of effort of will, your hand will unconsciously move the rule a trifle in the favored direction.

Use of the Combination Square

A common commercial variation of the steel rule with attachments is the combination square, Fig. 16. It consists of the steel rule, the center head, the sliding head or beam, the protractor head, and a scriber.

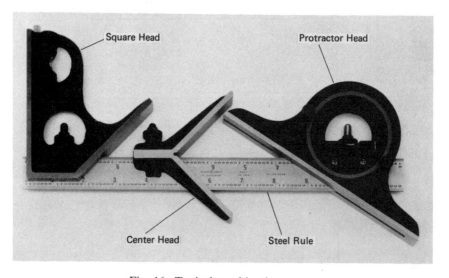

Fig. 16. Typical combination set.

A combination square is a typical multipurpose shop tool that enjoys a variety of uses and applications, including the ability to measure or discern across more than one standard (and increment) of measure.

Besides the more common or typical uses of a combination square, such as direct linear measurement, the tool can also be used to assess variable angles as a bevel protractor, the all-important right angle of 90°, an even "level" with respect to the horizon and its associated gravitational attraction, and the linear description of holes or situations displaying an element of "depth".

It is possible to use the steel rule and the sliding head to square a piece with a surface and at the same time determine whether one or the other is plumb. Also, by using the miter, it is possible to lay out 45° angles as well as 90° angles with the head. Inserted conveniently in the head is a scriber for this purpose. By setting the steel rule flush with the sliding head, it may be used as a height gage directly.

Also, by loosening the rule it is possible to use the combination as a depth gage where micrometer accuracy is not necessary.

The steel rule can be removed from the head, permitting the use of the rule and the sliding head separately. The head can be used as an ordinary level.

By substituting the center head for the sliding head, a center square is obtained for finding the center line of cylindrical objects. This center head is slotted in the center so that the rule, when inserted, bisects the 90° angle. In this way, the measuring surfaces become tangent to the circumference of cylindrical work. It is possible to locate the center of a bar.

The protractor can be inserted on the steel rule in the same manner as the sliding head and center head. The revolving turret can be graduated in degrees from 0 to 180 or to 90 in either direction. Also the head contains a spirit level to facilitate the measuring of angles in relation to the horizontal or vertical plane.

While it is not a precision instrument, it controls the accuracy of measuring and laying out angles within 1°.

The Calipers — Accessory to the Steel Rule

Calipers are also a handy and very traditionally satisfying means of assessing dimensions. These tools usually provide a measuring accuracy to within 1/64 inch, and are commonly used with steel rules and other measurement increments with compatible accuracy and discrimination.

There are four types of calipers, essentially characterized by the type of mechanical joint that connects the two sides of the unit:

1. Spring joint.
2. Firm joint.
3. Lock joint.
4. Hermaphrodite.

These types of shop and rough assessment calipers are mainstays of indirect or transfer-type measurements. They are often used with a rule in one hand and a calipers in the other — a visual joining of the indirect or transfer measuring tool with the scale or standard of measurement.

To an inspector, a pair of calipers is an accessory to the steel rule. A diameter may be measured with the steel rule as indicated in Fig. 9 but a more accurate diameter measurement will likely be gained by calipering and transferring the measurement to a steel rule as shown in Fig. 17.

 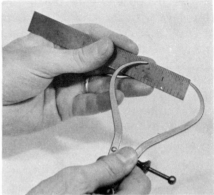

Fig. 17. Indirect measurement. (Left) Caliper being used to obtain an outside diameter measurement. (Right) Measuring a caliper setting with steel rule.

Likewise an inside diameter is measured after the manner shown in Fig. 18 and the reading is transferred to the rule by holding the rule vertically on a flat surface with the caliper ends against the rule, one caliper end resting on the flat surface.

Measure the True Diameter

Especially in taking an inside diameter (though for that matter an outside diameter, too, or any similar measurement) care must be used to measure the true diameter. Sketches A and B in Fig. 19 show incorrect measurements of an inside diameter. In each case a measurement greater than the true diameter would be obtained. One leg and point of the caliper should be set as the reference point — see sketch A in Fig. 19 — and the caliper should be rocked and adjusted until by "feel" the true diameter, the correct measurement, is secured as at C in Fig. 19.

At the same time, avoid letting the caliper legs become squeezed together slightly, thus allowing the shorter chord of the diameter circle

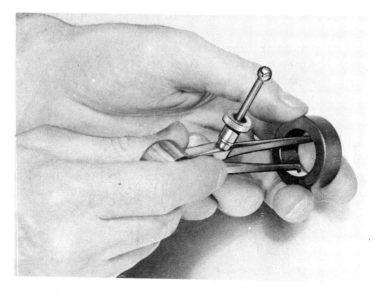

Fig. 18. Measuring an inside diameter with calipers.

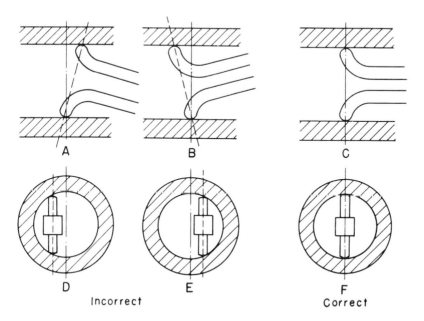

Fig. 19. (A), (B) Incorrect inside measurements are caused by tilting of calipers. (C) Correct method of measurement. (D), (E) Incorrect inside diameter measurements caused by excess calipering pressure which squeezes caliper legs together. (F) Correct method.

as at D and E in Fig. 19 to be measured rather than the true, full diameter as at F.

While calipers are ordinarily used to get diameter measurements, they can also be used to "explore" flat pieces to detect thickness, width, depth, taper, or parallelism.

Direct-Reading Calipers

Slide calipers, shown in Fig. 20, are the simplest of the direct measurement calipers available — literally a steel rule turned caliper, but are very handy for direct inside and outside readings where 1/64 inch discrimination is required.

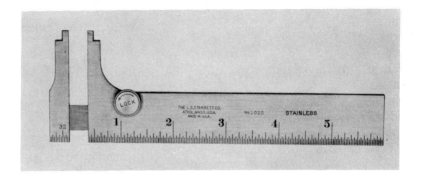

Fig. 20. Direct reading type of inside and outside caliper.

Other types of direct-measurement calipers are included in nearly all inspection and measurement work, because of the ease in which accurate, mid-range discrimination readings are attained. Unlike the types of calipers previously listed, these calipers are not compared on a separate incremental scale, but display a reading directly. These units display a wide range of discriminatory abilities, as well as varying sensitivity and resolution.

Mechanically Measuring Calipers

Dial calipers (Fig. 21) are the most typical of all hand-held calipers in use today, providing a typical direct-reading capability of .001 inch. These units are quite flexible, offering a typical measurement range of 6 inches with accessories available for range extensions, as well as specific measurement accessories.

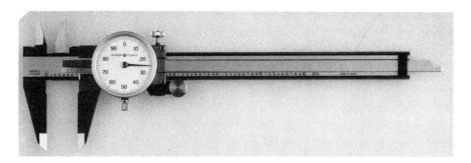

Courtesy of S-T Industries

Fig. 21. Dial caliper.

Dial calipers are actually one of the simplest forms of mechanical comparators — the translation of the linear form of the workpiece is accomplished by referencing the caliper jaws to a measurement scale by mechanical means. This mechanical translation uses principles of mechanical measurement covered in Chapter 8. The workpiece variation is assessed by a gear-train and lever (mechanical) translation technique that utilizes principles of amplification and sensitivity. Other modern calipers capitalize on electronic technology to assess and display dimensions.

Dial calipers, and other similar measurement units such as digital and electronic direct-read calipers typically offer a four function capacity for taking physical measurements: inside readings, outside measurements, depth assessment, and a step-function measurement capability. The advanced electronic versions of these calipers, as well as similarly designed micrometers, utilize basic mechanical principles of dimensional comparison and translation, with high-technology electronic enhancements through solid-state circuitry. (See Fig. 22.)

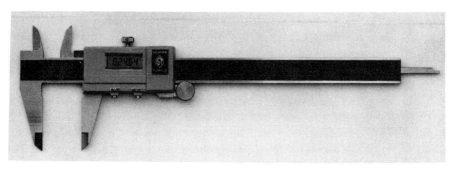

Courtesy of S-T Industries

Fig. 22. Electronic digital caliper.

Some of the most modern, microprocessor-based direct-reading calipers are also typically accurate to within .001 inch and operate without the use of gears and racks to provide the digital readout of the measured dimension, which, of course, simplifies the measurement task. These models are also automatically switchable from the English to the metric system, and are capable of direct hook-up or interface with computerized storage and display units, which chart and interpret assessed variation over time.

Gaging Pressure Should be Light

Measuring exercises with calipers furnish an excellent setting for discussing gaging pressure. Wherever gages of almost any sort are used, the pressure exerted either by the inspector's hand on the gage or by the mechanism of the gage itself will have a marked effect on the accuracy, the uniformity, or the repetition in measurements secured. Warnings concerning gaging pressure will be given in connection with measuring instruments mentioned hereafter.

As a general average, recognized gaging pressures vary between half a pound and two pounds. Some precision instruments are deliberately designed to exert measuring pressure of only a few grams. Occasionally, circumstances require a gaging or clamping pressure greater than two pounds.

An inspector using a variety of gages and measuring tools, such as calipers, micrometers, and snap and plug gages, should certainly practice and practice to develop, like a dentist or surgeon, a light, firm, consistent touch.

A good system to follow is to consider "pencil" pressure — the pressure you exert pressing the point of a pencil against a piece of paper as you write. Put a piece of paper on the platform of a postal scale, for instance. Then, resting the weight of the hand on some surface (like a book) that is about level with the postal scale platform, write with a pencil in normal fashion on the paper on the postal scale platform and read what the scale dial tells you in ounces or pounds.

From such an exercise try to establish between 12 and 18 ounces finger and hand pressure. Then, mentally, transfer the "feel," this sense of pressure, the firmness with which you grip a pencil, to your use of calipers, to your fingers turning a micrometer screw on to a piece of work, or to the insertion of a plug gage in a hole.

A gage, or any hand-measuring instrument, need not be lifted and moved the way a dowager raises her teacup from its saucer, her little finger daintily curled, but neither should it be applied like a snagging rasp.

Indicating gages and similar measuring mechanisms, as will be seen, have spring-controlled jaws and anvils, thus automatically establishing practically uniform gaging pressure independent of the inspector's actions.

Use of the Vernier Calipers

Studying common and conventional inspection tools from the viewpoint of discrimination, we come next to equipment having a vernier scale. The ordinary classes of this sort of apparatus are the vernier caliper, the vernier height gage, and the vernier depth gage. Where the steel rule has a discrimination of 1/64 inch or .0156 inch, standard vernier equipment can be used to measure to one-thousandth of an inch (.001 inch).

Vernier calipers are calipers that incorporate a vernier scale in order to obtain greater discrimination — usually to .001 inch and .05 mm. Vernier calipers, like the other calipers mentioned previously, have a large range of measuring abilities — commonly from 2 inches up to about 4 feet. They are manufactured as instruments for either internal or external measurements, depth and step increments, and, of course, they may display either metric or inch graduations, and sometimes both.

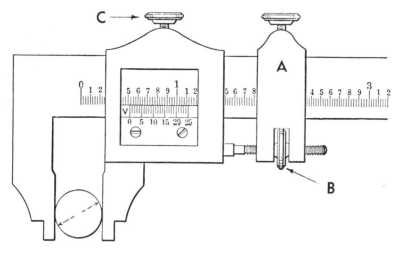

Fig. 23. Typical vernier caliper showing vernier scale V and fine adjustment nut B.

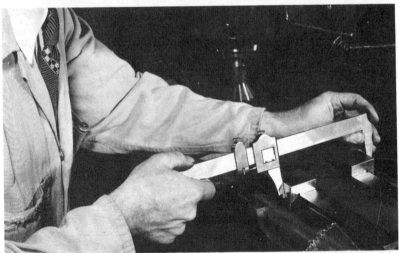

Fig. 24. (Upper) Vernier caliper being used to measure an inside diameter. (Lower) Making an outside measurement.

A typical vernier caliper is illustrated in Fig. 23. A couple of vernier calipers in action are shown in Fig. 24. A 50-division inch vernier scale and a 100-division millimeter vernier scale are shown in Fig. 25.

Vernier calipers, height gages, or depth gages are essentially steel rules. But the rule length, from 6 to 48 inches — depending on the size and model of instrument — has been not only accurately divided off or

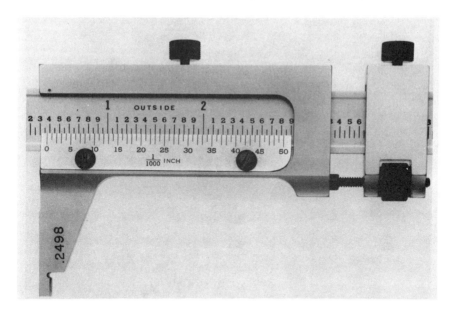

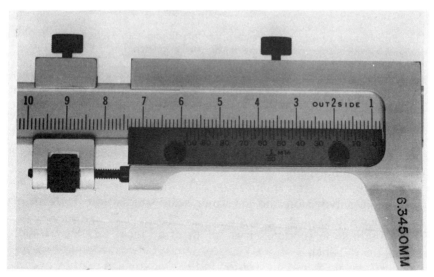

Fig. 25. (A) A 50-division inch vernier scale and (B) a 100-division millimeter vernier scale.

graduated, but a sliding jaw with a so-called vernier scale attachment has been added (see Fig. 23) as a visual aid, in subdividing the smallest graduation on the rule. Each inch of the main scale is subdivided into

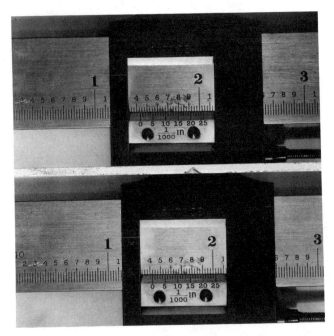

Fig. 26. Vernier caliper set to two different dimensions: (Upper) Set to 1.436 inches. (Lower) Set to 1.425 inches.

10 parts and each of these tenths of an inch again subdivided into quarters. (Study Fig. 26 carefully.) Hence, each *individual* graduation on the standard, commercial vernier caliper main scale represents 1/40th of an inch.

The "vernier scale," which slides along the main scale, see Fig. 26, contains 25 subdivisions in the same total length or span embraced by 24 subdivisions on the main scale. The difference, then, between a main scale subdivision and a vernier scale subdivision is 1/25th of a main scale subdivision.

A main scale subdivision is 1/40th of an inch; 1/25th *of* 1/40th is 1/1000 or .001 inch.

Looking at the top illustration in Fig. 26, you can read by eye the 1 inch, the 4/10ths of an inch (equalling 1.4 inch up to this point), and the 1/40-inch subdivision immediately to the left of the vernier scale zero division. In other words, you can read by eye: 1 inch plus .4 inch plus 1/40 inch or .025 inch, which totals 1.425 inches.

This is not the accurate reading, however, since the 0 index on the *vernier scale* shows it to be greater than 1.425 inches.

The additional, exact thousandths of an inch are secured by reading the "coincidence" of a line on the *vernier scale* with *any* line on the main scale. Read only the *vernier scale* coincident line. In the top view of Fig. 26 that *vernier scale* line is number 11, which is the only one that exactly coincides with a line on the main scale. Add .011 inch to 1.425 inches and get the correct instrument reading of 1.436 inches.

Had the vernier scale index of 0 and the vernier scale number 25 *both* been coincident with main scale subdivisions, then the true instrument reading would have been the *main scale* graduation marked by the vernier scale 0, or 1.425 inches, as illustrated in the bottom illustration of Fig. 26.

In routine shop inspection practice, it is best to find the coincidence of a specific vernier scale line to any main scale graduation with a glass in order to overcome any possible error of .001 inch, or so, by not determining the exact coincidence line on the vernier scale.

These direct-measurement devices are also available as 50-division vernier calipers with *one* inch graduations. In other words, the main scale or beam is graduated to .050 inch, and split by a 50-division vernier scale that shows each .001 inch, and has graduations in increments of .010 inch. English and metric divisions are simultaneously available on these units.

Proper Manipulation Avoids Errors

Experience with vernier calipers will show quickly that errors in measurement are not usually made by misreading the vernier scale or by incorrectly adding the vernier scale coincident reading to the relevant main scale reading but rather from manipulation of the vernier caliper and its jaws on the workpiece.

In measuring an outside diameter, be sure the caliper bar and the plane of the caliper jaws are truly perpendicular to the workpiece's longitudinal center line. In other words, be sure the caliper is not canted, tilted, or twisted. The warning needs to be reemphasized because the relatively long, extending main bar of the average vernier calipers so readily tips in one direction or the other.

If there is wear, spring or warp, you are most likely to see a knock-kneed condition like that illustrated in sketch A of Fig. 27. If in your opinion the condition would produce measurement errors greater than .0002 inch, the instrument should be sent back to the manufacturer for repair. (Remember, in sighting caliper jaws against a light, the smallest gap you can probably discern is half a ten-thousandth — .00005 inch.)

If the vernier scale is set at 0-0 and a strip of light appears between the ostensibly closed jaws, though no appreciable wear itself — or warp — shows, the jaws can then be closed tight and the vernier scale

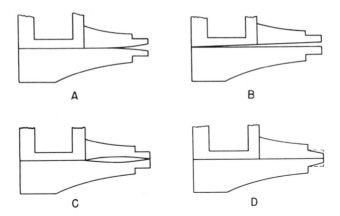

Fig. 27. Various jaw conditions that result in inaccurate caliper measurements.

readjusted to a new 0-0 by means of the two adjusting screws provided on the vernier scale (see Fig. 26). Also set the caliper jaws on a 1-inch gage block or suitable 1-inch master and in like manner check and reset, if necessary, the vernier scale 0 at the main scale 1-inch graduation. Or if the calipers are used mostly within a range of dimensions, close to, say, 3 inches, or 9 inches, or 18 inches, the vernier scale setting test should be made on master cylinders or blocks of corresponding size.

Returning to Fig. 27, a condition like that shown in sketch B is sometimes seen. This probably means that the sliding jaw frame has become worn or warped so that it does not slide squarely and snugly on the main caliper beam. The reconditioning job probably can be better executed by the gage manufacturer, if a simple adjustment of the gib and spring in the sliding member does not correct the trouble.

Where vernier calipers are used mostly for measuring inside diameters, the jaws may become bowlegged as in sketch C of Fig. 27, or their outside edges worn down as in sketch D.

Most vernier calipers are provided with two vernier scales and two main beam scales. One is marked for outside dimensions, the other for inside dimensions. The more skilled mechanics seldom use the "inside" scales. With caliper jaws in proper repair and closed tightly, measure across them with a micrometer as shown in Fig. 28, recording the dimension. Then, when measuring an I.D., read the "outside" scale of the instrument and add the known micrometer reading, just mentioned, to the outside scale reading to get an I.D. measurement which is usually more accurate than taking a reading directly on the instrument's inside scale. The above technique also obviates errors due to the wearing down of the outside surfaces of the caliper jaws.

Fig. 28. Measuring the jaws of a vernier caliper to determine the amount to add to the "outside" scale reading when making inside diameter measurements.

Vernier calipers should not be treated or used as a wrench or hammer. This is not to imply that a good mechanic or inspector would be so grossly careless, but to emphasize that vernier calipers are not rugged instruments. They should be set down gently — preferably in the box they came in — and not dropped or tossed aside. They must be wiped free from grit, chips, and oil. Bring vernier calipers to the workpiece; do not clamp the workpiece in the caliper jaws and wave them around in the air.

Vernier Depth Gages

Vernier depth gages are an interesting adaptation of the vernier principle — these gages are used to measure the depths of holes, recesses, slots, counterbores, and so forth. These instruments are similar to a common rule depth gage, with the addition of the vernier scale to increase discrimination.

A vernier depth gage is illustrated in Fig. 29. As an instrument, it is essentially a depth rule with vernier attachment and markings, and a base or anvil. In general, the base or anvil is rested on or against a reference surface and the scaled beam or tongue is pushed beyond the base to contact with the measured point as shown in Fig. 30. Readings taken at the vernier attachment show directly the length of beam or tongue protruding beyond the base.

The depth gage is carefully made so that the rule or beam is perpendicular to the base in both directions. The end of the beam is

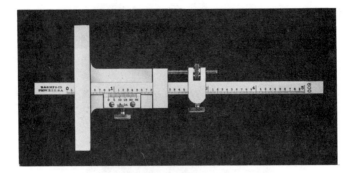

Fig. 29. Vernier depth gage.

Fig. 30. Vernier depth gage being used to measure a depth on a block.

square, and flat, like the end of a steel rule, and the base is flat and true, free from curves or waviness.

A depth gage itself then, because of its own careful construction, will give a true measurement when used properly, but it is easier to make errors with it, due to manipulation, than with almost any other form of measuring apparatus.

First of all, make sure the reference surface, on which the depth gage base is rested, is satisfactorily true, flat and square. Measuring depth is a little like measuring an inside diameter. The gage itself is true and square but it can be imperceptibly tipped or canted, because of the reference surface perhaps, and offer an erroneous reading as shown diagrammatically in exaggerated fashion in Fig. 31. Another type of error appears in sketch B of Fig. 31, illustrating how the gage may be unconsciously tipped forward or backward.

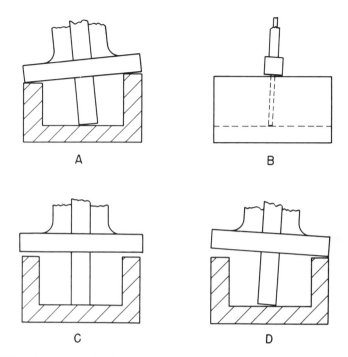

Fig. 31. Several depth measuring methods which produce erroneous measurements.

The most common error, however, is diagrammed at C in Fig. 31. In using the depth gage, the base or anvil may be at first firmly rested on the reference surface. The tendency then is to slide the beam or tongue against the measured point with so much pressure that the base is lifted as shown at C — perhaps only an imperceptible thousandth or so — or the point pressure cants the base a trifle as at D. In using a depth gage, press the anvil firmly on the reference surface and keep several pounds hand pressure on it. Then, in manipulating the gage beam to measure depth, be sure to apply only standard, light, measuring pressure — 8 to 16 ounces — like making a light dot on paper with a pencil.

Other Types of Vernier Equipment

Other typical pieces of vernier equipment are the vernier height gage and the vernier protractor. Since the height gage is so commonly allied with surface-plate practice, a discussion of its use is omitted here, to be included later, where its use in connection with surface plate measurement practice is described. In a similar vein, a discussion of the vernier protractor is included in the description of tools and methods of angular measurement.

Understanding the Micrometer

Micrometers are one of the standard instruments of industry and inspection, and a tool that is, like many other hand-held instruments of measurement, changing with fast-paced innovations in technological and electronic applications, including computerized connections to the measurement tools.

Traditional micrometers, or "mikes" as they are called, are distinguished by many styles and varieties. Basically, the instrument measures "inside" or "outside," or can be of the screw-thread variety, and even a depth micrometer; many specialized micrometers are also available. The micrometer measures by a fairly simple principle — how far does the end of a screw (the spindle) travel after completing one full revolution?

The principle of the micrometer, which is actually a special form of caliper, is purely that of a screw turning in a nut and of the point or end of the screw advancing toward or receding from the opposite anvil of a C frame. The common commercial names for the various parts of a micrometer are given in the sectional view of Fig. 32. The reader should become familiar with these parts, which include : *frame, anvil, spindle, barrel, thimble, screw, ratchet or stop, and clamp ring or locking screw.*

Frame sizes range from 0.5 in. to 48 in. The micrometer is considered to be one of the most versatile and accurate of the

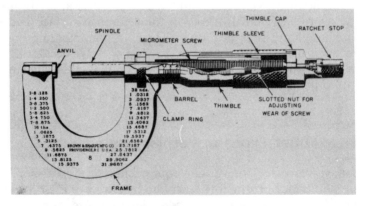

Fig. 32. Section through a micrometer showing principal parts.

hand-held measuring devices — commonly accurate to .001 inch. With the addition of a vernier scale, the accuracy of a mike can be increased to 0.0001 in. Metric micrometers are accurate to .01 mm, and to .002 mm with the additional assistance of the vernier scale.

In this book the fixed or reference measuring surface of the micrometer will be called consistently the anvil and the movable measuring surface, for securing the measured point, will be called the spindle.

The term micrometer, itself, or micrometers, is almost invariably applied to the standard, now conventional, 1-in. micrometer caliper. There are also inside micrometers thread micrometers and other special models such as are shown in Fig. 33. Micrometers come also in

Fig. 33. Various types and sizes of micrometer calipers commonly used.

ranges from 0.5 in., 2 in., 3 in., and up to 48 inches. So, as a matter of terminology, the plain word micrometers or "mikes" can be assumed in this text to refer to the 1-inch micrometer caliper, and the words 3-inch, or inside, or thread, or some other descriptive term will be used with the word micrometer when an instrument other than the 1-inch micrometer is being referred to. A few different kinds of micrometers are shown in use in Fig. 34.

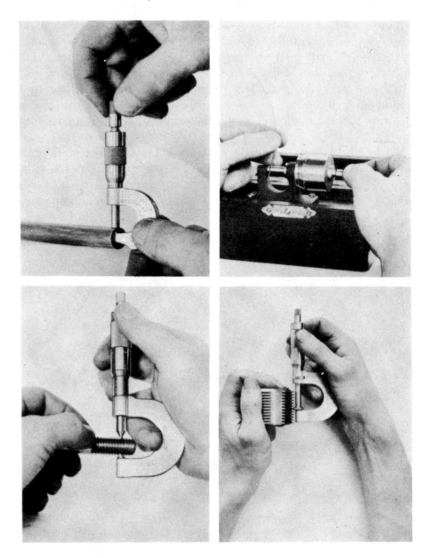

Fig. 34. Methods of using various types of micrometers.

Reading the Micrometer

Reading a micrometer scale is a skill that requires an understanding of the mechanical principles involved, as well as some practice. There are two primary scales on most units: the thimble scale and the sleeve scale. Measurement is accomplished by the spindle–anvil opposition. The spindle has a pitch of 40 threads per inch, or 1/40 in. Therefore,

1/40 inch equals a single revolution of the spindle screw; 40 turns equal 1 in. A clockwise motion sends the measuring device toward the anvil, and a counterclockwise motion sends it in the opposite direction.

The micrometer thimble, an enlarged sleeve or band, surrounding the screw and fastened to it, has 25 graduations engraved on its periphery in such a manner that the eye can readily register 1/25th of a turn of the screw. So, if the screw is turned in 1/25th of a full revolution, the point of the screw has advanced 1/1000 in., because 1/25 of 1/40 is 1/1000.

The 1-in. length of the *barrel* of the micrometer (reexamine Fig. 32) is divided into tenths of an inch — 0.1 inch, 0.2 inch, etc. And each tenth is graduated into four equal divisions of 0.025 inch. In other words — see Fig. 35 — each subdivision in succession, within each tenth division on the barrel, reads, respectively, 0.025 inch, 0.050 inch, 0.075 inch.

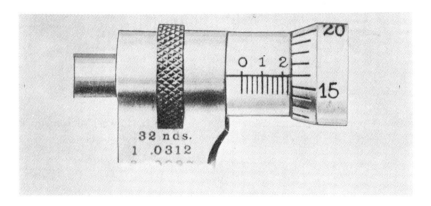

Fig. 35. Micrometer set to measure 0.241 inch.

Hence, to read the micrometer setting in Fig. 35, the edge of the thimble has uncovered first the "2" or 0.2 barrel division. Second, it has uncovered the 0.025 inch subdivision, making the reading up to this point 0.2 + 0.025 or 0.225. The part of the subdivision left registers on the thimble itself as "16" or 0.016. Adding 0.016 to 0.225 gives 0.241 as the micrometer reading. It means that the gap between the micrometer jaws is 0.241 inch.

Figure 36 shows micrometers reading 0.162 and 0.494 inches, respectively. However, the 0.494 reading in Fig. 36 does not fall squarely on a thimble divison — the actual reading is 0.494 in. and some ten-thousandths of an inch more. For reading the additional ten-thousandths, the micrometer is equipped with a vernier scale also engraved on the barrel. Note in Fig. 37 that a *thimble* engraving is

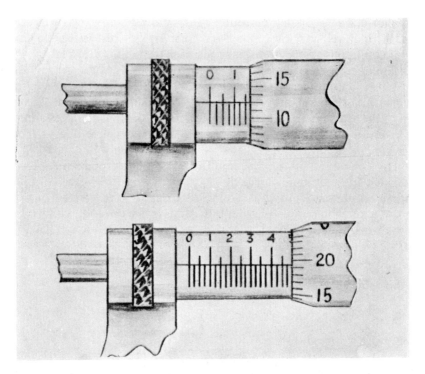

Fig. 36. (Upper) Micrometer set to measure 0.162 inch. (Lower) Micrometer set to measure somewhat more than 0.494 inch.

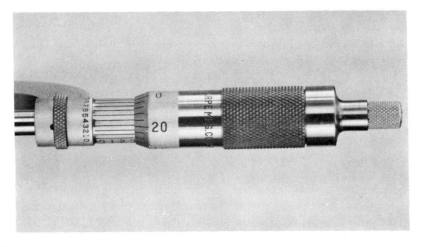

Fig. 37. Vernier giving reading in tenths of thousandths. Reading is 0.0004 greater than thimble reading (not shown).

coincident with the "4" vernier line, which means that 0.0004 inch is to be added to the thimble reading (which is not visible in this illustration).

Micrometers, today, like the range of mechanically and electronically enhanced calipers produced (Fig. 38), are available with many different capabilities, accuracies, and applications in mind. The newest digital micrometers have direct-reading capabilities to 0.00001, inch. Most micrometers are available not only with the typical ratchet stop, but also with a choice of operator-based thimble movement pressures, from a sliding feel to a friction feel. Extra features such as thumb stops and carbide measuring faces are available, depending on operator preference, and the application of the "mike."

Courtesy of S-T Industries

Fig. 38. Electronic micrometer.

Micrometers have many different shapes for different uses, especially in the area of the spindle–anvil opposition. For instance, these units are available with special configurations for measuring wire, grooves, or sheet metal; with spindle–anvils shaped like disks or blades; or three-pronged setups used to assess out-of-round conditions. Many uses and high technology combine to make this tool an ever-changing standard. These units are also commonly connected by electronics to recording devices that plot control charts and trends as readings are registered. Despite these changes and improvements, the basic principles of these measuring devices remain unchanged (Fig. 39).

Perhaps the best order in which to discuss the use of a micrometer for our purpose is to suggest a situation that is not entirely hypothetical

Courtesy of DoAll Co.

Fig. 39. Electronic micrometer printer readout with Z-bar and R capability.

in the average factory inspector's routine: that of coming into a shop area to measure certain product dimensions and using perhaps his own micrometer but, equally likely, using micrometers available in the area.

Cleaning the Micrometer

First, naturally, the micrometer should be wiped free from oil, dirt, dust, and grit. Nothing probably advertises a good inspector faster than the fact that he instinctively and consistently requires clean instruments. By this habit he inspires confidence on the part of others watching him or dependent on his decisions, and confidence in the measurements he takes.

Many times, when a micrometer feels gummy and dust ridden and the thimble fails to turn freely, inspectors are prone to dunk it bodily in kerosene or some similar solvent. Such practice, however, is not recommended. It is much better to have someone who knows what he is doing take the micrometer apart and thoroughly wash each component free from gum and dirt. Just soaking the assembled micrometer fails to float the dirt away; it may get softened up for a minute or so or transferred to another section of the mike. Besides, the apparent stickiness of the micrometer may not be due at all to grit and gum but to a damaged thread or to a warped and sprung frame or spindle.

Assuming reasonable cleanliness and a free-running instrument up to this point, the next step is to clean the measuring surfaces of the anvil and spindle. Technically, this function should be performed every time the micrometers are used. Screw the spindle lightly but firmly down on to a clean piece of paper held between spindle and anvil as shown in Fig. 40. Pull the piece of paper out from between the measuring surfaces. Then unscrew the spindle a few turns and blow out any fuzz or particles of paper that may have clung to the sharp edges of anvil and spindle.

Here is a good place to emphasize a condition that characteristically misguides an inspector into errors of precision measurement. A light film of oil or grease in itself may not disturb the accuracy of the reading, unless precision in something like a millionth of an inch is required, because the oil film is nearly impalpable. But oil is a collector of dust, dirt, grit, and particles. The product of human sweat glands is also a ready offender. A firm ridge of dirt several thousandths of an inch in elevation can readily collect — unnoticed — on the anvils of measuring equipment.

Testing for Parallelism and Flatness of Measuring Surfaces

The next test of a micrometer's reliability should be that of parallelism and flatness of the measuring surfaces, checking for conditions similar to those exaggerated in sketches A and B of Fig. 41.

For a quick test of parallelism use a pencil size steel rod, preferably with a very smooth ground or lapped surface. The test piece should be three or four inches long. Catch the rod lightly between anvil and spindle near one end of the rod. Tip the micrometer and rod into horizontal position as indicated at C in Fig. 29. If some condition like

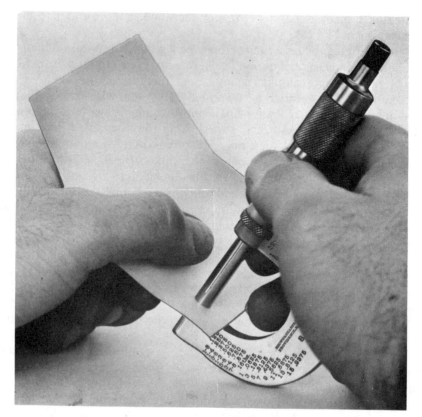

Fig. 40. Cleaning the measuring surfaces of a micrometer caliper.

that shown in sketch A exists, the rod will readily pivot on the high spots and turn or sag down if it is held in the mike as shown at C.

Lack of true flatness (sketch B of Fig. 41) can be checked by "exploring" the measuring surfaces with a precision ball as at D. The ball is moved from location to location around the measuring surfaces. Great care, however, must be used in applying uniform spindle pressure and in taking the reading for each location of the ball to detect minute errors in surface flatness.

The most foolproof test of micrometer measuring surface conditions is with an optical flat, an operation pictured in Fig. 42. This particular type of test and the reading and interpretation of light bands may be better understood after the study of optical inspection equipment in Chapter 12. The optical flat test has the advantage of disclosing readily and accurately all combinations of wear conditions — waviness, hollows, humps, and lack of parallelism.

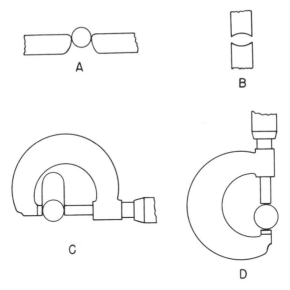

Fig. 41. (A) and (B) Two conditions of micrometer measuring surfaces which result in erroneous measurements. (C) and (D) Testing for conditions shown in (A) and (B), respectively.

Fig. 42. Use of an optical flat to determine the measuring surface condition of a micrometer caliper.

It is a good thing, too, to examine the sharp edges of the anvil and spindle faces with a magnifying glass or fingernail to be sure nicks and burrs have not been raised from accidentally dropping the mike or rapping those sharp square edges against metal.

Checking the Zero Reading

When you are reasonably confident that the condition of the micrometer measuring surfaces are free from dirt, nicks and surface irregularities that do not exceed .0001 inch, then check the zero reading of the micrometer. Screw the spindle down on to the anvil with standard, light, but firm pressure. Use the micrometer's ratchet device if it has one. Otherwise let the fingers slide a little over the thimble knurling, like a slipping clutch, as spindle and anvil contact. In other words translate 8 to 16 ounces pressure between your fingers to anvil and spindle faces. Above all, however, set the micrometer on zero with the same measuring pressure you intend to employ regularly with the mike in use. At that point the micrometer should read 0.

If the "zero" reading is not 0, if there is an error apparent of more than a ten-thousandth or two (.0001 to .0002 inch), when the measuring surfaces are closed together, then either the micrometer should be reset at its 0 station or the amount of error must be added to or subtracted from each actual measurement to be made with that micrometer — depending on which side of the 0 reading the test error falls. (Each micrometer manufacturer issues detailed, illustrated, and explicit directions for resetting and correcting his instrument. Hence the omission of such instruction here.)

When the 0 test is being made, observe whether or not the spindle turns freely, in your opinion. Try to wiggle it longitudinally to see if the micrometer screw is loose in the fixed nut (see Fig. 32). Too much freedom, play, or "end shake" signifies a worn instrument, usually. The micrometer manufacturer's instructions cover the techniques for overcoming looseness.

If and when any adjustments described by the micrometer manufacturer are made, the 0 setting must be checked again.

Calibrating the Micrometer

The next test is called calibration. Usually the test is made on a series of gage blocks of different sizes. If accurate cylinder masters — test plugs or wires — or precision balls are available, they are frequently used for calibration checks.

A calibration test is made after the 0 setting has been ensured by observing the comparison between the known size of gage block, master cylinder, or ball with the micrometer barrel, thimble, and vernier reading. The "mike" is calibrated at several stations. It is natural to think of testing at, say, the 1/4-inch, 1/2-inch, and 3/4-inch stations on similar size gage blocks. It is much better, however, to make up odd additions of gage blocks (or secure odd dimensions in cylinders or ball masters) such as, for instance, .195 inch, .390 inch, .585 inch, .780 inch, and 1.000 inch. By some such diversification of test gage block settings, a check is secured of the condition of the spindle and nut threads (that is, of the thread lead* of the instrument) at several different points on the screw, not only along its length, but also around its periphery. Where lead errors are discovered in this calibration test that exceed .0001 to .0002 inch, a record should be made of them. If the inaccurate "mike" cannot be immediately replaced or repaired, its actual readings on work pieces at the inaccurate stations can be compensated for by applying the relevant calibration correction.

Lead or calibration errors appear both plus and minus. For example, the test at the .195-inch station might disclose an actual micrometer reading of .1952 inch, a plus error, which means that .0002 inch is to be subtracted from any micrometer reading on a work piece close to .195 inch. Again, the calibration at the, say .780-inch station might read .7798 inch, a minus error of .0002 inch, and for workpiece readings in the vicinity of 3/4 inch the amount of .0002 inch should be added.

Do not neglect to test or calibrate a micrometer close to the 1-inch reading. If micrometers show any errors at all, they are likely to be most marked at this end of the micrometer screw range.

An understanding of the idea or technique of "wringing" is necessary before precise micrometer calibrations can be made with flat gage blocks or masters. Study and practice in correctly wringing gage blocks together brings out this basic concept. As the spindle is being finally tightened down on a gage block, with standard pressure, the flat micrometer measuring surfaces are carefully slid a trifle laterally along the gage block surface to secure the effect of wringing. Extra care must be used, of course, where the micrometer measuring surfaces are carbide tipped, for carbide can be unmerciful in cutting or shaving the surface of an accurate gage block.

---

*The lead of the micrometer screw is the distance it travels axially as it rotates through one turn in the nut. When the subject of screw thread lead errors is discussed in Chapter 13, the possibility of micrometer lead errors will be more apparent.

Proper Use and Care of Micrometers

The subjects of micrometer testing, adjustment, calibration, and wear brings up, of course, the need for using the micrometer properly and taking care of it when it is not in use. The idea of cleanliness has already been stressed. Micrometers should never be forced or sprung. If the principles of standard, correct gaging pressures already described are used in transferring the strength in your fingers to the pressure exerted by the micrometer spindle, there is practically no danger of springing a micrometer screw and little possibility of wearing the measuring surfaces unevenly.

When the micrometer is not in use or is to be put away, be sure it is wiped clean and free from oil, grit, and sweat, especially the measuring surfaces. And then, never leave a micrometer stored with the spindle clamped down on the empty anvil, not even over night. Salts, acids, and alkalis in oils, cutting fluids, and sweat induce corrosion of the measuring surfaces. Such corrosion may not be at all noticeable, in the sense of visible rusting, but it converts the metal on the measuring surfaces into an impalpable powder which readily scrapes off the first time the mikes are used. The loss each time may be less than a millionth of an inch, but after a while the tips of the anvil and spindle begin to resemble the conditions illustrated in Fig. 41, or the zero setting begins to show a discrepancy.

Leaving the "mike" closed seems to accentuate the corrosive action spoken of previously, probably because of an additional electrolytic action taking place when the measuring surfaces are left in contact.

Learn to hold and use the micrometers correctly; the 1-inch and 2-inch sizes in one hand so that the thumb, index finger, and third finger turn the spindle, while the fourth and fifth fingers clamp the frame against the palm of the hand (see Fig. 43). The larger sizes of micrometers are manipulated with two hands, also illustrated in Fig. 43, or sometimes there are special measuring conditions where, even with smaller range micrometers, it is more convenient to use both hands.

Remember the principle of reference point and measured point, especially in the two-handed manipulation of micrometers. It is better to hold the micrometer anvil, which is stationary, firmly against the work with one hand and take care of gaging pressure and finding the correct measured point with the other hand whose fingers turn the mike spindle.

Remember, too, that micrometers can be as readily tipped or canted as vernier calipers. This warning is emphasized here because micrometers have the knack of *seeming* to "home" readily, firmly, and squarely on the work being measured. Many workers get so used to

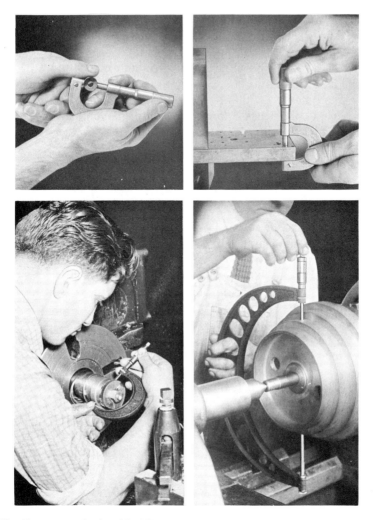

Fig. 43. Correct methods of holding various micrometer calipers when making a measurement.

their micrometers, and use them so rapidly, that they are not conscious of cramping them out of position on the work and of not getting, consequently, the true diameter or thickness.

One other word of warning should be issued here. While most micrometers are supplied with vernier scales, so that readings ostensibly to .0001 inch can be secured, the inspector ordinarily should not expect to measure with an accuracy or discrimination closer than .0002 inch in the ordinary use of micrometers.

A small variation from uniform finger pressure on the mike stem or the slightest — subconscious perhaps — canting or cramping of the mike or the workpiece can and will introduce an error of .0001 inch. The effect of hand temperature, too, is frequently forgotten, especially where the larger size micrometers are used. The purpose here is not to deny the micrometer's ability to measure to .0001 inch, but to warn that unless unusual care, finesse if you will, is displayed, the average user can readily mislead himself into believing he is measuring dimensions correctly to .0001 inch.

Misreading the Micrometer Scale

There is an error in reading micrometers that both the beginner and the experienced mechanic make, a form of error that seems a trifle preposterous at first thought. The neophyte makes his mistake usually in all innocence, but it is believed the expert's trouble arises more from a subconscious mental fixation than from carelessness. The mistake or error is that of misreading the micrometer barrel by .025 inch.

Look at Fig. 44. The one inch length of the barrel is primarily divided into tenths, as .1 inch and .2 inch, etc. The digit 1 shows as the primary division in Fig. 44. Then, as you know, each of the tenth divisions is again divided into four parts and each subdivision graduation represents .025 inch. Hence, in reading the micrometer setting of Fig. 44, we go through the mental arithmetic of adding two .025-inch subdivisions — which makes .050 inch — to the .1 inch main division and get .150 inch. To this we add the thimble reading of 12 or .012 inch, making .150 inch + .012 inch or .162 inch, the correct reading.

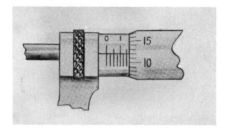

Fig. 44. Micrometer set to measure .162 inch.

However, it is easy to misread or miscount the number of .025-inch barrel divisions uncovered by the thimble. In the case of the reading shown in Fig. 44, the inspector might have read the barrel as .1 + .025 inch and, adding the .012-inch thimble reading, might have gotten a final, but inaccurate, reading or measurement of .137 inch. Again, he might have taken the initial barrel reading as .175 inch, rather than the correct .150 inch, and then, adding the .12-inch thimble reading, secured an incorrect result of .187 inch. In this example, Fig. 44, the latter error is probably the more likely, especially where the micrometer is dirty or the light poor, and more especially, for example, if the blue print or specification called for a dimension nearer to .187 inch than the actual workpiece measurement of .162 inch.

Mechanics have been known to start off with such an error, set their machines to the incorrect size, and persist for hours in reading their mikes .025 inch "off."

The intent of some of the foregoing detail is not only to present useful information to aid the inspector in securing accurate measurements, but also because little "tricks of the trade" give a man a professional air in the performance of his duties and thus inspire additional confidence in the results of his inspections.

Right here should be a good place to suggest, too, that an inspector devote considerable time, during his career, to studying carefully many of the catalogues, books, and leaflets furnished by the manufacturers of the various types of equipment he will be using in the normal course of his work. From such sources he will secure many valuable "professional" tips. In the same vein, the constant reading of gage and tool manufacturers' advertisements and literature keeps the inspector abreast of improvements, often revealing to him more efficient, less costly, and more accurate ways of performing his work.

All of the warnings covering misuse and inaccuracies — tipping, canting, excess gaging pressure — offered in preceding sections in connection with vernier equipment apply to the special forms and shapes of micrometer apparatus. The use of an inside micrometer with extensions offers an excellent object lesson in reference point, measured point, centralization, tipping, canting, and gaging pressure.

Micrometer Plug Gages

Thus far, in this chapter, the only devices described for measuring hole and bore diameters have been calipers and vernier calipers. During the years, several other means have been developed for this purpose, apparatus and instruments that are essentially variations of or accessories for basic measuring equipment.

So-called "inside" micrometers can be secured. One type is the micrometer plug gage shown in Fig. 45. Three blocks or guides make up the solid frame of the gage. Between the guides are three movable members or blades that are seated directly on a hardened cone ground

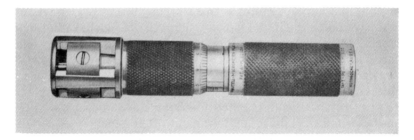

Fig. 45. Micrometer plug gage.

to an exact angle. A micrometer screw mechanism, with thimble and barrel, is attached to the solid framework of the guides, and the hardened cone, in turn, is the tip of the micrometer screw or spindle.

As the micrometer plug gage is introduced into a hole, the thimble and spindle are turned. The cone, traveling forward, then expands the blades against the sides of the hole. While barrel and thimble markings measure essentially the lateral travel of the cone, they are actually marked and calibrated so that the cone travel and, consequently, the expansion of the blades are literally translated into direct diameter readings in thousandths of an inch. Each such micrometer plug gage is nearly single-purpose in that its range of operation is confined to any 1/16-inch step in diameter. Gages of this type can be furnished in all sizes from 3/4 to 4 inches in 1/16-inch steps. In other words, one gage must be used, for example, for a 15/16-inch general hole size and another for 1-inch holes, etc.

Brief mention of this type of bore gage is made here in order to introduce a word of warning. In using it, an inspector may well *not* get the exact measurement to .001 inch he anticipates except with care. As the micrometer thimble is turned and the blades are forced against the hole sides, the blades may not "home" correctly. The gage should be rocked a little as the blades are homing. Then, a second or check reading should always be made to be sure of a repeat reading because the expanding blades also have a tendency to "creep."

Inside Micrometers

Considerably more versatile are the inside micrometers illustrated in the top illustration in Fig. 46. Anyone who has properly used micrometers and inside vernier calipers will see readily how the inside micrometer caliper of Fig. 46 is used. Needless to say, this sort of instrument must be checked regularly for 0 setting and for bent or worn caliper tips. The usual rules for rocking it and centering it on the true

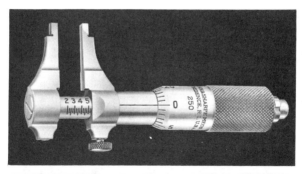

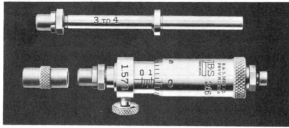

Fig. 46. (Upper) Inside micrometer caliper. (Lower) Inside micrometer with
extension spindles to increase range of measurement.

diameter, for standard gaging pressure, and for repetition (repeat reading) tests also apply, of course. These instruments have a range commonly to cover the measurement of internal diameters from .200 to 1 inch.

The inside micrometer sets like that in the bottom illustration in Fig. 46 can be secured in a variety of range groups so that bores anywhere between 1 and 36 inches can be measured to an accuracy of .001 inch. Closer study of this illustration will show that the instrument is basically a micrometer — screw, barrel, thimble — with a very short spindle. The spindle also contains a chuck in which the various extensions can be fastened. The No. 1 extension adds an inch in length to the micrometer and each successive extension in the series is 1 inch longer than its predecessor.

Figure 47 not only portrays an inside "mike" in use but emphasizes the ideas of reference point, measured point, and centralization. In addition to taking the usual measurement precautions, the inspector should watch for two causes of inaccuracy when using this form of inside mike. The extensions must be butted securely in the micrometer and extension sockets. Even the slightest inattention to this point can produce errors of several thousandths. He must also be careful not to use the extension inside mike too long. The temperature of the hands

Courtesy of Brown & Sharpe Mfg. Co.

Fig. 47. Inside micrometer shown in lower view of Fig. 46, being used to measure a bore.

can warm it up and expand the extension rapidly enough to produce significant errors.

Hole location micrometers use an extra arm to position or reference the location of holes and assess internal dimensions, as well as the locating dimensions themselves.

Split-Sphere and Telescope Gages

Ball-shaped small hole gages will be found in most gage cabinets. As Fig. 48 indicates, these gages consist primarily of split hollow spheres the halves of which can be expanded within a hole by means of a knurled knob which turns a screw thread that forces a spreader between the sphere halves. When the gage spheres have been set by applying standard gaging pressure to the knurled knob and by the feel of the split spheres within the hole, the gage is removed and the diameter of its spread sphere measured with a micrometer as in Fig. 48. This is an adaptation of measuring over the points of standard inside calipers as illustrated in Fig. 18.

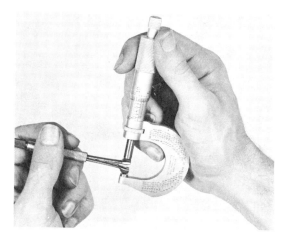

Fig. 48. One type of small hole gage being measured after setting to determine the hole size.

A "bigger brother" of the split-sphere, small-hole gage appears in Fig. 49, the telescope gage. One leg of the T is solid, the other leg telescopes into it. The telescoping leg is spring loaded. The knurled screw knob at the other end of the gage, in this case, turns a locking screw that will lock the telescoping leg in any position or length it may assume inside the hole being measured.

Courtesy of The L. S. Starrett Co.

Fig. 49. Telescope-type gage for measuring hole diameters.

Another design of gage for bore or hole measurement has three measuring points, each of which is designed to make line contact with the surface of the hole. Thus it can be readily centralized and aligned for accurate measurement of the true bore diameter. Extensions permit measurement of deep bores or holes and a micrometer-type scale permits direct reading. Various sizes are available to measure up to 4-inch internal diameters. Figure 50 shows this gage in use.

Enough has been said heretofore in regard to gaging pressure and the positioning of calipers, as well as the proper use of micrometers, so

Fig. 50. Three-point gage for measuring internal diameters. Design facilitates correct centralization and alignment for accurate measurement.

that the inspector should be able to get precise readings. In a sense, double caution is required because there is the need of properly using the T or ball gage and then of properly using the micrometer. In this type of measurement on close tolerances it is certainly best to repeat each measurement and reading. If you get repetition (repeat readings), you are probably measuring accurately; if not, you had better try again. Even so, with considerable care, measurements with a discrimination or accuracy better than .0005 inch are difficult to get.

Micrometer Depth Gage

Another common form of micrometer equipment is the micrometer depth gage illustrated in Fig. 51. One thing especially is to be noted concerning it as an instrument. The graduations on the standard micrometer caliper start at 0 when the thimble and spindle are screwed in to the fullest extent (when the micrometer jaws are closed) and as the micrometer is opened, as the thimble backs away, the readings of 1 (0.1 inch), 2 (0.2 inch), etc., are exposed. The graduation readings on the barrel of the depth micrometer are, however, reversed. When the

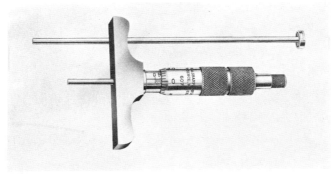

Courtesy of Brown & Sharp Mfg. Co.

Fig. 51. Micrometer type depth gage and auxiliary long-length spindle.

thimble is screwed way down and the graduations are fully concealed, the stem of the depth micrometer has been screwed *out* 1 inch, screwed out to its fullest extent. As the thimble is unscrewed the digits 9 (0.9 inch), 8 (0.8 inch), etc., see Fig. 51, become uncovered. The stem is receding toward the base, of course, and at the micrometer 0 reading the stem is fully receded, though the thimble and screw have been turned all the way out. Figure 52 shows a depth micrometer in use.

A simple way to check the accuracy of a depth micrometer appears in Fig. 53. Unscrew the spindle and set the base of the mike on a flat surface like a surface plate or toolmakers' flat as shown at the left in Fig. 53. Holding the base down firmly turn the thimble or screw in, or down, and when the tip of the mike depth stem contacts the flat firmly, with not more than two pounds gaging pressure, read the barrel. If the mike is accurate, it should read 0. Then rest the mike on a 1-inch gage block, as shown at the right in Fig. 53, and screw the stem all the way down to contact with the flat. There it should register 1 inch.

The warnings previously given for vernier depth gages also apply to micrometer depth gages.

Use of Hand-held Measuring Devices

Hand-held measuring devices, especially those with electronic enhancements, are well suited to the new directions of inspection, which focus on close process control through real-time sampling and reporting on product variation. Instead of being faced with a box or a pile of parts to inspect, modern inspection samples production pieces in-process, usually at the source or operation point, where critical

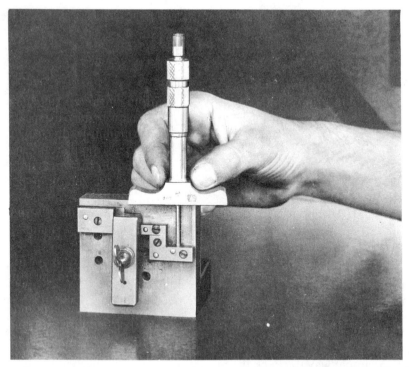

Courtesy of Brown & Sharp Mfg. Co.

Fig. 52. Depth gage being used to measure the location of a block.

variables are known and documented into the manufacturing process as surely as the production tools that created the part.

As pieces pass by the operator and/or inspector, many of these new calipers, micrometers, and other hand-held devices are extremely portable — they are well-suited for use outside the inspection lab, directly at the workplace. They are easy to use, requiring minimal training, and best of all, are reliable and easy to read, providing accurate feedback that the operator, management, and engineering groups can rely on.

Besides typical readouts in digital fashion, micrometers and other small measurement devices contain or are connected to small computer devices, which store and provide information to the operator — information that directly provides the critical "control" statement to the operator, which dictates whether or not it is acceptable to continue producing at that station. In addition to the measurement, these devices play back the average and range of the samples or entry measurements, and indicate whether these sample averages and ranges

Courtesy of The Lufkin Rule Co.

Fig. 53. Method of checking the accuracy of a depth gage: (Left) Zero reading being checked by using a surface plate. (Right) 1-inch reading being checked using a 1-inch gage block and surface plate.

have exceeded the control limits — a dependable statement directly and automatically issued to the operator that would allow the operator to stop if there is a problem causing excess variation.

CHAPTER 7

Comparison and Fixed Gages

Most important to the application and use of fixed gages is a comprehensive understanding of the principles of indirect readiness or measurements; that is, transfer and comparison to a calibrated standard as a means of assessing a dimension. Remember, the dimension — whatever that dimension may be or however the measurement is described or scaled — is not directly read. Rather, the dimension is *represented* by the measurement device, which is then directly calibrated or referenced according to the measurement standard that is applicable.

Ring gages, plug gages, and feeler gages are examples of this type of gage. In each case, the gage represents a calibrated or fixed dimension that is assessed against the area to be measured. As the plug gage is inserted into the hole, the relationship of the fixed measurement on the gage is compared to the hole. This type of comparison of a fixed gage to the dimension of the hole yields an assessment of the true measurement relative to the expectation — perhaps the specification.

Central to this entire concept of accurate gaging and reliable inspection and assessment through gaging are the principles of proper referencing to the standard or calibrated master, and the maintenance of that calibration. Without an awareness and an on-going effort to properly measure the work that these gages and fixtures are performing, the act of gaging may be useless, or worse, misleading.

Basically, fixed gages are designed and made to measure a single dimension, although many so-called conventional gages are equipped for adjustments over a limited range.

In general, there are several recognized types of fixed gages. For O.D.'s (outside diameters) and thicknesses, or similar measurements, there are snap gages and ring gages. For holes, I.D.'s (inside diameters), slots, and similar measurements, there are plug gages and feeler gages. Fixed gages take on special shapes such as taper gages, spline gages, length gages, and depth flush pin gages. Fixed gages are also classified as single purpose, progressive, double end, or reversible.

Using the Snap Gage

The snap gage is essentially a fixed caliper, although, as shown in Fig. 1, there may be provision for a limited range of adjustment. The rules that apply to calipering should be observed when using conventional snap gages, plus one or two extra admonitions that relate to the peculiarities of a snap gage.

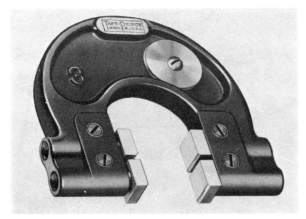

Courtesy of Taft-Peirce Mfg. Co.

Fig. 1. Snap gage with provision for adjustment over a limited range.

Probably the most common, but incorrect, tendency in using a snap gage is simply to slide it over the work like slipping the claws of a carpenter's hammer over the shank of a nail you want to pull. To slip a workpiece between the jaws or anvils of a snap gage the way you back a horse between the shafts of a carriage is not necessarily measuring. In using a snap gage never forget for a minute the conception of the reference point and the measured point.

First rest the upper anvil as a reference point on the workpiece and then swing the gage, the upper anvil acting as a hinge or pivot, and measure with the lower anvil as suggested in sketch A of Fig. 2. The complete and true check is considered technically completed when the anvils cover the true diameter of the workpiece equally as shown at B in Fig. 2. True measurement may not be obtained where one anvil has not been brought up fully, as sketch A in Fig. 2 shows, or an equal error may be made where one anvil of the gage is pushed too far — see C of Fig. 2.

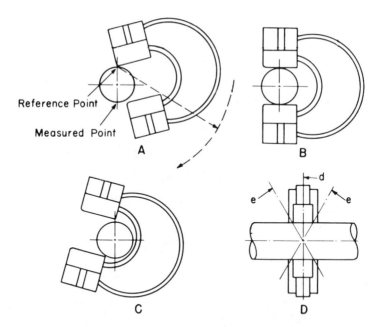

Fig. 2. Method of applying snap gage to workpiece: (A) First step. (B) Second step. (C) Result of pushing the gage too far onto the work. (D) Correct measurement is along line *d*, not along lines *e*.

Considerable care must be used, too, not to cant or tip the gage. The plane of the gage should be, of course, perpendicular to the center line of the work, after the manner shown at *d* in sketch D of Fig. 2, so as to prevent the possibility, shown somewhat exaggerated, of measuring false diameters like *e* and *e*. Such warnings seem superfluous, but it is surprising how casual, careless, and hurried the average worker becomes with a snap gage, especially under the pressure of commercial production.

Because the average commercial type of snap gage feels heavy, rugged, and solid, there is a natural tendency to use it as a sizing tool. The rigidity is built into this type gage purposely to prevent its warping or springing; to offer, in other words, error-proof measuring equipment. As a consequence, extra care or restraint must be practiced in order to apply no more than the standard gaging pressure of a pound or so when calipering with it. One system of standard gaging pressure recommended by some shops makes use of the natural dead weight of the average commercial snap gage, a system under which the inspector is instructed to let the gage slide over the work of its own weight without additional hand pressure as a criterion of dimensional conformance.

The snap gage may be applied to the work in the machine, or the gage may be gripped in one hand and the workpiece fed into it. A more efficient alternative may be where the snap gage is mounted in a bench stand and the workpieces can be applied alternately to it with both hands.

The Fixed Snap Gage

A most common use of comparison or fixed gages in the area of inspection is that of "Go/No Go" gages and conditions or assessments. "Go" gages are most accurately described as most material gages, (MMG) while "no go" gages are in essence the opposite: least material gages (LMG). Obviously, MMGs and LLGs are used to represent the top and bottom end of a specification for a part or workpiece size.

The end result of an inspection using this type of gage is certainly a "pass" or a "fail" distinction that is made on or near the production line. When a workpiece fails because of an excessive dimensional variation that's indicated by a Go/No Go gage, the question should be: what is the (dimensional) extent of the failure; how *much* out of spec is the workpiece? A failure, or a series of defectives, should be directly measured, so that a quantitative assessment of the variation can be communicated to those responsible for adjusting the production effort.

The tendency exists with go/no go gages to become absorbed in the act of defect detection, forgetting about the preventive aspects of the inspection. This is the reason for direct measurements, and their criticality — unless the extent of variation is known (and this is difficult with MMG and LMG gages), early detection of the trend toward unacceptable variation can continue to go unaddressed.

Therefore, the fixed gage offers two steps of measurement — one pair of calipers for the high limit of size and another pair for the low limit. (A very few fixed type of commercial gages offer only a single size limit, see A in Fig. 3.) The high and low steps may be two separate calipers at opposite ends of the gage, as at B in Fig. 3, although more usually the construction offers "progressive" measurement of succeeding calipers in line with each other as illustrated at C.

Sketch D, Fig. 3, illustrates the "Go — No Go" principle of the progressive type fixed gage, a design that has given the type of gage the general term of "Go — No Go" gage.

If the workpiece is of such size it can enter the "Go" end of the gage as at *a* in sketch D but not enter the "No Go" part of the gage as at *b*, it is within tolerances if the gage has been properly set. (Some refer to the "No Go" section of the gage as the "no go" limit.) In other words, work

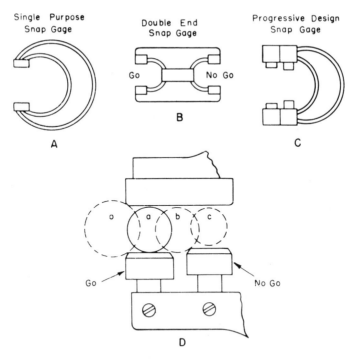

Fig. 3. Various types of snap gages: (A) Single-purpose snap gage. (B) Double-end snap gage. (C) Progressive snap gage. (D) Positions assumed by workpieces of various diameters.

that is within specifications must go in the "Go" section of the gage but it must not go in the "No Go" section.

Where the workpiece is too small it will enter the "No Go" caliper — see position *c* in sketch D of Fig. 3 — and is consequently rejectable.

Another equally relevant "No Go" condition is illustrated by the oversize diameter of the workpiece at position *o* (sketch D of Fig. 3), which therefore prevents its entering even the first or "*Go*" step of the gage.

It should be clear at this point that, for the worker or inspector who will use a fixed gage, the gage is "fixed" so far as size measurement is concerned. Hence, the first natural question should be: are the "Go — No Go" calipers set to the proper size?

Earlier in this book the inspector was warned to question the authority of any blue print or specification. A later section cautioned him to check the accuracy of micrometers before using them. He should be in the same frame of mind before he uses snap gages. Frequently the "Go" and "No Go" limits for a particular use of the snap gage are stamped on its frame. What authority is the source of these figures? Is the gage set exactly to them? Have the gaging surfaces become worn? Does the gage setting correspond with similar figures on the product blue print?

In many shops, the snap gage adjusting screws are covered with sealing wax, after the gage has been correctly set, and in those same shops the rule has been made that no snap gage is to be used if the sealing wax over any adjusting screw is missing.

The practice of setting the adjustment with sealing wax is effective in maintaining and monitoring a proper setting of the gage on the shop floor. Tampering and other sources of inaccuracies can be eliminated with this practice. But the true test of the tool is the ready ability to measure its settings directly.

Checking the Snap Gage for Accuracy

It is such a simple matter to check the accuracy of a snap gage and the parallelism of its jaws that either the occasional or prolonged use of an inaccurate gage is inexcusable.

In checking a fixed snap gage, as shown in Fig. 4, precision gage blocks are used, but master cylinders or disks would be equally successful. Have one set of blocks set up for the prescribed "No Go" limit of the gage (or a suitable diameter cylindrical master) and another set of blocks for the "Go" section. If the gage is properly set, the "No Go" gage blocks should just enter between the "No Go" jaws, perhaps with a slight wringing fit, but no more than two pounds finger pressure should be necessary. The same sort of test applies to the "Go" section. Be sure, of course, that the gage blocks are not unconsciously tipped, canted or cramped. Before using the gage blocks, see that the snap gage anvils are free from dirt, grit, edge nicks, or burrs that would either prevent the gage blocks from sliding between the anvils or scratch or cut the gage blocks.

If the snap gage anvils are not properly set — the gap between them may be too wide or too narrow, as shown by the gage block test — loosen the locking screws, as shown in Fig. 5, and unscrew the adjusting screws a turn or so. Push the anvils, the stems of which will slide in their sockets when the locking screws are loosened, against the adjusting screws thus opening the snap gage anvil gap a trifle. Put the gage blocks in between the anvils and screw down on the adjusting screw with not more than the standard two pounds gaging pressure. Tighten the locking screw and check the gap with the gage blocks.

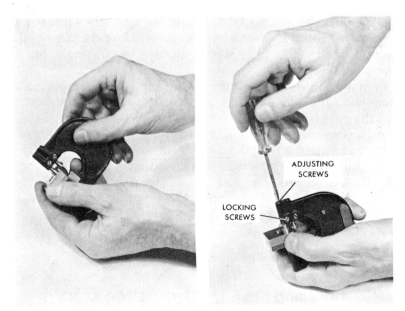

Fig. 4. (Left) Using gage blocks to check the "Go" and "No Go" settings of a snap gage. Fig. 5. (Right) Adjusting a snap gage to the required "Go" and "No Go" settings.

Factors to Consider in Setting the Adjustable Snap Gage

In the case of truly fixed snap gages, the nonadjustable, single-purpose gages like those illustrated in sketches A, B, and C in Fig. 3, gage maker's tolerances and gage surface wear allowances permit some variation in their size. Like any other worker, the gage maker cannot necessarily produce a gage to an exact size. He must have some latitude or tolerance. Usually the gage maker's tolerance is combined with the so-called wear allowance. This latter governs the allowable change in size due to wear of the gaging surfaces.

One common rule of thumb sets this combined tolerance at 1/10th, or 10% of the tolerance spread of the specification for the work the gage is to be used on. As an example, suppose the gage is bought for .837 inch ± .001 inch work. The tolerance spread of ± .001 inch is .002 inch. One-tenth of .002 inch is .0002 inch. Hence, likely, the new gage would not be made to measure .838 inch, the largest size piece allowable under .837 inch ± .001 inch tolerances, but at .838 inch minus .0002 inch or .8378 inch. Closing in the jaw gap by .0002 inch allows the gage to wear back in use to the full .838 inch. In some plants this sort of wear is permitted beyond the maximum, as, for instance, to .8382 inch — a total of .0004 inch wear. (In the case of adjustable snap gages, the so-called gage makers' tolerance is usually equal to the wear allowance. In this case, .0001 inch would be called gage makers' tolerance and .0001 inch wear allowance.) Because of the practice, and necessity, for recognizing and having gage makers' tolerances and wear allowances, the gage, in use, may at any time accept or reject work that is a tenth or two (.0001–0002 inch) either side of the exact tolerance or specification.

Hence, in setting, adjusting or calibrating the step or adjustable type snap gage (Fig. 1) the idea of wear tolerance should be kept in mind of course. The inspector can determine what his shop's standards are in connection with this somewhat controversial subject and apply the required tolerances at his gage setting.* If his shop has no established practice, the 10 per cent rule mentioned above will likely be found practical.

To reduce the error due to wear in adjustable and single-purpose snap gages, the normal hardened steel anvil surfaces are hard chrome plated, a very satisfactory addition. Worn surfaces of fixed, single-purpose gages can be built up with chrome plate and then fine ground and lapped to final precise size. Where snap gages get steady and prolonged service, and especially if they have to be used in the presence of grinding grit and the like, it pays to have the anvils tipped with carbide, Norbide, or sapphire. Chrome plating will increase wear life of gage anvils five times; carbide anvils last ten to a hundred times longer.

Fixed snap gages are made not only in the conventional, more or less universal shapes of portable hand gages illustrated thus far, but they can also be obtained in special shapes and forms for definite purposes in both portable and bench models. Figure 6 pictures a few special designs.

The C-shaped snap gages are useful too, of course, in checking flat stock or thickness and width of rectangular pieces, as well as on cylindrical work.

Ring Gages

Ring gages are an interesting and common use of fixed or comparison gages. Ring gages are of three major types: solid ring gages (for

---

*The suggestion has been made previously that the inspector secure relevant publications from the American National Standards Institute (ANSI), 1430 Broadway, New York. Information on American Gage Design (A.G.D.) standards, such as wear and makers' tolerances, can be secured from the same source.

Fig. 6. Snap gages of special design.

small diameters); flanged ring gages, which are used for assessment of "medium" sized diameters; and then there are larger types of ring gages, complete with handles, which can be used in the area of 10 in. measurements.

For limit measurements on cylindrical components, the ring gage is widely used. Several ring gages of standard design are shown in Fig. 7. The "Go / No Go" limit principle is obtained by using a pair of rings — the "No Go" ring distinguished from the "Go" ring by an annular groove cut in the former's outer knurled surface. For greater convenience, ring limit gages are frequently paired as hardened bushings inserted in a single steel plate.

In general, ring gages are used in the manner pictured by Fig. 8. The same universal rule for gaging pressure applies, the finger tips exerting from one to two pounds pressure. The strong human tendency, of course, is to force the workpiece through a ring gage, and this inclination must be sedulously overcome if anything like a true limit check is to be obtained.

For decades, a rather practical viewpoint has been held in many shops that if shafts, for instance, are to be turned to fit certain size bearings or holes, the best gage is a mating part. You will see this traditional practice prevailing in a few shops today — a sample bearing or a component with a hole in it hung on a nail or wired to the lathe. While ring gages carry out the same general mating part principle, they

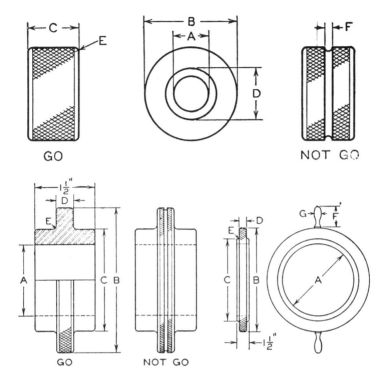

Fig. 7. Ring gages of standard design.

are much more accurate and reliable. Unlike the mating part hanging on the wall, they are made to precise tested sizes and have hardened surfaces, that prevent wearing oversize too quickly.

Gage Makers' Tolerance and Wear Allowance

Anyone manufacturing a solid nonadjustable instrument such as a ring gage or a single-purpose, fixed snap gage like that shown in Fig. 3, faces the question of gage makers' tolerance and wear allowance. The gage makers' tolerances shown in the following table were established by the American Gage Design (A.G.D.) Committee and have been widely used. It will be noted that there are five tolerance classes: XX, X, Y, Z, and ZZ.

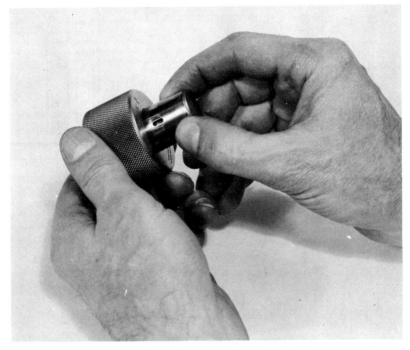

Courtesy of Taft-Peirce Mfg. Co.

Fig. 8. Using a ring gage to check the over-all outside diameter of a workpiece.

Plug and Ring Gage Tolerances (inches) — American Gage Design Standard

Nominal Size — Inches		(Male Gages Only)				(Ring Gages Only)
Above	To and Incl.	XX	X	Y	Z	ZZ
.029	—	.825	.00002	.00004	.00007	.00010	.00020
.825	—	1.510	.00003	.00006	.00009	.00012	.00024
1.510	—	2.510	.00004	.00008	.00012	.00016	.00032
2.510	—	4.510	.00005	.00010	.00015	.00020	.00040
4.510	—	6.510	.000065	.00013	.00019	.00025	.00050
6.510	—	9.010	.00008	.00016	.00024	.00032	.00064
9.010	—	12.010	.00010	.00020	.00030	.00040	.00080

XX—Precision lapped (plugs or male masters only)
X —Precision lapped plugs or rings
Y —Lapped plugs or rings
Z —Ground and polished (grinding marks may be in evidence)
ZZ—Ground only (for rings only)

The table can be best understood by studying one range of sizes like the 1.510- to 2.510-inch category for instance. In this range of ring gages, the gage makers' tolerance is to be confined to .00008 inch if a Class X gage, so-called, is purchased. This means that the hole in the brand new gage will usually be .00008 inch larger.

The tolerance for a Class ZZ gage is .00032 inch, an allowance four times greater than the Class X limit. The difference is a matter of cost and, consequently, price of the gage. If ordinary, quick measurements are to be made on fairly rough commercial work whose blue print tolerances range from ± .001 inch to ± .010 inch, say, there is no need of paying a premium for Class X gages. If, however, precise control to something like .0001 inch or .0002 inch is required on fine machine work, the Class X gages are necessary.

The tolerances listed in the table above cover the gage makers' tolerances. The usual custom, in ordering or making a gage or in setting an adjustable snap gage is to add a wear allowance. Suppose a basic dimension for a gage or gage step is to be .852 inch. The standard gage makers' tolerance would be .00006 inch (Class X). Hence, a ring gage might actually have an internal diameter of .85194 inch or .85206 inch. Normally, an additional .00005 inch or .0001 inch would be allowed for wear. Taking, say .0001 inch for wear allowance out of the ring gage internal diameter would give an actual, new ring gage size of either .85184 inch (.85194 inch — .0001 inch) instead of the basic .852 inch, or it would give .85196 inch (.85206 inch — .0001 inch), depending on which side of the *basic* dimension the gage maker happened to take his tolerance. Of course, it would be in between .85184 inch and .85196 inch if the gage maker only took part of his allowable tolerance.

In studying the foregoing, note, too, that the gage makers' tolerances increase materially as the size of the gage doubles, trebles, or quadruples. The reasoning behind this is that it is more difficult for the gage maker to maintain fine precision on the larger sizes; that, similarly, the lathe worker or grinder hand will have equal difficulty on larger diameter workpieces; that the larger rings, being heavier and bulkier, will wear faster; and, finally, that temperature has a more direct effect in measuring larger diameters.

In some shops, the various classes of gages are used in the following manner. The machinist, for instance, is supplied with a Class ZZ or Z gage. The inspector uses the somewhat more precise Y class gages and the Class X gages are kept as masters for final reference in case of dispute. Basically, all this goes back to using gages of finer discrimination where greater discrimination is needed.

The proper classification of gages (X, Y, or Z, etc.) to use can be determined by the 10% rule. If the workpiece dimension is 2.938 ± .002 inches so that the total tolerance or tolerance *spread* is

.004 inch, for instance, a Class ZZ gage would be adequate; 1/10th of .004 inch is .0004 inch. See 2.510-inch to 4.510-inch range in the table. On the other hand, a Class X gage would be needed where the specification allowed only .001-inch tolerance spread (1/10th of .001 inch is .0001 inch and in this table, size range 2.510 to 4.510 inches calls for .0001 inch gage tolerance in Class X).

Limitations of Ring Gages

Ringes gages indicate over-all size limits on cylindrical workpieces. They tell little else about the work. They will accept out-of-round work without a qualm provided the largest diameter of the piece is within bounds, see exaggerated case at A in Fig. 9, although such work may not help to make a satisfactory assembled product. Ring gages can detect excessive taper, as illustrated at B in Fig. 9, although where the largest diameter is still within limits the worker is very prone to "pass" it. These are some of the pitfalls in ring gaging. Ring gages present also the disadvantage that work must be taken out of the machine, ordinarily, and be deburred before an effective check can be made. To an extent then, the C-shaped snap gage is somewhat more useful.

One definite recommendation for ring gages must be mentioned, however. In a previous section the triangle effect on pieces coming from centerless grinders has been described. Unless the factory has available indicating, electric, or air gaging equipment specially designed for detecting the degree of triangle effect, ring gages are the only means of truly checking the limiting sizes of cylindrical pieces bearing the triangle error. Micrometers, snap gages, or any type of two-point, caliper measuring equipment will not reveal the "mating" or greatest

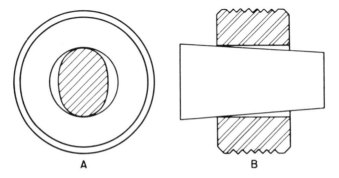

Fig. 9. (A) Ovality in the shape of a workpiece cannot be detected by a ring gage. (B) Excessive taper in a workpiece can be detected by a ring gage.

effective diameter of work in which the triangle effect appears. (The "mating" or effective diameter of a triangular or oval shaped workpiece is the diameter of the hole into which the piece will fit.)

Plug Gages

The idea of a mating piece has been mentioned previously, a ring gage being essentially the "mating" hole for cylindrical pieces. Plug gages are based on this same general concept. Is a hole the right size? One way to check it is to try in it the shaft or cylinder of the proper diameter that is to mate with it. Another way, of course, is to check the hole diameter with a more precise, nonwearing member, a cylinder called a plug gage.

For the sake of illustration, several varieties of commercial plug gages are pictured in Fig. 10. The average commercial plug gage is a precision ground, hardened cylinder somewhere around an inch long. Plug gages usually come in pairs and are gripped in hexagonal holders. Reversible plug gages, so-called, have a "Go" plug in one end of the holder and a "No Go" plug in the other. The "No Go" plug is frequently made shorter in length purposely to distinguish it more readily from the go plug. The specification limits are usually stamped on the holder. Figure 10 shows also a progressive, stepped type, which is faster to use than the reversible or double-end "Go / No Go" gage, but has the disadvantage that both sections must be replaced when the "Go" section, which usually gets the greater use, becomes worn down. Some plug gages are really steel wires, they are so small in diameter; others are large enough to require both hands to lift and manipulate them. As in the case of snap gages, plug gages frequently come chrome plated or in carbide or Norbide. Unlike conventional snap gages, they are inevitably single purpose — no adjustment whatsoever can be provided — a proper diameter plug being required for each tolerance or specification.

Plug gages are furnished in the X, Y, and Z classifications for precision and finish. If you will note the table on gage tolerances, given previously, you will see that tolerance specifications in the XX class are for plug gages only and that plug gages are not ordinarily furnished in the ZZ class.

Proper Use of Plug Gages

There is a trick to using plug gages if you want to get precise measurement. Too many people use them for hand reamers. Or

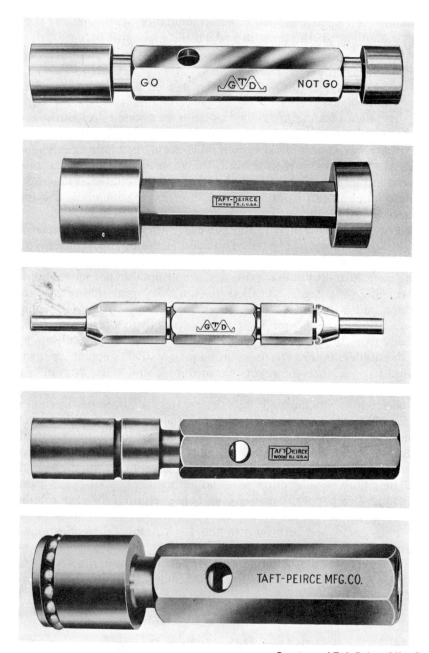

Courtesy of Taft-Peirce Mfg. Co.

Fig. 10. Various types of plug gages.

crowbars. (Workers have been seen pounding in nails or hammering up lathe dog clamps with them.) Let the punch, drill, reamer, broach, or boring tool cut the hole; use a plug gage only to measure the resulting size. If you will handle a plug gage in somewhat the same manner you do a pencil, you will come nearer to the true measurement than if you try to expand the hole to size.

It might be well to digress here and enlarge on a human foible implied in the last sentence of the paragraph above. What is involved is probably wishful thinking. Skilled mechanics and neophytes seem to be equally guilty. There is a strong tendency to "make" (using shop language) the workpiece "come" to the size you want it to. This, rather than to let the gage do its own work unhampered and measure the actual dimension of the workpiece. If the blue print, as an example, calls for .375 inch −.001 inch, + .002 inch and if the workpiece is actually, say, .373 inch, you are quite likely to manipulate micrometers, verniers, or fixed gages by either subconsciously increasing the gaging pressure or lightening up on it so that somehow you get the .374-inch reading or convince, "kid," yourself that you do. Indicating equipment, as will be seen from the subsequent description and discussion of it helps to eliminate this form of error or unintentional dishonesty.

Go back to the "measured point" conception. Consider a plug gage essentially as inside caliper points on opposite sides of a true diameter. The remainder of the plug gage cylinder can be considered as a skirt or guide. All the rules concerning manipulation discussed in previous sections apply to plug gages.

Many makes of plug gages are equipped with pilot tips (see Fig. 10) to facilitate inserting the plug and prevent the shaving action of sharp-edged gages. A plug gage, it goes without saying, should be kept clean, and it should be used only on clean work. Any small amount of dust, grit, or chips is especially merciless on plug gages. The plug surfaces should be coated with a film of oil.

One Difficulty to be Avoided

The "green" (inexperienced) inspector is liable to run into one embarrassing experience with plug gages. If he is gaging a fairly deep hole — something more than a .5 inch deep, perhaps; if the hole diameter happens to be right at specification size and, consequently, at the same size practically as the plug gage diameter; if both gage and workpiece happen to be dry — free from grease; then practically metal to metal contact is secured as the gage is forced into the hole and, if the inspector is stubborn enough to continue using undue pressure, the gage will "freeze" in the hole. The harder he pushes or pulls or twists to get it loose, the tighter it sticks. Dislodging the gage may mean rapping it, resorting to a vise, or even to heating the workpiece to expand it away from the gage. Where you are aware that the measurement with a plug gage might bring about this freezing, be sure the gage surface is oil coated and be sure especially, to use the words of experienced mechanics, "to keep it moving." Once you stop the motion, the workpiece may lock onto the gage. The danger of a gage freezing on the work also applies to ring gages.

Checking the Plug Gage

Before using a plug gage, the inspector should question first the authority for the "Go" and "No Go" sizes stamped on the gage. Do they correspond to the blue print tolerances? Then, of course, the actual diameter of the plugs should be carefully measured to be sure their actual diameters check with the size markings on the gage.

Note that, in contrast to a snap gage, the "Go" end of a plug gage has the smaller diameter. The "No Go" end is larger. If a hole is so large that not only the "Go" plug enters but also the "No Go" end, the hole is out of tolerance because it is oversize. If the hole is so small that the "Go" end plug will not enter it, the hole is out of tolerance in the sense of being undersize. The correct size hole will receive the "Go" plug but not the "No Go" plug.

If the end of the gage has a square edge — if it is not equipped with a pilot in other words — check the edge with your finger nail for fine nicks and raised burrs. Of all gages and measuring equipment, the plug gage is perhaps the most easily dropped. It has too easy a tendency to roll or fall off of bench or machine. To a somewhat greater degree than either micrometers or snap gages, the measuring surfaces and square edges on plug gages are unprotected, naked, if you will; consequently, plug gages must be handled, set down, and stored away more carefully.

Limitations of Plug Gages

Experienced mechanics claim they can detect out-of-round, tapered, and bell-mouthed holes readily with plug gages. The conditions being described are illustrated in the diagrams of Fig. 11. If the gage can be wiggled around in the hole, there is always a question of whether the hole is actually tapered, out-of-round, or bell-mouthed or whether the looseness is due only to a discrepancy between the plug diameter and the hole size. If the taper or bell mouth is slight — not exaggerated as in Fig. 11 — and if plug and hole size are close to each

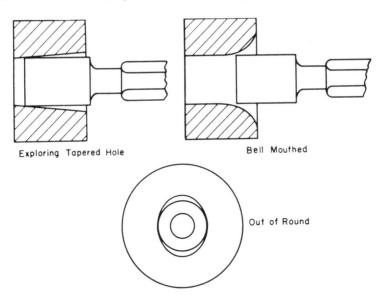

Exploring Tapered Hole   Bell Mouthed

Out of Round

Fig. 11. Various hole conditions that may be difficult to distinguish with a plug gage.

other, there can be some doubt as to the inspector's ability to discover the defects. Where the greatest diameter of an out-of-round hole corresponds very closely to plug size, there is considerable difficulty in diagnosing a condition of ovality.

When a plug gage is small enough to be in the "wire size" range, care must be used, of course, not to snap it off and, particularly, to make sure that the relatively long wire plug does not become imperceptibly bent.

Mention of wire-size plug gages brings up another fairly common practice used especially in connection with the smaller diameter plug gages. The butt end of the plug becomes worn down in diameter (bevelled or tapered) more rapidly than the main body of the plug because, almost inevitably, the end of the plug must fiddle around a little at the entrance of the hole. If the hole edges are burred, the butt end of the plug wears rapidly, shaving off such burrs. (Which, of course, is one big reason for the design and use of pilot features on plug gage members.) Hence, it is customary in many plants to regain the accuracy of plugs by grinding off or back 1/32, 1/16, or 1/8 inch from the end of the plug. This idea is followed even farther by having the inspector use only about the first 1/8 inch of the total length of the plug as a gaging, measuring, or setting area. By so doing, the rest of the diameter of the plug is not worn down, theoretically, and grinding off the 1/8-inch worn tip regains full accuracy of the plug.

Of the nonindicating and conventional and especially of the fixed type of gages, ring and plug gages come the nearest to being usable consistently to discriminations or accuracies of .0001 to .0002 inch. This statement must be limited of course to plug and ring gages made to sizes in tenths of a thousandth, as, for instance, a "Go" size of .3855 inch and a "No Go" size of .3857 inch on a plug gage.

Flush pin gages or cylindrical pin-type gages are also available in a wide range of sizes, typically in sets with increments of .001 in. and starting at about .01 in. gage size, to about a .5 in. in a set. These gages are available on the market for uses that approximate the regular cylindrical plug gages. They can be supplied with plus or minus tolerances, and are usually fitted with special interchangeable holders for ease of use.

Special Shapes — Spline Gages — Feeler Gages

The principle of plug gages is not confined to cylindrical plugs. Figure 12 illustrates a so-called flat plug gage with "Go" and "No Go" members designed to check the widths of slots, channels and rabbets.

One of the traditional methods for checking tapered holes is with taper plug gages of the sort also illustrated in Fig. 12. Usually, taper plugs show a pair of marks etched at the proper precalibrated location, as the sketch in Fig. 13 indicates. The tapered hole is too large if both marks on the gage sink down out of sight or too small if both marks are

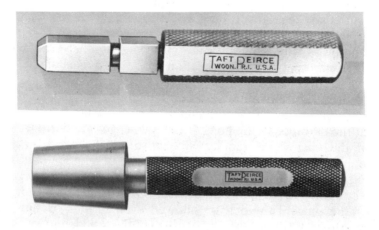

Courtesy of Taft-Peirce Mfg. Co.

Fig. 12. (Upper) Flat plug gage designed to check slots and channels. (Lower) Plug gage used to check a tapered hole.

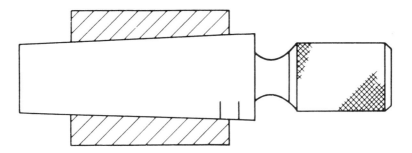

Fig. 13. Method of checking a tapered hole with a plug gage. If first gage mark
enters hole but second does not, taper is correct.

visible — the tapered hole being considered correct as to diameter and
depth if the edge of the tapered hole conceals only the lower gage
mark.

Gage pins are also a part of the plug gage family. Gage pins are
available in a variety of sizes, descriptions, sets, and accuracy. Essentially, these pins are small gage plugs and are used similarly.

Following also the mating part conception, and as an illustration of
the many special shapes commercial plug gages may take, a spline gage
is shown in Fig. 14. Gages of this general nature check a number of
conditions simultaneously. A spline gage, for example, may not enter
the workpiece (or fit it too loosely) not only because of workpiece
diameter variations but also because segments of the splined workpiece
are not evenly or properly spaced or because the splining is not parallel
to the axis of the workpiece.

The keyway gage, Fig. 15, belongs essentially in the class of flat
plug gages. In addition to confirming the width of a keyway, however,
it checks the depth and location.

Courtesy of Taft-Peirce Mfg. Co.

Fig. 14. (Left) Special plug gage for checking splines. Fig. 15. (Right) Double-
ended plug gage for checking the width of keyways.

Many movable-type gages such as calipers and micrometers can be specially adapted for use as a fixed or Go/No Go gage. Inside and outside diameters, with less precision, of course, and linear measurements and others, can be checked with a modified or fixed gage in a production environment, where a quick check to a known accuracy, can be obtained. Even gages commonly thought of as dynamic or movable direct measuring tools, such as functional hole location gages, can be adapted to the fixed gage approach. Such hole location gages are capable of multiple and simultaneous assessments, or may utilize movable and stopped settings for various locational assessments to be performed at one time. These types of fixed gages, adapted specially for unique designs, are very prevalent, and quite effective for use in volume production operations.

A close relative in the plug gage family is the thickness or feeler gage, so termed, of the general type illustrated in Fig. 16. For many of us, probably, the first time we saw a mechanic use a feeler gage was when we watched him establish the spark gap between the distributor points of an automobile. Feeler gage sets can be secured containing a fan of blades or leaves differing one from the other by one thousandth or several thousandths in thickness as Fig. 16 shows. The desired gaging thickness can be secured by folding together the selection of leaves that will build up to a thickness, width, height or clearance dimension that is to be checked. Heavier blades should of course be used in the outside positions in combination in order to protect the set and ensure accurate measurement.

One or two precautions in connection with using feeler gages should be observed. If width or height or space must be measured to a discrimination of less than about .002 inch, it would be wiser to try to secure a standard flat plug gage, see Fig. 12, or attempt a measuring

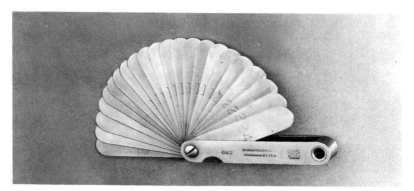

Courtesy of Brown & Sharpe Mfg. Co.

Fig. 16. Typical feeler gage set which has blades of various thicknesses.

setup where a dial indicator can be used. The leaves of feeler gages become rather readily burred, bent, or warped, or constant use shaves off a few ten-thousandths of thickness. The leaves of a thickness gage should be checked regularly and damaged blades replaced.

Radius and Template Gages

Most all parts and products are designed with rounded edges and with corners filled in. Appearance is not necessarily the main reason. Square sharp edges have a certain beauty too. But they become readily nicked and broken. The designer who deliberately introduces a fillet or rounded contour into what might have seemed more naturally or more ideally a sharp square corner between two sections of a shaft probably knows his business. Such a piece of metal — be it cylindrical or square — will often stand many times the torsion, tensile or breaking strain and, if its outer edges are rounded, will take a lot of rough handling without getting nicked or burred. Another practical consideration, in connection with fillets especially, is the common fact that the sharp corners or edges of cutting tools and wheels wear off readily; the mechanic has difficulty maintaining sharp corners when machining a series of pieces. (Radius gages are also called filet gages.)

So, the industrial inspector meets radii and fillets continually in his work. Usually a radius is not a precise affair; it may make little difference to the workings of a product whether a radius is 1/16 or 1/8 inch. However, there must be some specification limit for the amount of radius or fillet the machinist should cut and the inspector checks the evidence of control with radius gages. In occasional instances, the radius must be closely controlled because of clearance conditions and the like under which the piece is to be assembled.

Commercial radius gage sets consist of a series of thin steel leaves, such as are shown in Fig. 17, which are used as templates after the general fashion shown in Fig. 18. The radius gage, or most any template for that matter, corresponds in one way to the steel rule — do not expect to come closer than 1/64 inch (.015 inch) with your measurement. Using a radius gage without proper light is nearly useless. If possible, hold gage and workpiece between the light and the eye, but, if this is impossible, at all costs be sure that good strong light shines down on the junction of the radius gage with the radius or fillet being compared. Remember too, always, that the edges of radius gages can become readily nicked, burred, curled up, or worn back.

Template gages are not confined to the small sizes indicated in Fig. 17. Inspectors in aircraft work are familiar with the man-size sheets of metal with carefully calculated curvatures profiled in them to check the

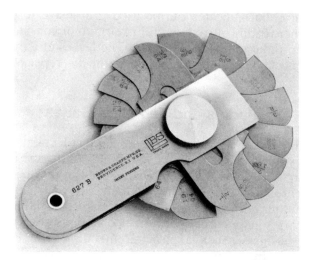

Courtesy of Brown & Sharpe Mfg. Co.

Fig. 17. Radius gage set for checking both internal and external radii.

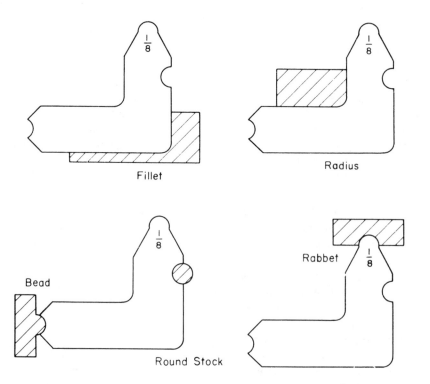

Fig. 18. Use of a radius gage to check various radii found on workpieces.

contours of wing surfaces, etc., or the more exacting templates used to shape, grind, balance, and polish propellor surfaces.

One of the common inspection tools, a straight edge, Fig. 19, is really a template. Likewise, the hardened steel square. The use of such tools seems evident, just from the photographs of them, but they also illustrate another principle of mechanical inspection work that is worth enlarging upon again. Assuming that either a straight edge or a try

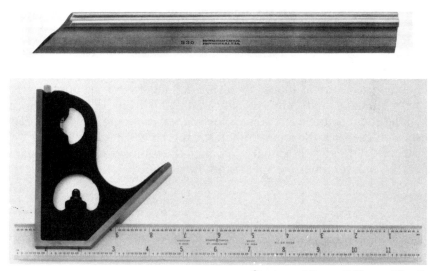

Courtesy of Brown & Sharpe Mfg. Co.

Fig. 19. (Upper) Precision straight edge. (Lower) Combination square.

square has been checked and found straight, true, and unwarped, with edges free from nicks, burrs, or worn spots, there are still limitations in their ability to measure and compare. With them, straightness or squareness can be determined to an accuracy of about 1/64 inch (.015 inch), ordinarily If the product specifications call for greater accuracy of measurement and control in manufacturing, the inspector would be wise to seek other means of measuring or comparing the squareness or straightness. This is the point to be emphasized. Know and admit the limitations of each type of measuring equipment and seek apparatus with finer discrimination where necessary. In other words do not guess. And do not let everyday laziness fool you into believing you are measuring correctly, establishing the degree of conformance, simply because you do not want to bother to secure better equipment.

Checking straightness or squareness with straight edge and square involves sighting between the instrument blade and the workpiece. Hence it is preferable to hold the work up to a light or arrange in some manner to have light behind it. While light can be distinguished through a gap .00005 inch wide between two perfectly smooth surfaces, it is still advisable for the inspector to assume that he is not guessing the degree of waviness, warp, or out-of-squareness closer than .010 inch. Even though you see light between a straight blade and the workpiece surface, you can never be sure of the difference between, say, .00005 inch and .0001 inch or between .001 inch and .005 inch. Few of us are gifted with such a basic sense of space.

Practical mechanics, however, do aid the sense of sight by inserting slips of paper in the gaps (ordinary typewriter or memorandum paper is .003 to .004 inch thick) or by using feeler gages where it is necessary to know fairly closely, for instance, how much more needs to be taken off the high spots of a workpiece in order to make it truly straight or square. This technique is illustrated in Fig. 20. Additionally, adjustable double squares are available for laying out and measuring applications. These versatile units are typically graduated from 1/32 to .01 inch, making them suitable for some shop and inspection applications. The use of these and the many other squares and squaring devices and instrumentation are essential in many inspection set-ups. The square-ness of measurement set-ups is often critical to the results of the

Courtesy of Taft-Peirce Mfg. Co.

Fig. 20. Proper method of checking the squareness of a workpiece by using a square and a paper or metal feeler gage.

measurements; this particular tool displays rather rough powers of discrimination. Many other such tools offer far greater accuracy and discrimination, and of course this function is available on many electronic instruments that will be discussed later.

Flush Pin Gages

Flush pin gages are practically always single-purpose gages designed for the control of a particular dimension on a particular component. Consequently, they are used mostly for gaging work produced in continuing operations and mass production, or where batches of the same sort of part are made every so often. Since they are made in so many different and special styles, no attempt will be made here to go into the subject of flush pin gages other than to describe the general principle and use of them.

Suppose we consider the shape of a flush pin gage outlined in the sketch of Fig. 21, a gage designed to check diameters. As can be seen,

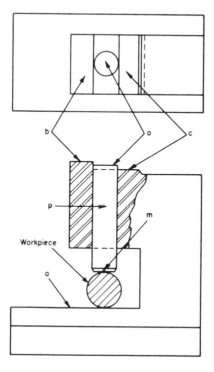

Fig. 21. Flush-pin gage of typical design.

the gage follows the standard snap gage pattern of a steel C-frame and a reference point or anvil, *a* in Fig. 21.

Rather than fixed (or slightly adjustable) Go / No Go upper anvils, however, the flush pin gage is provided with a hardened plunger or "pin," *p*, which, made to a snug fit in the hole in the C-frame, will slide up and down. The measuring surface, *m*, is carefully machined square to the axis of the pin and parallel to the reference anvil surface, *a*; it is also made flat to a few millionths. At the same time the opposite end of the plunger, *o* in Fig. 21, is likewise machined carefully smooth, flat, and square to the plunger axis.

The Go / No Go step idea is provided on this type of gage by two steps carefully planed on the upper surface of the C-frame as at *b* and *c* in Fig. 21. The difference in the heights of these two steps — usually anywhere from .002 to .010 inch — corresponds to the tolerance or specification limits. If the gage, for example, were to be used on diameters calling for .625 ± .002 inch, the upper step, *b*, would be marked .627″ and the lower step, *c*, would be stamped .623″.

The use of the gage is indicated by the sketch in Fig. 21. The workpiece is placed on the reference anvil *a* and the pin, *p*, is pushed down on to it with standard gaging pressure. If the flat top of the pin, *p* — its upper surface *o* — protrudes above surface *b*, the work is oversize. If it sinks below step *c*, the piece is undersize. In-specification work will position surface *o* somewhere between the two steps.

Where the workpiece is very close to the high or low tolerance, the amount the pin protrudes above either step, or sinks below, may not be discernible to the eye. The inspector then scratches across surfaces *b, o,* and *c* with his fingernail. The fingernail test cannot be depended on to a discrimination finer than .002 inch. If the blue print calls for a tolerance spread less than .002 inch or if the difference between steps *b* and *c* is less than .002 inch, the accurate use of the gage should be seriously questioned and other means of measurement (micrometer, vernier, or indicator equipment) should be used.

Tapered Parallels

Tapered or adjustable parallels can be used in many ways. For instance, they are often used to check internal diameters, as well as the more typical applications of assessing parallel slots or other parallel openings. These devices are also models of transfer, or indirect-comparison measurement tools. Parallels are typically 9/32 inch thick, and are available in sets with (measurement) ranges of from 3/8 to 2 1/4 inches.

Tapered parallels offer another means for measuring an internal diameter. The proper pair of parallels is selected and slipped into the hole with their flat tapering surfaces facing each other and their rounded parallel surfaces in position for contact with the inside surface of the hole. By sliding one tapered surface against the other, a wedging effect is obtained that brings the outside rounded surface of each block into contact with the hole surface as shown in Fig. 22. It should be mentioned that in obtaining this wedging contact only the standard gaging pressure of one or two pounds should be used.

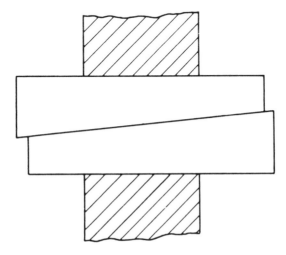

Fig. 22. Illustration showing how tapered parallels may be used to measure inside diameters.

The outside contact surfaces of the two blocks are rounded to some curvature *less* than that of the hole to permit contact of both on the hole diameter, but care must be used to make sure they are properly centered. It is possible to lock them to one side a few thousandths of an inch and thus measure a chord and not the true diameter. The internal diameter measurement is then made with a micrometer as shown in Fig. 23.

While the measurement of an internal diameter is illustrated, tapered parallels are also used to check the width and parallelism of slots and to "plug" holes for measuring hole center distances and layout.

Tapered parallels should be carefully examined at each use for edge nicks and burrs and worn, rough, or corroded spots; in other words, they must be used carefully and always kept in good shape — oiled or rust proofed before they are put away. They must, of course, be

Fig. 23. Using a micrometer to measure the setting of the tapered parallels after they have been wedged in place as shown in Fig. 22.

carefully cleaned to remove the rust proofing, dirt, and grit at each use. Again, there is a question as to the ability of tapered parallels, with micrometers, to measure to a precision finer than .0005 inch. Tapered parallels, of course, cannot be so readily used in a blind hole or where the hole is not accessible from both sides.

Planer Gages and Other Inside Measuring Gages

Another appropriate topic for the fixed gage section is a discussion of the planer gage. Similar to parallels, planer gages are comprised of two inclined planes. When these two planes are moved in coincidence or opposition, the distance between the parallel surfaces change[5]. These versatile units are assessed or directly measured with a micrometer.

A telescoping gage is another interesting device. This gage is used to measure the inside dimensions of grooves, slots, holes, and so forth. The T-shaped transer instrument is used in conjunction with a vernier caliper, regular calipers, or a micrometer.

"Small hole gages" are specialized instruments used to measure slots, grooves, and recesses of less than 0.5 inch.

Many other types of fixed gages exist, and this last category contains probably the largest number of fixed gages in industry today. This is the category of custom gages and fixtures that are designed as an integral *part* of the assembly operation, so that the operation *cannot proceed* if there is a critical dimensional problem in some area. It is now uncommon for these gages and fixtures to make several automatic Go/No Go assessments at once, often fairly invisibly to the operator, so that the act of assembling or manipulating and holding the pieces form the basis for the gage itself.

Often these assessments are progressive — that is, one operation is complete, and the assembly level piece is then gaged automatically at the next assembly level by the introduction of a new part to the gage. Much creativity is expressed in these gages; many of them are power-actuated, and can save assemblers and operators valuable time, but most importantly actually gage the operation as the work ensues.

CHAPTER 8

Surface Plate Methods and Equipment

Up to now, several fundamental methods of measuring have been discussed. The method illustrated by the steel rule — a length graduated or subdivided which, held up against an object, allows some pertinent dimension of that object to be determined by comparing by eye the dimension with the count of graduations on the measuring length or rule.

The sense of touch, aiding vision, has been brought in by suggesting the use of calipers to span a dimension and then comparing the caliper spread with a set of graduations, as on a steel rule.

Combining mechanical pressure and the sense of touch with a much finer discrimination in determining actual dimensions, the vernier principle was applied to the graduated length, such as the steel rule, with the equivalent of calipers attached to it.

Another addition to the methods of calipers, graduated lengths, and vernier devices was illustrated by the micrometer with its added basic mechanical principle of a screw thread which, when turned, advanced or withdrew caliper jaws 1/40th of an inch per turn.

The principle of the mating part plus the principle of a fixed measuring unit for comparison has been brought out in the discussion of fixed gages, such as snap and plug gages.

Precision Gage Blocks

The invention of gage blocks has resulted in another basic method being used in the field of precise measurement. If you were building a brick wall or column with bricks 3 inches thick, using 1/2 inch thick layers of mortar between the bricks, your wall seven bricks high, would measure 24 inches over all because seven 3-inch bricks stacked on each other equal 21 inches and the six mortar joints, each 1/2 inch thick, add 3 inches more. Of, if you took 12 1-inch kindergarten blocks and erected a tower for a child with them, the tower would be 12 inches high, more or less, depending on the uniformity of the blocks and the flatness and smoothness of their contacting surfaces.

Commercial gage blocks are special steel blocks, hardened, with carefully machined parallel and flat surfaces. They are used to build up various gaging lengths essentially as suggested in the simple brick and play block examples above. Commercially, too, gage blocks have been standardized so that in a cross section perpendicular to their length axes they usually are either a 1-inch square or a rectangle that is 3/8 × 1-3/16 inches. (One maker of gage blocks furnishes disks like coins or, in longer lengths, like rods.) While in the shop, terms like "Jo" blocks or "Hoke" blocks or just plain "blocks" have come into common use, the technical name for any individual gaging block is a *gage*. (The manufacturer of the round, coin- or rod-like gages, for instance, calls them as a trade name, "microgages.")

Gage Blocks Are Used for Linear Measurement

Figure 1 shows several sizes of these gages, and illustrates the concept of a gage block, which the new inspector should fix in his mind. The length of the gage or gage block is its fundamental feature. By joining a 1-inch block to a 2-inch block, 3 inches in length is obtained. To ensure the fidelity of the lengths of gage blocks the manufacturer makes sure, above all, that two of the surfaces or planes are strictly parallel to each other, truly perpendicular to the length axis, and that

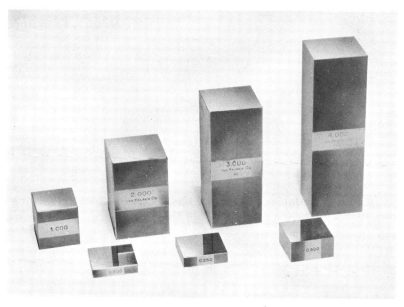

Courtesy of The Van Keuren Co.

Fig. 1. Gage blocks of various sizes.

these parallel surfaces are as perfectly flat as he can make them. Commercially, the flat parallel measuring surfaces are made square to the other or "side" surfaces of the blocks. Commercially also, the gage block manufacturer usually makes the contrast between the highly polished, flat and parallel measuring "ends" of the blocks and the "sides" more evident by putting a duller (sand blast) finish on the sides or at least by stamping the gage size on one of the sides.

Figure 2 illustrates these points; the "length" of the gage block, whether it is cut from .050-inch thick metal or .500-inch or 5.000-inch, is the dimension between the polished, flat, and parallel measuring

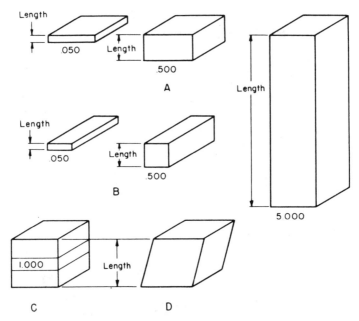

Fig. 2. (A through D) The "length" of the gage blocks is the distance between measuring surfaces.

ends as shown. In the case of the .300-inch long block of the rectangular type — series B in Fig. 2 — it might be difficult to distinguish which are the two "ends" because the block's other thickness dimension is 3/8 inch (.375 inch). Or, in the case of the 1-inch square type of block, since the 1.000-inch gage is, of course, practically a cube, unless the ends were distinguished by their high polish or, better, unless the nonmeasuring sides had a contrasting finish, pattern, or size marking as illustrated at C in Fig. 2, the precise measuring "length" or axis of the block might not be readily distinguishable.

While the manufacturers use care to put out "square" blocks, as a matter of good workmanship and for convenience in use, they do not guarantee this squareness. Some slight degree of condition D, Fig. 2, may exist. In other words, gage blocks should not be used to check squareness. The manufacturers do guarantee however a certain flatness of the measuring ends as well as their parallelism to each other and particularly the length between the ends.

There are several reasons for going into all this detail. One of the first rules a worker should adopt when using gage blocks is never to pick them up or grasp them by the ends. Take hold of them, handle them, by their sides. In the same vein, as far as possible set gage blocks down on the work bench resting on their sides. The ends are the precision surfaces and handling will damage them.

Gage blocks normally emerge from their manufacturing process with end surfaces that are flat, true planes within two millionths of an inch level. The two end planes are usually parallel to each other also within two millionths of an inch. The length of a gage block is usually guaranteed to be within about two, five, or ten millionths of an inch of the size stamped on it, depending on the commercial grade and price of the block.

As a matter of space or of length, a millionth of an inch is not much. It is about as hard to comprehend mentally as the astronomical light-year or the billions of dollars in federal government appropriations. Hence, gage blocks are not to be mishandled. (Actually they do withstand a great deal of unwarranted abuse without appreciably affecting the desired results in precision measurement.)

Grades of Gage Blocks

Commercial gage blocks are supplied as "master" blocks, "inspection" blocks, and "working" blocks; master blocks having almost invariably .000002-inch accuracy in specified length, inspection sets around .000005-inch accuracy, while working sets are liable to show errors in length up to 10 millionths or .000010 inch. The usual, and best, practice is to use master blocks very seldom and never directly on work or shop measurements. They are kept solely to compare with or to calibrate inspecting and working blocks. To a somewhat similar degree, the use of the inspection grade is limited as much as possible. An attempt is made to maintain their accuracy, precision, and serviceability in order that they may be used confidently to settle, perhaps, some dispute over the exact dimensions on workpieces, the precise setting of a comparator, or the accuracy of, say, micrometers or other gaging equipment.

It might be well right here to digress enough to emphasize that gage blocks, so far as the shop is concerned, are basic measuring equipment. In the standards laboratory it may be possible by optical means and spectroscopy, by means of light waves and interference bands, or by means of electronics, to be as basic and fundamental as the development of the science of measurement permits. Such delicate apparatus and techniques cannot be used of course in the shop. Fortunately, commercial gage blocks are so sturdy, reliable, and dependable, that it is possible to obtain with them basic accuracies and discriminations in the shop that compare very favorably with what the laboratory can do.

Stacking Gage Blocks for Measurement

To get a particular linear dimension, then, with gage blocks, they are built up, one on the other, using a suitable selection of various size blocks, until a "stack" like that shown in Fig. 3 is obtained. The gage block manufacturer uses of necessity, as has been stated, a tolerance of a few millionths in trying to obtain the precise length of each block he makes, at the same time trying to maintain parallelism and flatness. Some of these tolerances or gage allowances may fall a few millionths on the plus side; while as many others, it has been found, may cause the actual gage length to be minus. Gage blocks wear down to the minus side from use, of course, and certain size gage blocks usually wear faster than others because they happen to be used more. So, in practice, it is found that where several gage blocks are stacked together the pluses and minuses, to use an expression, usually add up algebraically, and the resulting or total length may show less error in millionths, either plus or minus, than the known error in any one of the individual blocks.

Unlike a child's building blocks, however, gage blocks are not simply rested one on the other. More like the bricks in a wall, they are actually cemented together. To use a shop term, they are "wrung" together. They are pressed together with a twisting motion that not only squeezes out the air between the almost geometrically flat surfaces of the gage block ends but gains additional adhesion from the phenomenon usually explained as a combination of molecular attraction and the "cementing action" of oil film or moisture on the adhering surfaces. Hence, there is no error in the total length of a stack of gage blocks due to any measurable spaces between the blocks' end surfaces.

Perfectly dry blocks will not stick together as, for instance, when blocks are carefully cleaned free from dust, grit, oil, or grease by bathing them in pure grain alcohol and then letting them dry completely moisture-free from any water absorbed in the alcohol by

Fig. 3. Gage blocks stacked to obtain an over-all dimension of 1.7545 inches.

evaporation in air. To put gage blocks together, then, for shop inspection and gaging purposes, the surfaces are cleaned free from grit and foreign matter but they are not cleaned in the sense of removing a final, impalpable oil film on the ends or wringing surfaces.

Wringing Gage Blocks Together

The steps in wringing blocks are shown in Figs. 4-A, -B, -C, and -D. These steps should always be taken exactly in the order shown — no step should be carelessly slighted. The two blocks, properly cleaned, are brought together flat as shown in Fig. 4-A and oscillated slightly with very light pressure. Be sure, in this step, that one block is not slightly tilted. If so, its edge may tend to shave, wear, or roughen the precisely flat, polished end surface of the mating block. This technique also minimizes the danger of one block scratching the other, and it will detect any foreign particles between the surfaces if the finger pressure is light but firm. Shift the blocks to the position of B in Fig. 4. If the blocks are clean, untilted — flat to each other — they will begin to take hold. (If any grit between the blocks has been felt, stop right there, of course, and clean them again.)

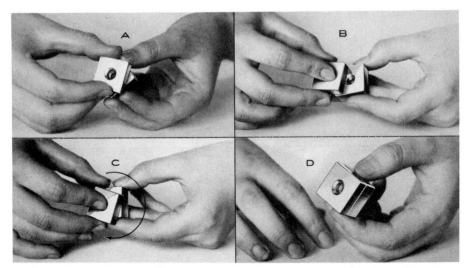

Courtesy of Colt Industries, Pratt & Whitney Machine Tool Div.

Fig. 4. (A through D) Four steps in wringing two gage blocks together.

With the top block half out of engagement, the two blocks are slid together, using standard gaging pressures and a rotary or spiral motion about as Fig. 4-C indicates. Maintain the motion until the blocks are professionally lined up as in Fig. 4-D.

A very common tendency in wringing blocks is to use unnecessary pressure to wring them together. Brute force is not needed to secure proper wringing or adhesion. And too much pressure not only makes it harder to separate the blocks but also sets up unnecessary wear. Furthermore, nothing is contributed to the accuracy of the setup. While it is possible to wring gage blocks together with an adhesion that will support several hundred pounds, all the inspector is after is just enough wringing pressure and just enough adhesion so that the blocks cannot inconveniently fall apart when they are moved or used.

Gage Block Measuring Technique

The proper manner of setting up a gage block stack to a given dimension can be considered as a series of steps. Ensure that the box of gage blocks is placed conveniently near the work area, to minimize the exposure and chance of dropping blocks and causing irreparable damage to the units. Handle the blocks separately as much as possible — combining them in one hand tends to scratch or damage the units. Any handling also starts the process of contamination and corrosion of these delicate blocks. Sweat and other acid-based emanations from the hands are always present, and begin a chemical reaction that violates the fine surface integrity of the units. Oil, grime, and even some airborne contaminants can be extremely damaging to the texture of the metal blocks, and often times a destructive condition is only understood when it is too late.*

Gage Blocks Must be Kept Clean

The second step should inevitably be that of putting down a clean sheet of paper, a clean cloth, or chamois between the gage block box and the work. The idea is to have a clean location prepared on which to set the gage blocks after they are selected from the box. Never, as has been said, handle gage blocks by their precision lapped ends, if it can be avoided. Take hold of their sides and set them down on their sides. Never put gage blocks down on a bare bench top or machine bed. Set them on the clean paper, cloth, or chamois prepared for them.

As a general rule a clean paper towel or a clean piece of typewriter paper is used. (Never a newspaper; it is probably dirty.) Some mechanics prefer the softer facial tissues, but tissue, like cloth, felt, or velvet, bothers sometimes, because it sheds lint or fuzz which may actually interfere with the accuracy of a precise measurement. Perhaps the best material is chamois. Chamois is soft and fuzzless. If a gage block is accidentally fumbled and dropped, the chamois cushions the fall. The one danger in the use of chamois (also cloth) is that it readily collects dust and grit and most of us, being lazy and inattentive, fail to wash it out thoroughly just about every time we use it. By and large the best solution seems to be to line the bench top area near the gage block box with a clean piece of cloth or felt as a soft mat and top the cloth with a clean piece of paper such as a fresh paper towel.

(It should be unnecessary to suggest at this point that perhaps the inspector needs to interrupt his routine to wash his own hands, especially if he has been recently handling extra dirty components or parts. In general — sloppy work place, sloppy worker, sloppy work.)

---

*Several makes of gage blocks are chrome plated. Chrome plating obviates the danger of corrosion, of course. Chrome, being a so-called slippery metal, also seems to withstand better the frictional wear of wringing. Where chrome-plated blocks finally wear minus, they can be "built up" by replating and then sized and lapped. But it is nearly impossible to get chrome-plated blocks to stay wrung together. For this reason they are disliked by experienced workers and the accuracy of stacks of chrome plated blocks is also suspected. Add to all this, the tendency of chrome plating to flake off, especially at the edges.

Selecting the Proper Blocks

The third step is the selection of the proper blocks for the particular measurement needed. Before proceeding, it might be wise to re-examine the blue print or specification tolerances and review the nature of the work, the measurement of which seems to involve the use of gage blocks. If the workpiece tolerances are less than .001 inch, if it is felt that the decision or setup to be reached depends on extra precise measurement, then gage blocks can be the proper tools. But if the limits are broader, the work and the measurement less particular, some alternative measuring method or medium probably should be chosen. The point is that too many times an inspector is prone to get out gage blocks when micrometers or vernier equipment would be easier and faster to use and adequate as to accuracy or discrimination. Use gage blocks as little as possible. They are perishable and expensive. It costs time and money to keep on constantly rechecking them for wear and inaccuracy. As a rule of thumb, it has been said that gage blocks wear down a full millionth of an inch, even with very careful use, in the vicinity of 200 wringings. Use other measuring techniques and apparatus if possible.

Let us suppose, however, that the set-up calls for a gage block stack for 2.6437 inches.

Examine the box or set of gage blocks available A popular size box contains 81 blocks, see Fig. 5. Another size is the 36-piece set. Gage

Courtesy of Brown & Sharpe Mfg. Co.

Fig. 5. A set of gage blocks consisting of 81 different sizes

makers' catalogues list other assortments and sets including blocks in metric measurement. The inventory of block lengths in an 81 piece set is listed in Table 1.

Table 1. Sizes of Gage Blocks in 81-Block Set

First: One Ten-Thousandth Series — 9 Blocks

.1001″	.1002″	.1003″	.1004″	.1005″	.1006″	.1007″	.1008″	.1009″	

Second: One Thousandth Series — 49 Blocks

.101″	.102″	.103″	.104″	.105″	.106″	.107″	.108″	.109″	.110″
.111″	.112″	.113″	.114″	.115″	.116″	.117″	.118″	.119″	.120″
.121″	.122″	.123″	.124″	.125″	.126″	.127″	.128″	.129″	.130″
.131″	.132″	.133″	.134″	.135″	.136″	.137″	.138″	.139″	.140″
.141″	.142″	.143″	.144″	.145″	.146″	.147″	.148″	.149″	

Third: Fifty Thousandth Series — 19 Blocks

.050″	.100″	.150″	.200″	.250″	.300″	.350″	.400″	.450″	.500″
.550″	.600″	.650″	.700″	.750″	.800″	.850″	.900″	.950″	

Fourth: Inch Series — 5 Blocks

1.000″	2.000″	3.000″	4.000″

The trick in selecting blocks from the box is to eliminate the right-hand figure or digit of the specification. In this example the dimension is 2.6437 inches. To eliminate the final digit 7, select the .1007-inch block from the box. Using a piece of scratch paper and a pencil (or a calculator), subtract .1007 from 2.6437, getting 2.543, as below:

$$\begin{array}{r} 2.6437 \\ -.1007 \\ \hline 2.543 \end{array}$$

(This procedure of selecting a block to eliminate the right-hand digit is followed in each succeeding step.)

Now the digit 3 is the right-hand digit and the selection of the .103-inch block, see Table 1, will eliminate it as the next step, below.

$$\begin{array}{r} 2.6437 \\ -.1007 \\ \hline 2.543 \\ -.103 \\ \hline 2.440 \end{array}$$

Again, the digit 4 is the essential right-hand figure and the 140-inch block is the next choice.

$$
\begin{array}{r}
2.6437 \\
-.1007 \\
\hline
2.543 \\
-1.103 \\
\hline
2.440 \\
-.140 \\
\hline
2.300
\end{array}
$$

Rather obviously, now, the next choice is the .300-inch block and, finally, the 2.000-inch block.

The selected blocks should now be resting on the paper covered mat ready for work. But first read the sizes from the selected blocks themselves and list them down on the scratch pad. Add the list up to check the correspondence of the total to the original specification size.

$$
\begin{array}{r}
.1007 \\
.103 \\
.140 \\
.300 \\
2.000 \\
\hline
2.6437
\end{array}
$$

Always perform this last mentioned "rite." Many times, unwittingly, the wrong size block has been picked out of the set; occasionally a mistake has been made in the paper-and-pencil operation of selection by elimination of digits. This final check is a safeguard against such errors.

Several Combinations are Usually Possible

As an inspector gets used to selecting blocks, he will find other combinations designed to allow the use of a minimum number of blocks. In the example above, for instance, the .103-inch, the .140-inch and the .300-inch blocks were selected. But the gage block set contains a .143-inch block and a .400-inch block. The use of the .103-inch block was not strictly necessary. The professional trick is to use as few blocks as possible to build up the total stack.

On the other hand, some measuring problems call for the simultaneous use of two equal stacks of blocks or for the use of two stacks the building up of which might call for the same size block in each stack at the same time. (The use of the sine bar, to be described, very frequently brings up this contingency.)

The following example points out the use in Stack A of the .143-inch block in place of the .140-inch and .103-inch blocks just mentioned, and also the use of combinations of substitutions in Stack B for blocks already in use. The specifications in this example called for the 2.6437-inch stack and another simultaneous stack for 1.6437 inches is also to be made up.

2.6437		1.6437	
−.1007	.1007	−.1003	.1003
2.5430		1.5434	
−.143	.143	−.1004	.1004
2.400		1.4430	
−.400	.400	−.103	.103
2.000	2.000	1.340	
	2.6437	−.140	.140
		1.200	
Stack A — 4 blocks		−.200	.200
		1.000	1.000
			1.6437

Stack B — 6 blocks

As can be seen, in order to build the 1.6437-inch stack, B, the first substitution came up in the combination of the .1003-inch plus .1004-inch blocks in place of the single .1007-inch block already in use in stack A, and the second substitution involved .103-inch plus .140-inch blocks instead of the already used .143-inch block. In other words, the 1.6437-inch stack B used two more blocks than the 2.6437-inch stack A.

Cleaning the Gage Blocks

The fourth step in the use of gage blocks is to clean them ready for wringing and stacking. The cleaning tools are a soft clean cloth, chamois, or tissue, and cleaning solution such as carbon tetrachloride or Stodsol, plus clean hands and the clean paper topped bench location to set the cleaned blocks on. (Benzene and gasoline are not taboo as cleaning fluids if the fire and explosion hazard are properly recognized. Some favor the use of kerosene. It is wise to filter kerosene clean first through a chamois, however.)

Checking the Gage Blocks for Injury

Connected with the cleaning step in the use of gage blocks are several items of surveillance. Note from the appearance of the blocks whether or not the blocks were properly cleaned and oiled or rust proofed the last time they were used. If blocks in poor shape in this respect are discovered, a bit of detective work directed toward correcting somebody's careless habits should be performed. As each block is cleaned, examine the polished ends for scratches, worn spots, and corrosion. Check the edges with your eyes and fingernail (perhaps even with magnifying glass) for nicks. *Don't* use scratched, nicked, dirty, or rusted blocks.

Where gage blocks that have apparently been properly cleaned and rust proofed reveal rust and corrosion spots, more detective work should be in the offing. Look for (a) an employee (it might be yourself) whose sweat pores exude the type of acid which will start up a corroded spot simply from touching a block or (b) look for oil, grease, or rust proofing compounds that contain excess acid or alkali* or (c) check to be sure the blocks were properly cleaned free from corrosion at the last using.

Scratches and nicks have been mentioned. At first thought, it seems as if a slight scratch on the polished measuring surface of a gage block should be harmless; however, if a scratch on metal is examined under a powerful glass, it will be seen that the metal has also been forced into minute ridges projecting above the rim of the scratch. A scratch looks like a furrow in a plowed field. The plow not only scrapes out a V trench but it also piles dirt up on either side. A scratched block can scratch another block.

Nicks and digs along the edges do the same thing. It is very easy to drop a gage block or unconsciously rap it against a bench or piece of metal and never realize that a tiny nick has been raised on an edge. While the edges of commercial gage blocks are deliberately "broken" a trifle in the sense of being slightly dulled, they are not definitely bevelled or rounded.

In wringing the cleaned blocks together to form the stack, following the instructions previously given and illustrated in Fig. 4, it is customary to wring them together in the order of size. Using the example of stack A, previously mentioned, for 2.6437 inches, which required four blocks, the .400-inch block would be first wrung on to the 2.000-inch block. Then the .143-inch block would be added to the .400-inch block and, finally, the 1.007-inch block would top the stack.

---

*Gage block manufacturers will recommend suitably tested oils, greases, or rust proofing compounds. Many industrial oil and grease suppliers catalog suitable compounds. Or see your laboratory, engineering, or purchasing departments.

At this juncture, the stack of blocks is stood end up on the clean paper, usually with the largest size block at the base of the stack and resting on the polished measuring end of this largest block. The blocks, as a unit stack, are to be handled, as always, by gripping the sides.

The Use of Wear Blocks

For the purpose of protection, particularly when the block stack is being used as a gage for direct measurement, some gage manufacturers supply so-called wear blocks. Several wear blocks come with each such set. They are usually .050-inch or .100-inch thick and they are wrung on each end of the stack being made up. (Of course, in calculating the total height of a gage-block stack, with wear blocks in use, the calculation takes into account the thickness of the wear blocks.) The idea of wear blocks is implied in the name — they are to take the wear and tear of handling, setting down, and measuring with a gage block stack, and it is expected that they will be checked regularly and perhaps replaced rather often. Obvious care is usually taken, too, in the use of wear blocks, to face the same end out, thus keeping the opposite surface always free from scratches, wear, and dirt. This method preserves one face of a wear block always smooth for wringing against the regular measuring blocks.

Provisions for Clamping and Use of Attachments

The type of gage blocks illustrated in Fig. 6 come with countersunk holes which allow the blocks to be clamped together with extension rods and flat head screws, in addition to their being wrung together. By this device a gage block stack cannot, of course, fall apart in use.

Another purpose behind this type of block design is the provision for adding caliper and scriber point attachment blocks to the main stack so that for certain occasions the gage block stack becomes a complete fixed snap, caliper, plug or height gage in itself. Figure 7 shows types of clamping and holding devices for both types of gage blocks illustrated in Figs. 5 and 6.

In an early section describing the steel rule, it may be recalled, the "hook" attachment or hook rule was discussed. The statement was made that "a general difficulty in connection with various helpful attachments offered with measuring equipment, to make its use easier and quicker, is the fact that the accuracy of the attachment itself is too seldom questioned. The attachment . . . becomes relied on and yet by

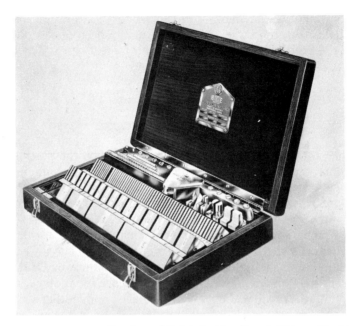

Courtesy of Colt Industries, Pratt & Whitney Machine Tool Div.

Fig. 6. Commercial gage blocks with countersunk holes that facilitate clamping the blocks together by means of a rod through the center.

its very nature, design and use proves to be the item that becomes bent, worn, loose, and inaccurate . . ."

Because gage blocks are basic measuring equipment, because their accuracy may be relied on down to a few millionths of an inch, the inspector should give whatever attachments he uses priority in the attention he pays to checking and calibration. He should also learn his own shop's standards and preferences as to the types of attachments to be used or, failing to get clear-cut answers on this subject, he can consult manufacturers' catalogues and instruction books.

Guarding Against Injury to Measuring Surfaces

One quite common shop practice represents a paradox. It has been said that perfectly clean, dry, grease free blocks will not wring together. It has also been said that the acids in natural sweat will start up corrosion. Yet many time inspectors and mechanics will be seen carefully degreasing and cleaning gage blocks only then to wipe the measuring ends quickly on their wrists or forearms or even against the sides of their noses. The idea, of course, is to get a film of oil on the

Courtesy of Dearborn Gage Co.
Courtesy of Colt Industries, Pratt & Whitney Machine Tool Div.

Fig. 7. (Upper) Gage blocks of type shown in Fig. 5 clamped together to form a snap gage. (Lower) Gage blocks of Fig. 6 clamped together to form a snap gage.

polished gage block surface for wringing purposes. Perhaps it is assumed that the sweat oozing only from finger tips contains acid, that the oily skin of wrist, forearm, nose, or scalp is singularly pure. The machining trades display a number of quaint, traditional inconsistencies like this. It is safer to give each cleaned block a dab with a clean soft rag slightly moist with a strictly neutral, pure and light mineral or vegetable oil. If the blocks have been cleaned with petroleum products, enough resident grease usually will remain for wringing purposes.

The effect of corrosion on hardened steel gage surfaces is deceptive, because many times, unless a high-powered glass is used, it is not seen. The metal, as it first corrodes, frequently does not noticeably discolor. The corrosion, itself, is substantially the formation of a molecular fineness of powder or a disintegration and crumbling of the surface to an impalpable depth, all of which, nevertheless, scrapes right off when the blocks are together.* Add this to the natural burnishing

---

*Try not to leave gage blocks wrung together any longer than necessary and never leave them wrung over night. This same idea was mentioned during the discussion of the maintenance and care of micrometers. An electrolytic effect transpiring between two metal surfaces in close contact speeds up the chemistry of corrosion or adds a special pitting effect of its own.

and shaving action, the mechanical wear from wringing, and the blocks may lose length an unnoticed millionth of an inch at a time.

The use of gage blocks in measuring procedures will be discussed further in connection with descriptions of other measuring apparatus and methods farther on. Two fundamental points should be taken up here, however.

Cleanliness is an Important Factor

It would be inconsistent to clean, check, and handle gage blocks meticulously and then pay no attention to the condition of other gages, surface plates, machines, or workpieces the gage blocks are to be used with and on. Any surface a gage block or stack is to rest against should also be as carefully cleaned free from grit, oil, or corrosion. Examine the work surfaces also for scratches, nicks, and burrs, which will damage the polished gage block ends or attachments. As far as pressure is concerned, gage blocks or stacks should be applied delicately, though firmly. A gaging pressure not exceeding eight ounces (half a pound) should be sufficient. Never ram a gage block stack into position; the use of gage blocks is always a slow-motion operation.

Temperature Changes Can Affect Accuracy

Another factor too commonly forgotten or carelessly ignored when gage block stacks are used is the effect of temperature. This question is to be taken up again in detail farther on, but right here it should be pointed out that a 1-inch gage block will expand or lengthen about six millionths of an inch (.000006″) for every degree Fahrenheit its temperature increases. (In like manner it will shrink .000006 inch per degree drop in temperature.) A gage block's temperature can increase from normal room temperature some 10 degrees in something less than 10 minutes if it is simply held in the hand.

To illustrate the effect of ordinary inaccuracies in measurements occurring from the everyday use of gage blocks, suppose a situation (not uncommon) where a gage block set-up (single or a stack) is made for 10 inches in length. The ordinary cleaning, wiping, wringing, and handling of the blocks, plus the handling of the completed stack in the measuring or comparing operation will usually mean contact with the inspector's hands for about 10 minutes, perhaps, and the block's or stack's temperature will have increased probably about 10 degrees F. Multiply the .000006-inch expansion per inch per degree by the 10-inch length and the 10-degree temperature rise and the stack will actually measure .0006 inch more than it should:

$$.000006 \times 10 \times 10 = .0006$$

To put it as a sort of rule of thumb, ordinary handling of a 1-inch length of block might produce an expansion error of nearly .0001 inch and ordinary handling of a 10-inch gage block stack can readily develop more than half-a-thousandth inch error.

The professional custom is to make up the gage block stack and set it on, near, or against the surface plate, gage, machine, or workpiece with which it is to be used for about 20 minutes. In this manner, the gage block stack temperature equalizes to or with the apparatus or work it is to be used on. Don't expect accuracy from gage block stacks where they and the apparatus or work are in direct sunlight, near steam radiators, or in line with the draft from doors, windows, and fans.

The handling of gage blocks has been gone into at considerable length because most of the principles and techniques described and the warnings issued apply to the everyday use of practically all precision measuring apparatus.

Surface Plate Provides Basic Reference Surface

In several previous sections, considerable attention has been directed to the reference or position point and to reference surfaces like anvils and jaws on gaging equipment. The surface plate, in one sense, is the epitome of the reference surface. Or, to put it another way, if the inspector, when using a surface plate, always keeps the geometrical idea that it is a reference surface in mind, he will complete and use surface plate setups to the greatest advantage.

A surface plate, primarily or geometrically, is a truly flat, level plane. Practically or physically, it may be a cast iron, granite, or glass block, set level on bench, stand, or table, with its one flat, polished surface facing upward.

Types of Surface Plates

The surface plate is the horizon, the horizontal plane from which *both* direct and indirect measurements can be made. The concept of reference or position point is crucial — this is the reason for the heavy and reliable construction and care of the surface plate — the x and y axes are the plate or table, and this beginning reference point must always be in the same place. Remember, accurate measurements and descriptions of geometric form must begin with an established relationship to some known position — in many cases, the surface plate is that reference position.

Iron surface plates are *not* used as much in industry or inspection applications today — corrosion and stability, or the lack of corrosion resistance, and a tendency to change shape are some reasons why granite blocks are almost exclusively used. Use, care and maintenance, and reliability are all enhanced with the use of the granite units. Granite blocks (Fig. 8) are readily available in many sizes, shapes, and grades, and, of course, they are typically used in today's sophisticated measurement systems such as CMMs and laser interferometers.

Courtesy of S-T Industries

Fig. 8. Granite surface plates.

During the last few years true, flat, smooth surfaces have been polished on granite blocks and granite surface plates have increased in popularity. Occasionally a cast, polished glass plate is seen and ceramic surface plates are also manufactured. The iron plate has the advantage of allowing a certain amount of wringing (see farther on) and holes are not readily chipped out of its surface or edges from something dropped on it. The glass plate will shatter, craze or break and be destroyed if heavy pieces happen to be dropped on it.

Granite (and glass) surface plates present some advantage over the iron plates. When a dig or scratch is accidentally made in iron, there is not only the below-surface depression but metal is forced up the sides of the scratch above the true plane of the surface in the same manner as a ploughed garden furrow. When a granite surface is scratched, fine particles of granite are routed out. There is the depression of the scratch itself but no accompanying ridge because the loosened particles of granite readily brush away. A scratch on an iron surface must be suspected; it may elevate or tip, by a few ten-thousandths or thousandths of an inch, whatever apparatus is set on it. Scratched granite can be disregarded in this respect.

Manufacturers claim that the scraped surface of an iron surface plate is free from warp, waviness, or chatter — that it is a true flat plane within .0001 inch between any two points on its surface. Considering the possibilities of scraping errors, or of some possible settling or warping due especially to local sunlight, drafts, vibration, and temperature conditions surrounding the plate in use, and to the presence of scratches on the surface, it is probably safer to presume a possible surface error of .0002 inch or even .0005 inch. Granite surface plates are claimed to be more reliable in this respect.

Surface plates are of three primary varieties, in terms of shape: rectangular, square, and square with a center hole for the addition of accessories.

On the other hand, because of the crystalline or molecular structure of the surface of the granite block, any degree of wringing is practically impossible. Where wringing is perceptible, as in the case of an iron surface plate, errors of manipulation are reduced if not eliminated. Hence, because a height gage or other apparatus slides or slips along the granite block so readily, the inspector should take somewhat extra precautions to be sure the apparatus does not tip or lift. An unconscious error up to .0001 inch or .0002 inch is possible under these circumstances.

Either type of plate will wear down. A gradual hollow may form in an area commonly and regularly used, and when the plate is checked, a wear error of several ten-thousandths is often found. Plates should be checked at least every few months — their surfaces explored with a .0001-inch discrimination indicator or with an optical flat. Either type of plate can be returned to the manufacturer for reconditioning.

It is important to emphasize that surface blocks, as primary reference or position points, are used in conjunction with gage blocks, height gages, a variety of dial indicators, sine bars and plates, V-blocks, parallel bars, and other extenders of the vertical and horizontal plane.

Direct and indirect measurement techniques are both centered around the physical aspect of the surface plate as starting point of reference position. But most of the surface plate work tends to be of the transfer and comparison variety — known standards are compared to the workpiece by many different transfer tools and methods, and then the differences between the workpiece and the standard is measured through the comparison tools.

Flatness is, of course, a main criterion or assumption in the use and reliability of measurements on surface plates. With a distorted horizon, or horizontal plane, all measurements, which compound themselves from the surface plate, will be corrupted.

There are several methods of determining and maintaining a flat surface on a surface plate. Many of these methods are time-honored ways of assessing surface plate flatness, and, as such, they occupy a secure place in industry and inspection. But there are also many new and more accurate and sensitive ways to assess surface plate flatness. As advances in technology allow greater discrimination and accuracy on these plates, that same technology provides more discrete techniques of flatness measurement, mostly incorporating electronic gaging and circuitry.

Care and Use of Surface Plates

Reference to manufacturers' catalogues will show that surface plates can be obtained in many sizes, from 4 × 4 inches area to 48 × 144 inches. As a consequence, in many inspection areas, they are the inspectors' work benches, especially where the inspection demand requires larger sizes. And as a consequence also, many inspectors are inclined to keep them littered with micrometers, calipers, gage block accessories, pencils, workpieces, screwdrivers, pliers, clamps, finger rings, rags and what not.

Put on the surface plate at any one time only what is needed for the measurement immediately required. Don't drop everything on it because it's a handy surface; use the bench top beside it. The measurement may require a dozen pieces of apparatus anyway; there's no need of adding to the confusion. And, remembering that the surface is supposed to be flat and true to about .0001 inch, use slow, careful motions in setting apparatus down on it.

Any surface plate should be provided with a felt lined, wooden cover, like an inverted shallow box or the equivalent, to set down over the surface of the plate every minute it is not in use. A surface plate left uncovered, exposed, on bench or stand, and bare of any routine apparatus, becomes a magnet for anything anyone wants to put down. In no time at all, in the average shop, its surface will be loaded with packages, pieces of stock, somebody's shoes, or almost anything else imaginable. All this though the bench it rests on is entirely empty. Keep the cover on the plate to take the gaff.

In general, the same precautions in regard to maintenance and care described previously for gage blocks apply to surface plates. Try to prevent the edges of a surface plate from becoming nicked. Watch for corrosion and rust spots on iron surface plates. Clean the surface at each use and rust proof the surface for storage. Don't leave apparatus or metal objects on the surface longer than necessary and certainly not overnight.

Surface Plate Accessories

Mention of surface plates always brings to mind the conventional types of apparatus and accessories more commonly used on and with them — knees, cubes or box parallels, V-blocks, parallels, squares, planer gages, height gages, test sets and many other articles. Some of these items appear in Figs. 9 and 10. They are all intended to make the job easier and the measurement more accurate or possible. The setups in Fig. 11 show the practical use of some of these accessories.

Using a surface plate as an inspection tool means defining the relationship of a surface plate to the primary tools used on that plate. The position of the plate's top as the horizon is the primary reference point involved. Through the use of gage blocks, and mechanical surface gages with dial indicators, transfer and comparison, or indirect, meas-

Fig. 9. Surface plate and various surface plate accessories.

Fig. 10. Surface plate accessories: (Left) V-blocks; (Right) Toolmaker's surface-plate square.

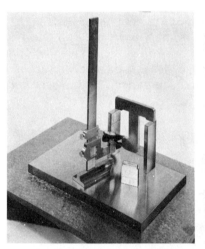

Fig. 11. Measuring devices being used. The surface plate provides a reference plane.

urements can be made on the surface plate. Direct measurements can also be made on the surface plate through the use of height gages and other direct measurement devices that actually incorporate the plate itself as part of the direct measurement scale.

What is a horizon without the ability to look "up"? The use of the "x" and "y" axes is somewhat limiting without the component of the third dimension, the measurement of height. Surface plate measurement tools and accessories unlock all three dimensions, whether by direct or indirect measurement methods. The ability to recognize and use a perpendicular or "z" axis is critical to successful measurement and comparison on a surface plate.

It would take an entire book to discuss surface plate techniques in detail and to educate fully a person in their use. Every shop has different problems; there are a multitude of shapes and sizes of components and parts, and about as many clever ways of handling their measurement. In regard to surface plate methods, there are as many diverse, sometimes sharp, opinions as there are experienced inspectors, engineers, tool-makers, and seasoned mechanics. Perhaps, in the face of a problem of such magnitude, the best a book of this sort can do is to demonstrate and impress a few basic principles — a little geometry — and the best the new inspector can do is to observe and learn from experience and to use all the ingenuity and common sense he possesses when he is setting up at a surface plate.

Basic Principles of Surface Plate Use

Always remember, in using a surface plate, that the basic intention is to secure, geometrically, vertical and horizontal axes. The workpiece to be measured and the instruments and accessories used are manipulated, placed, or clamped to effect this up-and-down and lying-flat idea. In general, too, all measurements are read or taken from the surface plate up.

An attempt will be made to illustrate these general principles simply with a few diagrammatic sketches. As an example, study the workpiece sketched in Fig. 12 — a shaft with two threaded ends. The

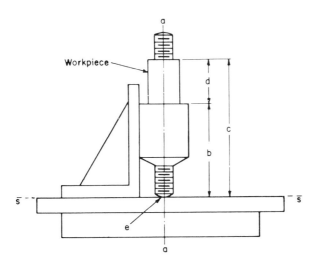

Fig. 12. Dimension *d* of the workpiece is obtained by measuring the distances *b* and *c* with respect to the surface plate and then subtracting *b* from *c*.

inspector is to check the dimension *d* which is to be machined, according to the blue print, to a length tolerance of ± .001 inch.

Dimension *d* might be checked with a steel rule, except that the tolerance is ± .001 inch, which is below the discrimination (.015 inch) of a steel rule. There seems to be no combination of standard commercial vernier caliper or micrometer — no mongrel combination of inside and outside jaw — with which *d* could be measured directly to any such accuracy as ± .001 inch.* A depth vernier or micrometer might be considered but, if the shaft shoulders were more shallow than half the width of the depth instrument anvil or base, a true measurement could not be readily secured. Besides, ordinary clumsiness in trying to use a depth gage on such a combination would make accuracies within ± .001 inch doubtful. To measure from the shaft end *e* to both shaft shoulders with vernier calipers sounds feasible except that the rough, rounded-off, threaded end of *e* is too insecure as a reference surface for ± .001 inch tolerances. And, again, if the shaft shoulders between which the measurement is to be taken are shallow and if some portion of the shaft between these shoulders has a radius greater than the throat depth of the caliper, then the resulting interference would prevent a true measurement.

When the shaft is set up on a surface plate, however, as shown in Fig. 12, the measurement becomes simpler. The trouble with the irregular spherical end *e* is overcome by holding or clamping the shaft vertically against a machinist's knee. Thus, we get true perpendicularity between the shaft axis, *a-a*, and the flat plane of the surface plate, *s-s*, making sure, with a try square, that the shaft is not tipped forward or backward from true perpendicularity in the other plane. The next step, then, is to get measurements *b* and *c* (most likely with a height gage — see farther on) and, by subtraction, obtaining the desired dimension *d*.

Typical Surface Plate Inspection Problems

Suppose the inspector's next problem is to check the machining on a casting, typical of automotive transmission castings, shown diagrammatically in Fig. 13 where the top surface *a-a* has been planed off, likewise the under surface of the lugs, *b-b*. No other machining has been done; the exterior and interior of the casting, other than the

---

*Where the production quantities of pieces of this general type warrant the expense of a specially designed indicating gage, the direct, accurate, and quick measurement of a dimension like *d* is readily secured as will be seen in the descriptions of modern indicating gage designs found farther on.

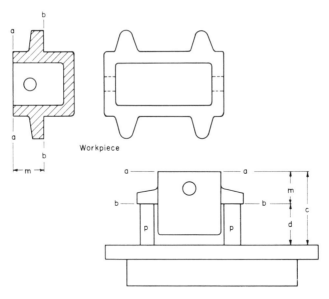

Fig. 13. This use of a surface plate facilitates measuring dimension *m* of the casting.

surface *a-a* and *b-b* are as rough and unfinished as when they emerged from the foundry mold. Dimension *m* is "critical" because the cored holes in the casting must be bored out eventually to line up exactly with and be a certain distance above another set of holes in another casting to which the workpiece lugs (*b-b*) will be bolted when it reaches the assembly line. The tolerance is again ± .001 inch for dimension *m*.

As Fig. 13 shows, the surface plate plus a pair of steel parallels, *p-p*, make measuring dimension *m* easy. In fact, after the casting is rested on the parallels, only dimension *c* need be taken (again, likely, with a height gage) because dimension *d* is known — the height of the parallels, *p-p* stamped on them.* Dimension *m*, of course, is dimension *d* subtracted from measurement *c*.

To illustrate the facility of surface plate techniques, try to figure out how you would get measurement *m* on an S-shaped piece, like the workpiece sketched in Fig. 14, by means of any of the sort of measuring apparatus described previously. Figure 14 shows how simple the

---

*This assumption brings up another word of warning. All surface plate accessories should be checked regularly to be sure their measurements correspond to the legends stamped on them. If the parallels in this particular example were stamped, for instance, $1\frac{1}{2}'' \times 3''$, the 3-inch height, and the parallelism itself, of each of the steel parallels should be checked independently before they are used on the surface plate. Any inaccuracies should not exceed .0005 inch.

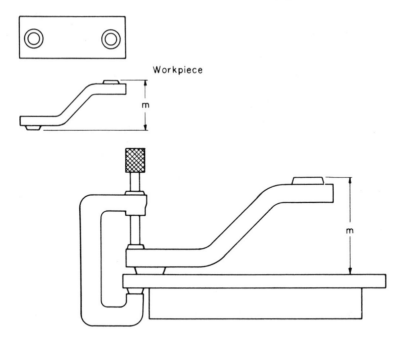

Fig. 14. Dimension *m* is easily measured if one leg of the S-shaped workpiece is clamped against a surface plate.

measurement becomes if one leg of the S-shaped piece is held down firmly on a surface plate.

The ease with which surface plate setups can many times pick difficult measurements "out of the air" is illustrated even in the rather simple problem of Fig. 15. The relevant dimensions supplied for this workpiece layout by the draftsmen are shown as letters *a* to *h*, plus *k*, along with the 30-degree angle. The question is: Did the mechanic lay out the hole centers and the slot correctly to tolerances of ± .001 inch?

Checking Hole and Slot Positions

Figure 15 shows the workpiece in the position it appears in on the blue print. Note dimensions *f* and *g* in the upper right-hand corner. The assumption is made that the machinist, in laying out, drilling, or jigging the workpiece plate followed the draftsman's suggestion with *f* and *g* to the extent that sides *P* and *R* were the reference surfaces from which he worked.

The professional touch in surface plate work is illustrated in the diagrams of Fig. 16. Set up the work so that a number of dimensions are vertical to the surface plate plane. After measuring these, the work-

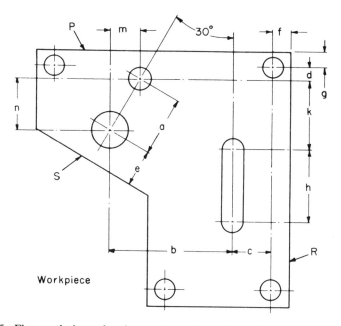

Fig. 15. Flat workpiece showing various dimensions that must be measured or calculated from other measurements.

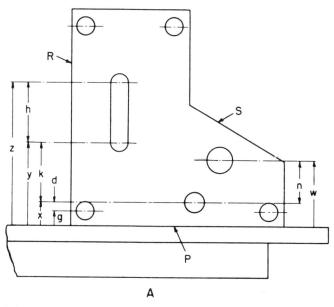

A

Fig. 16. (A) Workpiece of Fig. 16 with reference surface *P* placed on surface plate. The various dimensions indicated are then determined by measurement and calculation.

piece is then located so that the normally horizontal dimensions are vertical, etc. In this case, since surface *P* is the reference surface for the dimensions showing vertically on the blue print and since we measure always from the bottom upon a surface plate, the workpiece is turned over or completely around so reference surface *P* rests on the surface plate as Fig. 16-A indicates.

In setting up and measuring, maintain the usual precautions of freeing the workpiece of dirt, chips, and burrs that might prevent its being squarely clamped.

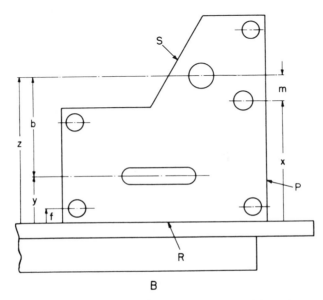

Fig. 16. (B) Workpiece rotated 90 degrees from position shown in Fig. 17 (A). Dimensions indicated are obtained by measurement and calculation.

Only the centerlines relevant to the vertical measurements to be taken in this position are shown in Fig. 16-A. The diagram is thus made, purposely, to point out how the inspector's mind unravels from the blue print (which in Fig. 15 is represented by the sketch of the workpiece) only the dimensions he wants to measure in any one setting. So, he proceeds to measure *g, w, x, y,* and *z*. From these, with pencil and paper and subtraction, he can determine dimensions *d, h,* and *k.*

But note dimension *n* both on Fig. 16-A and on the workpiece sketch in Fig. 13. The thorough inspector would get out his trig book (or calculator) and compute from the dimension *a* and the 30-degree angle what dimensions *m* and *n* in Fig. 14 should be and check them also. Dimensions *m* and *n* did not appear, of course, on the blue print.

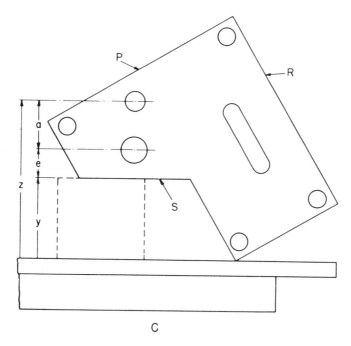

Fig. 16. (C) Workpiece positioned so that the remaining dimensions may be measured and calculated.

They were added to the workpiece sketch of Fig. 15 to bring up the point that hole coordinates, so-called, should be checked two ways, especially where an angle is involved.

Incidentally, also, the inspector assumes that the centers of the other three holes in the corners of the workpiece, for which no dimensioning appears on the blue print, bear the f and g relationship to their respective sides even though the draftsman failed to include specific f's and g's for the other three holes.

In a similar manner, Fig. 16-B, the workpiece is turned again so that what were horizontal dimensions on the workpiece sketch are now arranged vertically for ready surface plate measurement. In this position, dimension m is also checked, following the thinking in connection with n mentioned in a previous paragraph.

Finally the workpiece is turned at a 30-degree angle, Fig. 16-C, so that surface S is now horizontal. This position permits the ready measurement of dimensions a and e that on the workpiece sketch of the actual blue print, appeared as diagonals.

In the preceding examples, it has been shown that the basic surfaceplate technique of measuring is always perpendicularly upward from the horizontal plane of the surface plate as a general reference surface.

Accessories for Checking Hole Location

The discussion of Fig. 16, thus far, has taken into account only the simple geometry involved. The next question, of course, is how, actually and physically, the several measurements are to be secured. When it is necessary to measure the location or coordinates of a hole, the customary technique is to "plug" the hole. The idea of plugging a hole and measuring hole centers by getting, actually, the location of the circumference of the hole is illustrated in Figs. 17 and 18. Figure 17 shows how a simple measurement could be secured with micrometers or vernier calipers and Fig. 18 shows the use of a height gage equipped with a dial indicator.

Fig. 17. Measuring the center distance of two holes by "plugging" the holes and measuring over the plugs.

These illustrations also show a favorite device — the use of standard plug gages for plugging holes. Another possibility, not thought of often enough in this connection and, consequently, not used for plugging holes as much as it could be, is the correct combination of tapered parallels. For many measurement conditions where tolerance requirements are not too precise, drill rod or drill blanks form satisfactory and frequently readily available plugs. A number of commercial tool and gage manufacturers supply wire plug gages, taper inserts, and measuring rolls. Sometimes the measuring stint is suffi-

Fig. 18. Measuring the position of a hole by using a stack of gage blocks, a vernier
height gage, and a plug.

ciently worth having the inspector's own tool room or machine shop turn and grind the necessary sizes and sets of hole plugs.

Whatever is used as an auxiliary for securing hole location or hole-center coordinates in a surface plate setup, there are several routine precautions in the direction of greater precision that the inspector should observe. Item number one, the inspector should measure accurately the diameter (the O.D.) of any plug he is about to use. His problem is to measure to the center line of the hole. Actually he measures to the outside diameter of the close fitting plug and *subtracts* from this measurement half the diameter of the plug to get the distance to the hole centerline.

The surface of the plug should be smooth. When the edges or interior of the hole to be plugged for coordinate measurement are unnecessarily burred and rough, it is frequently wise to smooth off excess metal with crocus cloth.

The plug should be selected and measured so that its diameter in relation to the hole diameter will yield a firm fit — something akin to a wringing fit — when the plug is inserted. The inspector should be unable to rock, shake, or turn the plug, after it is inserted in the hole, with normal gaging pressure. If either the plug or the hole is out-of-round, as shown at A or B in Fig. 19 errors in measurement will result. The condition shown at C in Fig. 19, where a true plug contacts an oval hole along its vertical axis is not as undesirable as A or B.

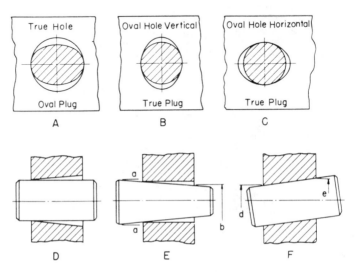

Fig. 19. (A through F) Various conditions of fit between a hole and measuring plug that result in erroneous measurements.

In the same general manner, the use of tapered plugs and the presence of tapered holes can produce errors in securing the coordinates of holes. When measuring a plug for its diameter (O.D. size), check its periphery also for out-of-round and taper. Check the hole likewise. The conditions implied are exaggerated in diagrams D and E of Fig. 19.

Many inspection departments allow — even establish as a routine — the definite use of tapered plugs. If a plug with a slight, uniform taper is used, it will of course wedge in the hole and give the effect of a wringing fit because the metal around the hole is slightly and probably evenly expanded. Properly done, such a procedure will not affect the accurate securing of the hole-center location enough to make material difference. Nevertheless, the condition pictured in E of Fig. 19 remains present. Suppose the diameter of the plug itself is measured as lines *a-a* in sketch E of Fig. 19 indicate and suppose the surface plate measurement for the hole location happens to be made as *b*. Subtracting half of *a-a* from surface plate measurement *b*, it can be readily seen, will yield an error in recording the actual hole axis vertical coordinate. The practical rule does not forbid the use of uniformly tapered inserts, but the total taper of the plug itself should not be greater than, say, one-tenth of the tolerance or specification for the hole-center location being measured.

Where the centerline of the hole is not parallel to the plane of the surface plate, another, but generally similar, error can be introduced. The inserted plug will be tipped as portrayed in the exaggerated diagram of F in Fig. 19. If half the diameter of the plug is subtracted from either measurement *d* or *e*, sketch F as taken from a surface plate, it is quite apparent that the true location of the hole center will not be obtained.

Sometimes it is necessary to establish the distances between hole centers or locate their coordinates under conditions where the axes of the holes are deliberately, by design and blue print, at an angle with the horizontal, perpendicular, or normal. If precision in terms of, say, .0005 inch is required, or if the angles of the hole axes exceed several degrees, the geometry of measuring tapered plugs with measuring rolls can be applied — with a little ingenuity and trigonometry.

The Height Gage

As a rough analogy, the height gage is to the surface plate what butter is to bread — you seldom think of one without the other. Like the other members of its family — the vernier caliper and the vernier depth gage — it has primarily a graduated beam and a sliding frame with a vernier scale as shown in Fig. 20. It is mainly different from its family because of the foot block or base that supports the vertical beam and that adapts the height gage to direct use with a surface plate or on any smooth, truly flat surface.

Even though the height gage seems almost entirely different from the vernier caliper, the basic resemblance, the basic principle, is present nevertheless. In the case of the height gage, its base *plus* the surface plate are the equivalent of the fixed or solid jaw of the vernier caliper. The movable or sliding arm of the height gage differs from the similar member on the caliper only in a few mechanical details, which include accessories for making surface plate measurements easier. In other words, between the surface plate on which the height gage foot block rests, and the sliding arm of the height gage, there is the general C-frame principle of the vernier caliper with a reference jaw, literally the surface plate, and the adjustable measuring jaw on the gage itself.

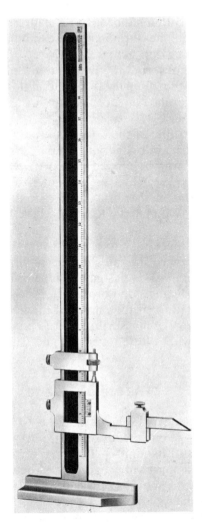

Fig. 20. Height gage equipped with a vernier scale.

In the case of the height gage, however, the vertical beam is so graduated and calibrated, in relation to the under side of the base or foot block, that the correct vernier reading records the actual height to surface *s* of the extension arm *a* as shown in Fig. 22. To put this another way, the height gage's scale and vernier give a direct reading (correct to a possible .001-inch instrument error) of height above the surface plate, such as measurement *c* in Fig. 22, where the standard, so-called scriber point attachment is used on the extension arm *a*. Figure 21 shows this sort of measurement which is, in the larger sense, "calipering."

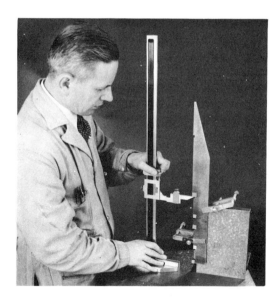

Fig. 22. Diagram showing the various components of a scriber point attachment.

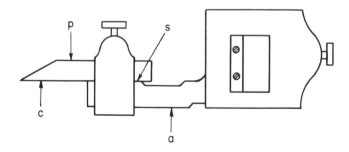

Fig.22. Diagram showing the various components of a scriber point attachment.

Height gages have also changed dramatically over the years. The vernier scale height gage is still very much an integral part of inspection and measurement methodology around the world. Like handheld vernier calipers, some manufacturers have altered the vernier scale relationships on their height gages, producing a long vernier relationship with 50 divisions in English and metric units that provide the .001 inch definition and its metric equivalent on easy to read height gages with varying ranges.

Dial or mechanically driven height gages are also available (Fig. 23). These units are similar to a vernier height gage in appearance, but have a mechanically driven dial that indicates the referenced dimension, making the unit easier to use in some respects.

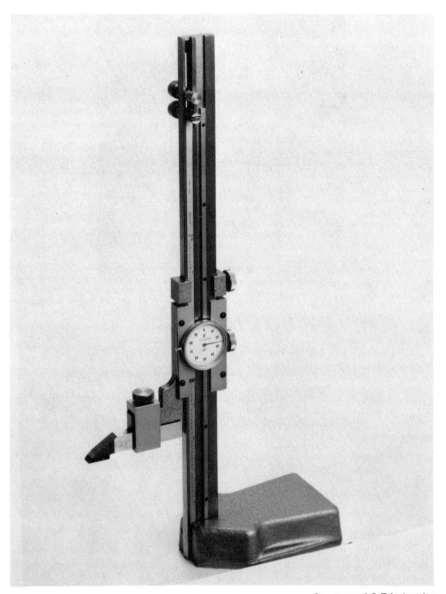

Fig. 23. Dial height gage.

More dramatically, the world of the height gage on the surface block has been revolutionized by electronics. Height gages that physically resemble vernier height gages are available with a digital readout, typically with "LCD," or liquid crystal display, battery power, memory, and computer recording capabilities, and ever-increasing accuracy and resolution (Fig. 24). Like the handheld varieties, they are switchable from inches to millimeters, and like the handheld varieties, their accuracy and reliability depend heavily on the operator or inspector, and that person's ability to understand basic concepts and principles of measurement, in this case, measurement on a surface plate.

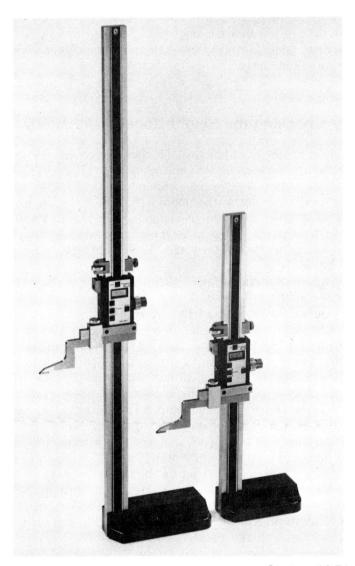

Courtesy of S-T Industries

Fig. 24. Electronic height gages.

As these gages become more sophisticated, and equipped with options such as electronic gaging heads, memory, air bearings, and a host of other technological advancements, they become more accurate, versatile, and effective as both a (direct) measurement apparatus and a transfer instrument. More expensive models are even keyboard-programmable and perform a multitude of direct functions, such as internal and external, depth measurements, and speciality routines, and then translate the measurements into the desired format or deviation description that has been asked for by the inspector. Many units can carry on numerous measurement and transfer and comparison activities simultaneously.

Checking the Height Gage for Accuracy

Before a height gage is used, the common checks for instrument accuracy, wear, and looseness should be made as generally described in the previous section for vernier calipers. Certainly, arm *a*, Fig. 22, should be perpendicular to the main vertical beam of the height gage when the sliding vernier frame is clamped tight against the beam. An experienced inspector, using a height gage strange to him, will first check for any looseness or play where the vertical beam is fastened into the foot block.

The important section of the height gage, from the point of view of instrument condition, is the under side of the foot block or base. This surface originally is machined flat, true, smooth, and unwarped to within a few millionths of an inch of a perfect plane surface.

As with gage block surfaces, examine the height gage base measuring surface for scratches, nicks, corrosion or any other damage. (Once in a while examine the solid junction of the vertical beam with the base for strain cracks, especially if the gage has been dropped or if some amateur has used the gage for a hammer.) After setting the height gage on the surface plate, press down on the base with light but firm fingertip pressure and try to rock the gage. Try rocking it from several different angles. There should be no perceptible feeling of instability; in fact, sliding the base a little bit on the test flat should yield a slight sensation of wringing. Some mechanics occasionally tip the height gage upside down and indicate across the measuring surface or underside of the foot block (a principle described more in detail in a section farther on) or, more rarely, make a surface check with an optical flat. Where it can be seen that the condition of the under surface of a height gage might produce a measuring error greater than .0002-inch, the instrument should be turned in for reconditioning.

A height gage's full scale accuracy can be calibrated step-by-step by setting it up and checking vernier and beam graduations at several different stations along the vertical beam against the heights of gage-block stacks.

The height gage 0 check is made by sliding the arm *a*, Fig. 22, and the vernier bracket down to the base. In this position the vertical distance between the surface plate and the surface *c*, the under side of the scriber point attachment *p*, is supposed to be 1 inch. Check this space with a 1-inch gage block and, simultaneously, the vernier reading on the beam which should be, of course, 1 inch. The graduations on the height gage read directly the height of surface *s*, above the base or the surface plate the gage rests on. Where elevations less than 1 inch must be measured, the height gage is provided with an offset attachment, such as is illustrated in Fig. 25. Using the offset attachment in the position shown, the readings for height are made between the 1-inch and 2-inch graduations on the height gage beam and 1 inch is subtracted from each such reading in order to secure the correct figure for the actual height measurement between 0 inch to 1 inch. All this, because the offset attachment's measuring surface *c*, Fig. 25, is purposely made 1 inch below the clamping surface *s* in contact with arm *a* of the gage.

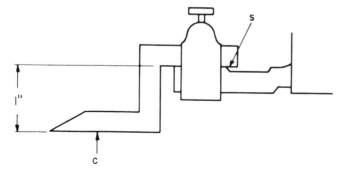

Fig. 25. Offset attachment used with a vernier height gage for height measurements of less than 1 inch.

Accessories Should Also be Checked

A height gage is probably never used without having some accessory or attachment clamped to its measuring arm — a scriber point or offset, a depth attachment, or an indicator. These accessories must be examined regularly, as independent units, for such errors as lack of parallelism, taper, wear, corrosion, warp, or curvature. All clamps, springs, gibs and screws should be kept in shape and replaced when worn or loose. When a calibration check on a height gage shows an appreciable error, suspect the condition of attachments, accessories, and clamps first before attempting to adjust for calibration errors at the vernier scale.

It is difficult within the limitations of static print and photographs to describe to the neophyte the actual, practical use of such a generally mobile instrument as a height gage. Some of its applications will appear farther on, either directly or incidental to descriptions of other instruments, apparatus, measuring techniques, and inspection methods. The best school for height gage instruction is at a surface plate, machine, or bench with workpieces requiring measurement. The coaching of an experienced inspector or skilled mechanic helps, of course. Every instructor, it will be found, has certain ingenious methods of his own plus likely sharp disagreements with the techniques of others. A few pointers, however, can be offered here that will help the amateur when he first uses a height gage.

Care and Use of the Height Gage

In the ordinary shop there is too much tendency to leave a height gage kicking around. When it's not in use, put it back in its box. Too many times we take it off the surface plate and stand it upright on a gritty, dirty bench top. At least it could be rested gently on its side and thus protect the smooth flat under surface of the base.

Naturally, be sure this latter surface is carefully cleaned each time the height gage is used. Also clean the surface of the plate or machine where it is to be used. This is only reemphasizing all the general rules for instrument care which have been expressed and described thus far. Then, always move a height gage slowly, gently; don't slam it around. Don't let the base of a height gage inadvertently dig, scratch, or shave the surface of a surface plate.

Theoretically, the base of the height gage should wring to the surface plate. Practically, of course, this is impossible because the most useful feature of a height gage is its sliding along on a plane, level, flat surface. Skill and practice in its use does produce a slight suction or wringing sensation, a slight drag, discernible at the fingertips. If these effects are not present at least to a small degree, look for dust, grit, scratches, or excess oil on the gage and plate surfaces. Perhaps the gage base or the surface plate is warped, curved, or worn. Perhaps, too, the gage is being imperceptibly, and unconsciously, tipped, tilted or rocked.

Standard gaging pressure, 8 ounces to 2 pounds, is used in sliding a height gage along a surface plate and especially as an arm extension slides over the workpiece surface or, say, the outside diameter of a plug when measuring for hole coordinates. Again review the warnings concerning the unconscious and imperceptible tipping, canting, and lifting of a depth gage base. The same principle applies here. Always remember not only to slide the height gage forward and back, but also to keep pushing downward on the base with firm though light and constant pressure. The exerienced mechanic rechecks every reading and measurement made with a height gage by sliding the measuring arm over the workpiece surface at least twice. If he gets repetition (a repeated reading), he feels that his setting and manipulation are correct.

Accuracy Expected in Height Gage Measurements

An accuracy in reading can be expected of a height gage to .001 inch, probably not much better. (Where an indicator with a discrimination to .0001 inch is used on the height gage, accuracies greater than .001 inch are, of course, obtained.) Height gages come, commercially, 10 inches, 18 inches, and 24 inches in height. The taller the gage, the greater must be the care taken concerning tilting, and in pressures and wringing effect to secure accurate readings. The repetition test is a safe test of manipulation. However, the manufacturers of height gages do not guarantee the extension of .001-inch accuracy to the upper ends of the 18-inch and 24-inch gages; an allowance up to .002 inch may have to be expected in the longer gages.

The question of instrument temperature and expansion probably enters less in the use of a height gage than with most other measuring instruments. This is because, professionally, the height gage is always manipulated by its base as generally indicated in Fig. 21. Only the amateur grasps the height gage up along the beam. The only reason for holding the beam, and expanding its length by heat transfer from the hands to the beam, is generally when the vernier slider is being set and clamped for a measurement or when the height gage is picked up in order to read the graduations on beam and vernier. If the measurement involved is fairly precise, and if you realize you are handling the gage rather steadily, it is wise to let it rest untouched on the surface plate for a few minutes to let the effects of temperature subside. Like other measuring instruments, a height gage and surface plate should not be used in direct sunlight, too close to a hot radiator, nor in a direct draft.

V-Blocks

V-blocks are essentially tool steel blocks that are very precisely 4-square. As Fig. 26 indicates, they usually have two V's, one some-

Courtesy of Brown & Sharpe Mfg. Co.

Fig. 26. (Left) A pair of precision V-blocks with work-holding clamps.
Fig. 27. (Right) A pair of V-blocks being used to facilitate measurement of a workpiece dimension.

times deeper and wider than the other. Channels are cut in either side to accommodate special holding down clamps as illustrated. Standard V-blocks come as 45-degree blocks, i.e., the V-sides slope 45 degrees from horizontal or vertical, the included angle of the V being, of course, 90 degrees. Because they are made with sides and ends parallel and/or square to each other, they may be used lying flat, as shown in Fig. 27, or turned over on their sides, or up-ended vertically. For special purposes such as checking triangle effect or for taps and other three-fluted tools, 60-degree V-blocks can be secured. The included angle of the V then is 120 degrees.

The major purpose of V-blocks is to hold cylindrical pieces or, more to the point, to establish precisely the centerline or axis of a cylindrical piece. Understanding the V-block is a matter of a little geometry.

When a cylindrical piece rests in a V-block it takes the position shown diagrammatically in Fig. 28. From the point of view of geometry, the figure is a circle touching or tangent to two sides of an angle. Therefore, radii of the circle, lines *a-c* and *b-c*, extended to the point of tangency, are perpendicular to the sides of the included angle and angle *d* will equal angle *e* or angle *f* will equal angle *g*. Conversely, then, where a circle touches two sides of an angle, lines drawn perpendicular to the sides of the angle from the point of tangency will intersect at the center of the circle. A line drawn through the center of the circle and the vertex of the angle will bisect the angle included between the tangents.

In mechanical terms, the V-block "finds the centerline" of the circular section or, to put it another way, it locates the cylinder so that its vertical diameter passes through the vertex of the V as shown in Fig.

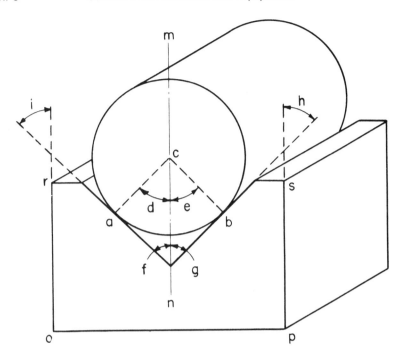

Fig. 28. Illustration of the geometry involved when a workpiece rests in a V-block.

28. If the V-block of Fig. 28 is on a surface plate, the center line (*m-n*) of the cylinder is perpendicular to the surface plate plane. At the same time, of course, the cylinder's center line, *m-n*, is perpendicular always to the base of the V-block, *o-p*, and always parallel to its sides *o-r* and *p-s*, provided the V-block is made not only 4-square, as it should be, but also with angles *f, g, h,* and *i* all equal.

Checking the V-Block for Accuracy

In ordinarily using a V-block, first check it visually, or with the fingernail, for dents, edge nicks, scratches, and burrs that might either damage the surface plate or cause the V-block to tilt imperceptibly from true flat or vertical. Be careful not to let it drop on the surface plate. V-blocks are heavy, solid and sharp cornered; they readily slip out of the fingers.

Any V-block should be checked periodically for basic accuracy. If it has rusted, if it is worn, if it has warped a little, if it were made inaccurately in the first place, its four sides and two ends may not be truly parallel or square to each other. The V-channel may be out of parallel. Constant use with the same size cylindrical workpieces wears hollows in the V-sides. Some of the possible types of inaccuracies discoverable in a V-block are indicated by the light dotted lines in the sketch of Fig. 29. Probably the best way to check a V-block's accuracy is to "explore" with an indicator test set, a technique that is to be described in a section farther on devoted to the subject.

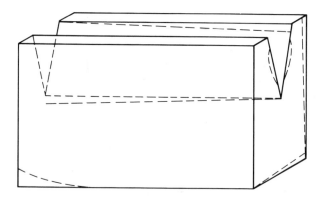

Fig. 29. Various inaccuracies that may exist in a V-block.

The same general principle of watchfulness in connection with nicks, burrs, scratches, corrosion, etc., applies, of course, to parallels, cubes, knees, cylinder squares, and all other surface plate accessories.

Earlier sections have discussed the conditions to be found on cylindrical workpieces that spoil their true geometrical shape, conditions, such as out-of-round or taper. The eccentricity or lack of concentricity between two adjacent cylindrical sections has been described. The V-block is most useful probably for discovering inherent defects of this general type.

Figure 30 shows the general technique for using a height gage over a piston resting in a V-block. If taper is suspected, run the height gage arm over the workpiece as Fig. 30 suggests. Let the vernier slider down until the height gage arm attachment touches the workpiece surface and tighten up the fine adjustment clamp. Then, using the fine adjustment screw, move the arm up or down, at the same time sliding the height gage back and forth, with firm downward pressure on the height gage base, until the height gage arm seems to be at an adjustment where about a half pound of lateral gaging pressure is necessary to "rub" the arm over the workpiece. Clamp the vernier slide and check for repeat reading, and read the scale. Then move the height gage along the length of the workpiece cylinder so that two separate measurements are taken as shown in Fig. 31.

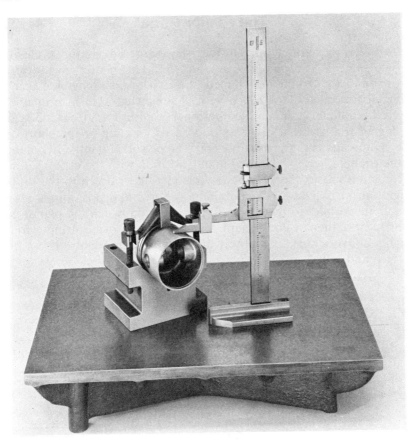

Fig. 30. Checking the taper of a piston that is resting in a V-block.

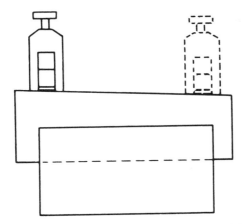

Fig. 31. Two separate measurements taken on workpiece of Fig. 30 will indicate
the amount of taper.

The Planer Gage

Another piece of equipment that may be used in conjunction with a surface plate is the planer gage, Fig. 32. The planer gage consists primarily of two triangular shaped blocks whose sloping sides can be clamped together. Geometrically, it is somewhat akin in principle to tapered parallels. Its main member or base, *a* in Fig. 33, is a 30-, 60-, 90-degree triangle that can be stood on its long leg or the short one. The sliding and clamping member, *b*, can be slid along the hypotenuse of the triangle, *a*, and clamped in any position with the clamp screw *c*. Such action moves the surfaces *s* vertically up or down to the desired height, surfaces *s* always maintaining a horizontal position — always remaining parallel to the surface plate plane on which the planer gage stands. It is then a convenient and quick way of securing variable heights above a surface plate as illustrated diagrammatically in positions *m* and *n* in Fig. 33.

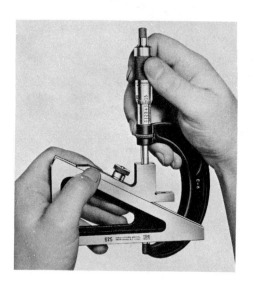 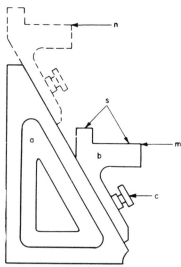

Courtesy of Brown & Sharpe Mfg. Co.

Fig. 32. (Left) Planer gage, a tool often used with a surface plate. Fig. 33. (Right) Component parts of a planer gage.

The Toolmakers' Flat

A small "cousin" of the surface plate is called the toolmakers' flat, machinist's flat, or the optical flat. Commercially it is ordinarily a disc of special aged steel 5 or 6 inches in diameter, 3/4 to 1 inch thick, whose upper and lower surfaces are machined, ground, and lapped smooth, polished, flat to .000010 inch, almost the accuracy of gage blocks. The two finished surfaces are usually guaranteed to be parallel to each other within .000010 inch. It is precision equipment and should always be kept in the velvet lined box it comes in. In fact, frequently, the measuring setup is accomplished on a toolmaker's flat while it is still in its box. Fig. 34 illustrates this apparatus.

Courtesy of The Van Keuren Co.

Fig. 34. Toolmakers' or optical flat in use. This is a high-precision surface plate.

If you will turn back to Fig. 6, which shows an 81-piece gage-block set with attachments, you will see a triangle base, so-called. This is essentially a gage block attachment but can be used after the fashion of the toolmaker's flat. The triangle base is made to the precision described above for the toolmakers' flat. It is smaller, however, each side of the triangle shape being about 3 inches.

A toolmakers' flat or triangle base should receive constantly the same care and maintenance recommended for gage blocks. They should always be carefully cleaned for use, and thoroughly rust proofed for storage. They require more care than gage blocks in handling because they are larger and heavier. If a toolmakers' flat is dropped, it is probably more or less ruined and should be returned to the manufacturer for resurfacing. Somehow, too, it seems easier to gouge or scratch the surface of a toolmakers' flat with apparatus than almost any other piece of equipment.

Where a measuring setup calls for precisions down to .0001 inch or less, the toolmakers' flat should be selected. In other words, don't use it indiscriminately. If the blue print tolerances are greater, say, than .0005 inch for any measurement you need, don't be getting the triangle base or toolmakers' flat out. It is useless and inconsistent, for instance, to use a flat under a height gage unless an indicator with .0001 inch discrimination is also used and needed. If one is available, there is a temptation to use it oftener than the practical circumstance requires.

While flats are used directly as if they were miniature surface plates for certain precision setups where gage blocks, V-blocks, and height gages may be involved, probably they are best used on a surface plate to present local areas of extra precise flatness on which to rest a gage-block stack, V-block, or indicating height gage. If an extra precise indicator and gage block measurement is required and if the face of the surface plate is suspect for some reason, then the triangle base or toolmakers' flat forms a useful, even necessary, accessory.

These micrometer height gages are now common in shops where surface plates are used. Used properly, they contribute excellent accuracy and repeatability owing to the physical interface of the micrometer heads and the built-in gage blocks. Many new models offer the advantage of a digital readout or display of the measurement. They are also capable of extended and adaptable range, making them an appropriate all-around tool for use on the surface plate, with many different comparison applications.

An interesting height gage design utilizes a number of precision blocks stacked alternately and supported by a carrier which is moved vertically by a micrometer head. Settings can be taken from either the top or bottom of any block (the blocks are exactly 1.0000 inch high) and the micrometer head permits settings in ten-thousandths of an inch over the entire range of the instrument. Thus, with a minimum of adjustment, a wide range of settings can be quickly and accurately obtained. Figure 35 shows this type of gage being used to check the height of the surface of a casting that has been ground.

Other surface plate accessories include parallel bars or blocks, and solid parallels, which can be made from steel, granite, or cast iron, and machined for opposing surfaces. These devices are used to supplement set-ups and enhance the abilities of the surface plate to reach out into the extended realms of the three dimensions (Fig. 36).

Sine bars and sine plates are used for angles and their complementary relationships.

Universal right angle plates are commonly used to assess surface square, right angle, and parallel relationships.

Toolmaker's vice and clamps are used in cooperation with other holding and measuring devices, such as the V-block. Proper and steady holding of both the workpiece and the measuring tools is essential in

Courtesy of Brown & Sharpe Mfg. Co.

Figure 35. With this height gage, settings can be taken from top or bottom of 1.0000-inch gage blocks spaced one inch apart.

Courtesy of Starrett Co.

Fig. 36. Surface plate setups.

accurate readings. The flexibility of these measuring tools often determines whether or not the inspection is carried out on a surface plate or by some other means.

CHAPTER 9

Mechanical Indicating Equipment

Mechanical indicators are designed to provide more flexibility, as well as increased accuracy, in measurement approaches, often in a surface plate measurement setup or other specialized dimensional inspection. This chapter is designed to provide a useful explanation of common mechanical measurement devices: comparators and indicating gages.

In most shop applications where precision tooling is used and fine discriminatory measurements are required, it is necessary not just to measure a condition that is "in spec" or "out of spec," but to assess actual dimensions, with an appointed degree of precision.

In fact, much more production and control information can be gleaned from direct measurements, where trends and dimensional changes are apparent, rather than go/no go assessments. Control charting and timely reaction to excessive variation is obviously enhanced by having such direct measurements at hand.

What is a part or component's dimension? What is the amount of *variation*, plus or minus, from the specification or control limit cutoff point? Basic to inspection and gaging is this ability to discriminate dimensional deviation — deviation directly indicated or measured, as well as the assessment of a measurement or deviation within a certain range.

Much of the basic measuring that is done in industry, whether in the inspection shop or through on-line applications, is performed with the mechanical comparator or adjustable gage, on a setup that either makes use of a surface plate or some special setup directly in the manufacturing flow. The dial test indicator is typically used on a test set or height transfer gage; other mechanical contact gages are set up on special fixtures that meet the particular needs of the application.

Mechanical measuring equipment provides the opportunity to measure anything in three dimensions, to the necessary level of accuracy and repeatability. Some of these measurements can be accomplished directly with a height gage or similar device, but many of these crucial assessments must be conducted with the aid of a variety of mechanical equipment, in order to define the geometry of the workpiece.

Perhaps the inspector's first step connected with indicating equipment would be the use of a test indicator on a height gage as shown in Fig. 1. Test indicators are provided with clamp arms and attachments especially designed to fit the height gage measuring arm and clamp. Figure 2 shows a typical commercial test indicator and the universal swivel with which these instruments are usually equipped so that a test indicator may be cocked at almost any convenient angle or extended position.

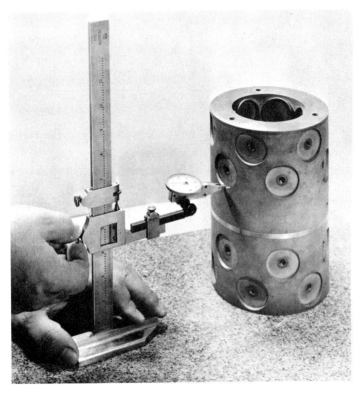

Fig. 1. Dial test indicator being used in conjunction with a height gage.

Before going into detail in the discussion of comparator instruments, it would be helpful to distinguish general categories among the types of (mechanical) comparator devices that comprise this broad category of measuring instruments. Technically, test indicators are instruments for in-process checking, not intended to be used for measuring. Even though test indicators may hold a graduated dial (face), they are intended to be used for comparison of standards by showing change and the direction of change. They are specifically indicate: on size or off size, and in which direction.

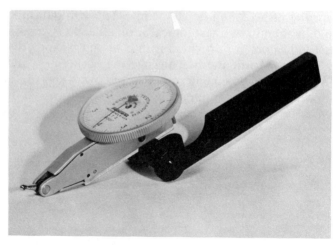

Fig. 2. Typical commercial test indicator equipped with a universal swivel.

By contrast, dial indicators are refinements of test indicators that are to offer greater precision, accuracy, and reliability. These units hold discriminatory abilities that range from half a mil to one mil. Further along the scale, true comparators, as contrasted with the previously described indicators, use principles of amplification to obtain greater discrimination and range. These are termed true mechanical comparators and, the newer and rapidly improving, electronic comparators. Obviously, the choice of which instrument to use depends on the measurement requirements.

Use of Test Indicator on Height Gage

Test indicators on height gages are especially useful in testing for out-of-round and taper. Look back at Fig. 31 in Chapter 8 and think of a test indicator clamped in the height gage rather than the scriber point. If the amount of taper is slight, the inspector might have trouble detecting it by pressure and "feel" with the solid scriber point, but a test indicator will translate a minute variation into visible readings on a dial. The indicator quickly and automatically displays the amount of taper in .001-inch or .0001-inch graduations, when it is desirable to know the degree of such errors, as compared to the much more tedious process of clamping and unclamping, adjusting and reading the conventional vernier device on the height gage, to say nothing of the accuracy gained. When an inspector uses a test indicator on a height gage, he is spared the dual chore of controlling the correct amount of gaging pressure applied to the object being measured plus the manipulation of the gage. He needs only to be sure the height gage base is in full flat contact with the surface plate — wringing it, as it were. The indicator does the rest, practically.

Checking for Various Shaft Conditions

Quite likely the most important service performed by the indicating height gage is the check of concentricity. In Chapter 8, see Fig. 13, the measurement of shaft shoulder lengths was described and illustrated. Suppose now it was desirable to know that one shaft section is concentric with the other (within definitely expressed specifications or, lacking them, within half the diameter tolerance.)

The first step is to indicate for ovality and taper; the second step is to search for the evidence of a bent or bowed shaft. Any of these elements — ovality, taper, crookedness, or curvature, or combinations of them — might give the same general indication as eccentricity or lack of concentricity. In other words, the mechanic who turned the shaft might hunt fruitlessly for reasons why his machine, chuck, or setup produced eccentricity when all the time the real trouble with the shaft was that it had become bent, bowed, or oval. Machine conditions, the way a workpiece is handled (it might have been dropped), or the condition of the original cold-rolled stock all contribute to spoiling the true geometrical form of the workpiece, and in a manner different from the way eccentricity is formed.

As a matter of review, half of Fig. 3 presents diagrams of crooked, curved, tapered, oval, and eccentric shafts, all of course in exaggerated form. It also shows diagrammatically the diagnosis of these several ills by means of V-block, height gage, and indicator setups.

So, as a matter of definite procedure when checking concentricity, the inspector tests the shaft first for ovality. One end of the stepped shaft is rested in the V-block, as demonstrated in sketch A in the right-hand side of Fig. 3, the height gage with its indicator is moved in over the shaft section as in Fig. 1 and the shaft section is revolved in the V-block to allow the indicator to register any ovality present in that section of the shaft. If an accurate measurement or reading of ovality is required, the indicator should be placed in a horizontal position (possibly by holding its base against the vertical surface of an angle plate) and then moved up and down in contact with the shaft until a maximum reading is obtained. This procedure is repeated each time the shaft is rotated to a new position. The sides of the V-block shown at A in Fig. 3 are too high to permit this arrangement, except where the

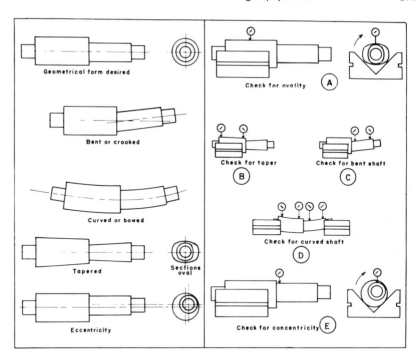

Fig. 3. Several undesirable conditions which may be present in a workpiece and methods of checking for them.

shaft extends beyond the end of the V-block. In many cases, however, a notch or hole is provided in one or both sides of the V-block so that the indicator can be used in this way. Since a sideways pressure is exerted on the shaft, the supporting V-block should be held firmly in place on the surface plate. A record of the ovality should be made. In an exactly similar manner the opposite, the smaller section, of the shaft is rested in the V-block and its degree of out-of-round measured and recorded.

While the larger end of the shaft is still in the V-block, after the out-of-round check has been completed, the indicator is moved lengthwise along the shaft section to "explore" for taper as diagrammed at B in Fig. 3. Similarly, after changing the shaft's position in the V-block, the smaller diameter section is checked for taper. The kind or direction of taper and its degree should be recorded. The workpiece is not, of course, revolved in the V-block when taper conditions are being explored.

One reason for observing any taper in the workpiece, especially back taper, appears when the check for a bent shaft is tried as illustrated in sketch C of Fig. 3. To determine whether or not the shaft is simply bent (in contrast to being curved), it should be indicated only at the shoulder between the two shaft sections and once again at the far end of the shaft. Do not explore in this particular operation. Probably, it is better not to revolve the shaft in the V-block in the sense in which it was turned when checking for ovality. Make the two indications on the shaft at rest in the V-block, as C of Fig. 3 shows, then turn the shaft only 90 degrees, a quarter turn, and again indicate at the shoulder and shaft end. By noting the indicator readings secured at this "bent shaft" examination and comparing any increase over or discrepancy with the readings observed at the ovality and taper tests, the inspector should be able to separate any noteworthy degree of crookedness from the defects of ovality or taper.

Distinguishing One Shaft Condition from Another

The matter of separating the effects of curvature or a bowed shaft is more complex, but probably the best step is to suspend the shaft in two V-blocks as illustrated in sketch D, Fig. 3. Have each end of the shaft make contact only, say, 1/8 inch into the V-block. (Another way of checking a curved shaft is between precision centers.) The shaft is then "indicated" at several points along its periphery, preferably at locations not previously chosen for checking ovality, taper, or crookedness. At each indicator position, the shaft is rotated in the V-blocks. Care must be used in differentiating between these several indicator readings and the degrees of ovality, taper, or crookedness previously recorded if the single element of curvature is to be diagnosed and its own value noted or recorded.

The final step is the measurement of eccentricity between one shaft section and the other. Place the larger diameter section of the shaft in the V-block and run the indicator over and onto the smaller diameter section, preferably fairly close to the shoulder. Revolve the shaft in the V-block. If the shaft sections are not concentric, the condition will show up under the indicator if the indicator is placed close to the shoulder as sketch E in Fig. 3 suggests.

When noting the degree, in thousandths of an inch or fraction of a thousandth, of ovality, crookedness, or curvature, the sector of the periphery of the shaft where the condition is maximum should have been noted or marked also, as well as the direction of taper (front or back), if any taper appeared. The tendencies toward curvature, crookedness, or ovality may tend to "subtract each other out" or counteract each other, in the sense shown in Fig. 4, or one or several of the degrees of defectiveness may add themselves to the true eccentricity reading — all depending on the peripheral locations of maximum

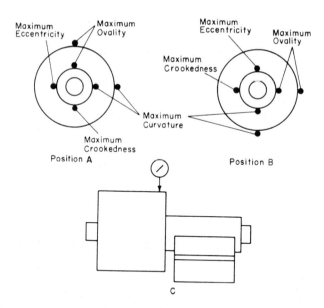

Fig. 4. Defects in a workpiece may be so located as to cancel each other out thereby giving an erroneous impression of the workpiece condition.

indicator readings for the several faults. If the actual shaft conditions — maximum readings — are located as diagrammed in Fig. 4, crookedness and ovality would tend to cancel each other out in their effect toward falsifying a true reading for eccentricity, as sketch A shows. Similarly, as the shaft is rotated in the V-block to position B, during the check for concentricity, the maximum curvature indicator reading could tend to cancel or reduce the maximum eccentricity reading.

Where maximum readings for ovality, curvature, or crookedness occur on the same "side" of a cylindrical workpiece as the maximum reading for eccentricity, then, as can be seen, their values should be subtracted from the reading for eccentricity in order to secure the true, so-called concentricity figure. Another double-check in effect may be made by reversing the position of the shouldered shaft in the V-block and obtaining an eccentricity reading as at C in Fig. 4.

In normal, everyday, shop inspection work, however, the mixture of conditions is frequently ignored. The one reading for maximum eccentricity is made, as at E in Fig. 3 — and the shaft is rejected, if need be, on the simple count of "concentricity" not meeting specifications, because of the practical effect of eccentricity, ovality, curvature, and crookedness when the shaft is assembled in a hole is about the same.

The discussion concerning concentricity versus ovality, curvature, etc., has been presented in considerable detail in the hope that the new inspector will, by poring over it several times and by making some pencil sketches of his own, correctly analyze the geometry of the several situations. If a little mental preparation along this line is not taken beforehand, the inspector will probably be puzzled over what an indicator on a height gage registers when it is working on a shaft in a V-block.

Measuring Parts with Bench Centers

Bench centers are pointed precision ground devices designed to clamp and hold a workpiece, particularly a shaft or cylindrical workpiece, for purposes of measuring concentricity and other similar conditions. When the part to be measured contains tooling holes on each end, it is possible to engage these holes with the bench centers, and arrange the workpiece in a line that is perpendicular to the test setup. With this relationship established, measurements may be taken in the suspended position, in much the same way as the V-block was used to assist the set up. Bench centers can be obtained in a variety of sizes and shapes for specialized holding operations.

The Test Set

The apparatus known in shop parlance as a "test set" is a complete instrument made up, in general, of a base, a rodlike stand, an arm, and an indicator. Several varieties of test sets commercially available are shown in Fig. 5, which also includes an illustration of a test set in use. More commonly, perhaps, than height gages, test sets will be found not only in inspection areas but also in machine shop setup and control practices.

The main components of a typical test set include a stand, mounted on a base, the contact or tip, the measuring system itself, to transmit the workpiece variation as mechanical motion, and the display, which exists to interpret the transmitted mechanical motion.

Comparing a test set with a height gage, again, the general rule can be stated that test sets are not moved or slid around a surface plate. One reason is that test sets are heavier, bulkier, and cover greater surface plate area than height gages. The under side of the base of a test set is not ordinarily machined, ground, and lapped to the precise flatness of the similar height gage foot block surface. An inspector should not attempt to use a test set like the more mobile height gage

Courtesy of Brown & Sharpe Mfg. Co.
Courtesy of Federal Products Corp.

Fig. 5. Commercial test sets.

unless he has made certain that the test set base has unusually good characteristics in regard to flatness, warp or level, and smoothness. The height gage may be used, and is used, consistently as a test set, but there are few situations where a test set may be used literally as a height gage.

The upright post in a commercial test set may not be truly vertical — square — to the base. Any good height gage should withstand a squareness test. The uselessness of attempting too much precision in making a test set appears in the illustrations in Fig. 5; not only is there a clamp device that will move up and down the post, that can be turned and twisted to any angle, but this clamp holds an arm that can be extended or retracted, elevated or lowered, turned or twisted to almost any conceivable angle or location in space.

No, the test set is not moved; it is set up where required and the work is brought *to* the test set. In fact, one model of test set, see Fig. 5, has a so-called magnetic base to anchor it firmly to the surface plate or to a machine bed. A test set offers no graduated, calibrated scale with a vernier, in the sense of the height gage's beam; its measurement is made with an indicator.

Because a test set is purposely made either with a heavy base, a magnetic base, or a base that can be readily clamped down, it is perhaps more useful, more reliable, and more accurate for measuring diameters, heights, offsets, and shoulders than a height gage, which is prone to tip or cant a little unless held firmly by one hand. A test set has, too, a "production" value in that it is frequently the instrument selected when a succession or a large quantity of similar pieces is to be inspected. Once set up, the test set allows the inspector the use of both hands for manipulating the pieces to be measured, as Fig. 5 illustrates.

These test set stands have also changed and varied drastically over the years. New designs that help to ensure greater stability and flexibility of the indicator are on the market. Improvements in the stability and precision of the base, finer and more controlled vertical movement, and 360 degree orientation of the gage head are foremost among these improvements.

Using the Test Set Correctly

In addition to the sort of routine apparatus maintenance and examination already discussed many times thus far in this book, one or two precautions relating particularly to the test set should be mentioned. As a test set is being set up, be sure all clamps are tight. Be sure it sets firmly, without any tendency to wobble or rock, on the surface plate — clamp it down if necessary.

Make every reasonable effort with a test set to have the extension arm as low down toward the base on the upright post as possible. Then try not to let the extension arm holding the indicator extend, overhang, any further than necessary beyond the base and the clamp holding it to the upright post. Where the extension arm holding the indicator is high up near the top of the post and/or where it is out well toward the limit of its length beyond its clamp and the base, both a powerful leverage and a cantilever effect are obtained that will produce measurement errors. Under such setup conditions, the base will invariably tilt a trifle, heavy as it is, or, if the base is clamped down or magnetic, the upright post and especially the extension arm will bend or deflect a little. At all costs, the effect to be avoided is illustrated in Fig. 6.

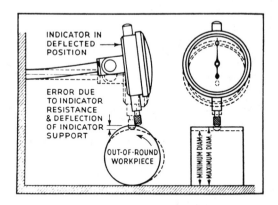

Fig. 6. Measurement errors caused by the deflection of the indicator support arm in a test set.

The sort of test set difficulty just described can be ingeniously offset almost completely by using the model of test set supplied with a channel iron base and by using an extra post with an extra clamp. The extension arm is threaded through the pair of clamps, as in Fig. 7, and gains the extra support, steadiness, and stability of the parallel upright posts.

A thousand excellent examples of parts or pieces might be selected to illustrate surface plate practice and the use of test sets and other indicating equipment in modern industrial inspection and gaging methods. To choose just one example that would supply the answer to

Fig. 7. Test set equipped with extra supporting post to minimize deflection errors.

every surface plate problem any new inspector anywhere might come up against is, of course, impossible.

Inspection of an Automobile Engine Piston

Realizing this, we have chosen to talk about the inspection of a piston like that pictured in Fig. 8, having in mind that even the most amateur mechanic doubtless has some conception of an automobile engine piston. Pistons are used, of course, in aircraft engines and diesel locomotives, in motorcycles and outboard motors. Pistons of this general type appear in tiny compressors, which keep your refrigerator cold. The choice of an ordinary piston for this example does offer a fair variety of tolerances and precisions and a reasonable choice of ordinary surface plate techniques.

The use of the surface plate and other measuring tools and apparatus has already been described, in test and illustrations used thus far, and surface plate techniques will appear casually, at least, in the

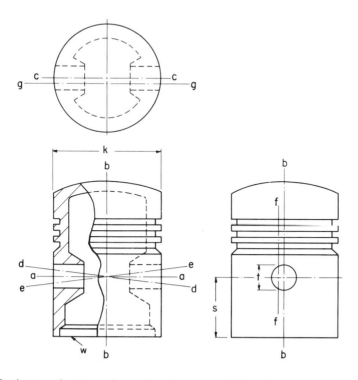

Fig. 8. Automotive-type piston showing some of the dimensions that are to be checked.

discussions to come on special measuring problems. Out of it all, it is to be hoped that the new inspector will get enough of the basic principles and geometry involved to take the raw edges off, at least. The rest will have to come from practice, experience, observation, and the use of native mechanical ability, wherever he works at inspection and uses a surface plate in a factory.

Figure 9 shows not only the piston in question but also a collection of surface plate apparatus to be used to check the dimensions on the sketch in Fig. 8. This sketch does not, purposely, show all the dimensions to be checked — all of the dimensions appearing on the original blue print — but enough typical measurements have been selected to illustrate the general procedure of surface plate work. Also, any experienced inspector can quite properly disagree with the general order of taking the measurements as suggested below and even with some of the particular methods or instrumentation about to be described. Perhaps more than in any other general system of measurement, surface plate practices reflect personal preferences.

Fig. 9. Some of the surface plate apparatus that is used to check the piston of Fig. 8.

Checking for Squareness of End and Measuring Outside Diameter

So then, the first step in checking the conformance of the sample piston to specifications would be to rest it in a V-block and check the squareness of the machining of end surface *w* with the axis of the piston. Since this element is not too critical, it can be checked with a try square as illustrated in Fig. 10, turning the piston, once, 90 degrees in the V-block.

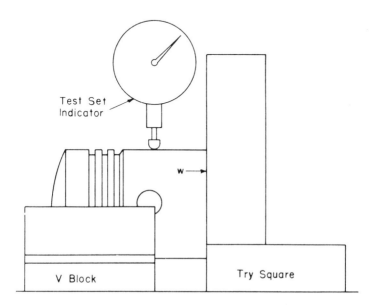

Fig. 10. Checking for squareness of end and ovality of the outside diameter of the piston shown in Fig. 8.

Almost simultaneously the test set indicator can be set over the piston and its ovality or out-of-round can be checked as Fig. 10 also indicates diagrammatically. Since several of the blue print tolerances for some of the piston-ring groove and skirt diameters are in the neighborhood of .001 inch, any out-of-round should not exceed .0005 inch.

Remember the operation above does not measure the diameter itself. Remove the V-block and locate the piston, on its side, directly under the test-set indicator to make the O.D. (outside diameter) checks.

For this purpose make up a gage-block stack the exact equivalent of the blue print specification for diameter *k*, Fig. 8, and set the test set indicator to its 0 by manipulating the clamps, post, and arm of the test set and turning the indicator dial. Then slide the gage-block stack away from the test set and roll diameter k of the piston under the indicator point, as in Fig. 11.

If the sample piston diameter is less than the specific diameter, as represented by the gage block stack "master," it will show in tenths or thousandths as a minus reading on the indicator. Similarly, an oversize piston diameter will appear as a plus indicator reading.

Fig. 11. Measuring the outside diameter of the piston shown in Fig. 8.

Of course O.D.'s like k of Fig. 8 could be checked with micrometers or vernier calipers, but the method above was offered to illustrate one surface plate technique and also to lead up to the use of "comparators," a type of instrumentation to be described farther on.

Checking Direction of Wrist-pin Hole Axis

The location of the wrist-pin hole, dimension s in Fig. 8, is also important but, before the actual measurement is taken, conditions like those shown by lines d-d, e-e, f-f, and g-g must be checked. Or, shall we say, the geometrical position of the hole in relation to the piston axis — whether it slants up or down, whether it's twisted, whether it's off center — is more important to the correct operation of the piston in the cylinder, perhaps, than the actual location of the hole, dimension s, in relation to the bottom of the piston skirt, surface w in Fig. 8.

The wrist-pin holes should have been bored on the center line a-a, at 90 degrees to the axis, b-b, of the piston. The bore should not slant as lines d-d or e-e suggest, Fig. 8. Such a relationship can be verified by first plugging the wrist-pin holes. In this case, a long plug should be selected — long enough to go through both wrist-pin holes and over-hang a little on each side. Its diameter should be so carefully dressed that virtually a wringing fit is secured. The blueprint allows a tolerance of only .0001 inch for diameter t of the wrist-pin holes, Fig. 8. For this reason, plugging these holes with tapered parallels or selected plug gages will not be accurate enough. A through plug with the suitable lapped diameter is the answer.

Next clamp the plugged piston upright in a vertical V-block setup as illustrated in Fig. 12. Using the height gage with a test indicator, move

Fig. 12. Checking to determine whether or not the wrist-pin bearing holes are perpendicular to the axis of the piston.

the point of the indicator over the top of the plug projecting out of one side of the piston as shown in Fig. 12 and also diagrammatically at *n* in Fig. 13.

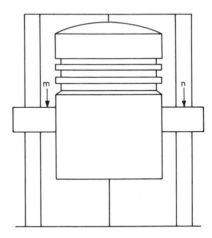

Fig. 13. Diagram showing the points at which measurements are taken when using the set-up of Fig. 12.

The particular technique involves loosening the height gage slider and moving it up and down until the test indicator point just touches the top of the plug in the wrist-pin holes. As the indicator point touches, the indicator hand will move or register. Clamp the height-gage arm and slider tight. Move the height gage back and forth so that the test indicator point rides up to the high point of the plug periphery. The indicator reading should be taken at position *a*, Fig. 14, and not at *b* or *c*. Position *a* can be accurately determined by moving the indicator point back and forth — literally by sliding the whole height gage back and forth — and as the indicator point progresses from, say, position *b* in Fig. 14 to position *c* the indicator hand will move in one direction around the indicator dial to a "high point" and then start to recede or reverse. With the pointer at this high point, the indicator point is at *a*, Fig. 14. Having located the height gage and indicator so that the indicator point is at *a*, set the indicator to its 0 by using the height gage fine adjustment or by turning the indicator dial to 0. Many inspectors do not bother with "zeroing" the indicator; they simply note or record the actual indicator dial reading at the high point. However, memory is slippery and there is always the chance either of misreading the indicator dial or making an arithmetical miscalculation.

Slide the height gage around and over to position *m*, Fig. 13, in other words to the other end of the plug locating the wrist-pin hole.

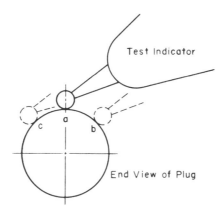

Fig. 14. When indicating over a plug as shown in Fig. 12, the reading at the high point, position *a*, should be taken, not that at position *b* or *c*.

Again, move indicator and height gage back and forth as previously directed until it comes to rest on the high reading (until it rests on the true vertical diameter of the plug). This time, however, do not touch the height gage fine adjustment nor the indicator dial. Gain the high point for *m* simply by locating the height gage on the surface plate so that a maximum indicator reading is obtained.

The indicator reading at *n*, Fig. 13, was set at 0. The other indicator reading, in "tenths" or thousandths, at point *m*, then, tells how much the centerline of the wrist-pin holes tips up or down. If the wrist-pin hole centerline is not parallel to the surface plate, it is not perpendicular to the vertical axis of the piston. In other words, the wrist-pin bores slant like *d-d* or *e-e* in Fig. 8. If the test indicator reading at *m*, Fig. 13, is the same as at *n* (or within, say, .00005 inch), the wrist-pin bores can be considered on the true centerline *a-a*, Fig. 8.

The possibility of noting a high point indicator reading at position *n* rather than setting the indicator to 0 was mentioned. It can be seen now that there is more likelihood of error in getting the difference between readings *m* and *n*, if the 0 setting of the indicator is not used.

Checking for Stepped or Offset Wrist-pin Holes

If, for some reason, the wrist-pin bores are perpendicular to the piston axis but are "stepped" or not on the same centerline, as exaggerated in Fig. 15, another, though similar, type of test is needed. In the first place, if the two bores are stepped, the longer through-plug probably cannot be forced through both holes. In the second place, to analyze a stepped condition as illustrated in Fig. 15 in contrast to the

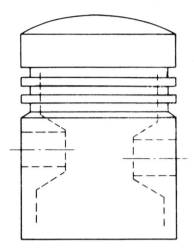

Fig. 15. Possible misaligned condition of wrist-pin bearing holes; misalignment shown exaggeraged.

slanting bores of *d-d* and *e-e* of Fig. 8, each wrist-pin hole must be plugged with its own separate, shorter plug. Then the height-gage indicator test is made as in Fig. 13.

Once more, referring to Fig. 8, we need for obvious reasons to be sure that the wrist-pin holes have not been offset — bored off center — as lines *g-g* or *f-f* would suggest — that the axes of the wrist pin holes are on the true diameter of the piston itself.

This time, the plugged piston is rested on a horizontal V-block as diagrammed in Fig. 16. With the height gage indicator at points *a* and

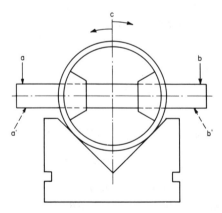

Fig. 16. Checking to determine whether or not the centerline of the wrist-pin bearing holes and the center line of the piston coincide.

b, rotate the piston a trifle in the V-block, as indicated at c, until the indicator reading at a corresponds to the reading at b. By this means, the plug is made truly horizontal and parallel to the surface plate. Now turn the plugged piston completely end over end in the V-block, opposite side up in other words, so that points a' and b' on the plug come up under the height gage indicator. Again slightly revolve the piston as at c until the plug is level and indicator readings a' and b' are alike.

If readings a' and b', which are in themselves alike, differ from the previous readings a and b, then the plug centerline is not concentric with the piston centerline. It is offset by one-half the number of tenths or thousandths difference between the pairs of readings.

The same general technique is used where slots are milled in shafts. Is the slot on the centerline where it belongs as at A in Fig. 17 or is it at one side as at B? By plugging the slot with a flat strip of metal and using a V-block as at C, the true location of the slot can be determined. "Level off" the flat metal plug and note the final, level indicator reading. Reverse the shaft and slot as shown at D, level off again, and compare the latter final reading with the indicator reading at position C. (Before using the plug, be sure that the two surfaces which the indicator is to contact are parallel with each other.) One-half of the difference, if any, is the amount the slot is located to one side of the true geometric shaft center.

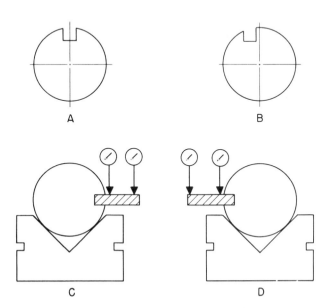

Fig. 17. Checking to determine whether or not the centerline of the milled slot is radial. (A) Correct alignment. (B) Incorrect alignment.

Measuring Vertical Location of Wrist-pin Holes

Returning to the piston measurements, Fig. 8, we are now in better position to measure the vertical location of the wrist-pin holes, dimension s in Fig. 18, because we are aware of any possible slanted, stepped, offset or non-concentric conditions.

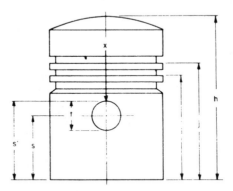

Fig. 18. Dimension s' and t are measured and used to calculate dimension s.

Ordinarily the tolerance allowance for a dimension like s in Fig. 18 is of the nature of .002 inch or greater. Hence, dimension s could be secured accurately enough with the solid height gage measuring arm (using the scriber point and not the indicator) and reading the dimension by the graduations on the height gage beam. (Of course, the hole is first plugged and measurement s' over the top of the plug is taken. One-half the diameter of the plug is then subtracted from s' to obtain s.) Similarly, dimensions i, j, and h of Fig. 18 could be taken directly with the height gage.

But suppose the blue print gave a closer tolerance (.001 inch or less) for dimension s, Fig. 18. In that event, a test indicator would be clamped on the height gage that would be set to its 0 over a gage block stack equal to dimension s', Fig. 18. Dimension s' is of course the sum of the blue print specification for s plus one-half of t. Whatever the indicator reads, plus or minus at x, Fig. 19, would then be the amount the actual dimension s of the sample piston varies from the blue print requirement.

Efficient Way to Use Gage Block Master

The exercises described above that suggest the use of a height gage, indicator, and gage block stack bring up another professional touch

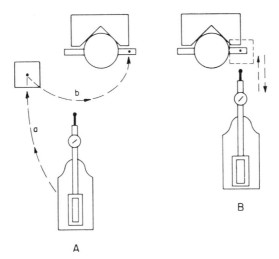

Fig. 19. (A) Possible positions of workpiece, gage block stack and height gage. (B) For more accurate measurements, gage block should be placed close to the work as shown so that height gage movements are short.

that the beginner can adopt, especially where accurate measurements to .0001 inch or even .00005 inch are required.

The general tendency is to stand the gage block stack on the surface plate at one location and the workpiece in another location. The height gage occupies a third location. This relationship is illustrated diagrammatically at A in Fig. 19. The height gage is then slid over path *a* and the indicator is zeroed over the gage-block stack master. The height gage is then moved in a path somewhat like *b*, until the indicator registers for the measurement on the workpiece.

Now remember the measurement is wanted to an accuracy of .0001 inch or .00005 inch. Also remember the face of a surface plate is not guaranteed to be utterly free from warp or waviness — at best a tolerance of at least .0002 inch must be considered for surface unevenness or surface plate irregularities. There is always the possibility of dirt, chips, scratches, and burrs on the plate, instruments, and accessories any of which can introduce an error of .0002 inch or more.

To reduce the chance of error to a minimum, it is suggested that a height gage be moved always in as narrow, restricted, and short a path as possible. It should be slid back and forth over the *same* path if possible. This is one situation where "staying in a rut" pays off.

The professional way is illustrated at B in Fig. 19. The gage-block stack and height gage are located close to each other. The height gage is "mastered" on the gage blocks. The gage-block stack is then moved away and the workpiece is placed on the spot occupied by the gage blocks. As shown at B in Fig. 19, the original gage block stack position appears in dotted lines. Where this technique is followed, the height gage can be moved back and forth, perhaps only a fraction of an inch, and always in the same straight narrow path, as the arrows at B in Fig. 19 suggest.

Where the accuracy of the surface plate is questionable, and where extra close measurements are needed, the especially accurate machinist's flat is frequently used. In the example shown in Fig. 19 it would be entirely possible to place a flat on the surface plate and locate the gage-block stack and then the workpiece on the flat. Furthermore, the height gage could be stood on another, adjacent machinist's flat.

Exploring with Height Gage and Test Indicator

The idea of "exploring" with a height gage and test indicator has been mentioned. If, somehow, a short motion picture could be readily introduced into the reading of a book, the description of exploring would need few words.

If it is a matter of simple taper or lack of parallelism on a simple oblong, flat, or cylindrical piece, the conception of exploring can be shown geometrically as at A in Fig. 20. The workpiece can be laid on a surface plate and the height gage moved along its upper surface as at *a*,

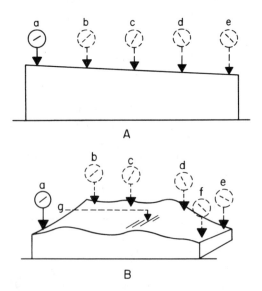

Fig. 20. Methods of exploring the contour of a workpiece. (A) Tapered workpiece. (B) Workpiece of irregular shape.

b, c, d, and *e*. If the piece is cylindrical, it can be located in a V-block for the same sort of test.

In exploring with an indicator, however, the principle of the reference point comes into use. In Fig. 20-A, for instance, the height gage indicator might best be zeroed at position *a* (or at position *e*). As the height gage and indicator are trailed along the piece, the indicator will register the changes in height as it moves from *a* to *e* (or *e* to *a*) through *b, c,* and *d*. Usually, the mechanic will zero at *a,* say, and then take the reading at *e* without bothering with readings *b, c,* and *d*. But one position must be the reference point, or else exploring will not mean much.

Surface irregularities, waviness, hollows, humps, warp, any lack of parallelism or flatness on flat stock can be explored, for example, with an indicator-equipped height gage. An attempt has been made to sketch this conception at B in Fig. 20. Here it would be better probably to zero the indicator at position *a* as a reference point. Then the degree of waviness or warp, either plus or minus as compared to *a,* can be determined by checking at various locations such as *b, c, d, e,* or *f,* or even to an extent into the general area of the workpiece as at *g*.

Checking Inside Diameters with Test Set

The surface plate techniques described thus far have at least implied measurements and comparisons made on outside diameters, cylindrical workpieces, or flat work. Many of the suggestions apply also to inside diameters. Figure 21 shows a set-up for checking an inside diameter.

Under the conditions shown in Fig. 21, a tapered or bell-mouthed situation in the bored hole could be explored with the height gage, but one or two sharp warnings are in order. Where large-diameter bores are involved, such as are pictured in Fig. 21, it is difficult, many times, to plug them for the purpose of securing coordinates or center locations. The coordinates can be secured by taking measurements at the circumferences, as Fig. 21 intimates, but several precautions, based on obvious, common-sense geometry, must be rigidly observed.

In the first place, what might be called the axis of the height gage and indicator should be strictly parallel with the axis of the hole or bore. The indicator should approach the circumference, as shown in position *e,* Fig. 22, and not at an angle as suggested by the dotted line *d-d*. The height gage is then moved slowly from right to left a trifle — motions strictly perpendicular to the hole axis suggested by the arrows at *f* in Fig. 22 — from positions *b* to *c*. The point where the test indicator hand reaches a static position, hesitates, and starts in the

Fig. 21. Surface plate setup for checking an inside diameter. Fig. 22. (Inset) Technique used in locating point on the true vertical diameter of a hole.

reverse direction is location *a*, very close to being on the true vertical diameter of the hole.

Where hole locations must be checked for blue print tolerances finer than .0005 inch, the preceding method very probably should be discarded in favor of somehow plugging the hole.

Mechanical Principle of the Dial Indicator

Thus far, test indicators and dial indicators have been mentioned as if the new inspector had full-grown acquaintance with them. The indicator is, however, a special measuring tool in itself. It represents another basic method of securing measurement in contrast to the linear system of the steel rule and vernier equipment, the 40-threads-to-the-inch lead of a micrometer screw, or the building-block idea of gage blocks. An indicator has many attributes, but also peculiarities and some shortcomings all its own. Even experienced mechanics may forget or ignore some of the intrinsic values and shortcomings in dial indicators as measuring instruments.

One basic mechanical principle of the dial indicator can be quickly understood by considering the geometry of the elementary mechanism appearing in diagram A of Fig. 23 illustrating a 10-to-1 lever. It is obvious that the tip of the indicating hand will move ten times as far (and ten times as fast, incidentally) as the contact point. Some commercial test indicators directly use this principle of the ratio of lengths either side of a pivot.

If the distance from contact to pivot, Fig. 23-A were $\frac{1}{4}$ inch, for instance, and the length of the pointer beyond the pivot were 4 inches — a 16 to 1 ratio — then a .001-inch movement of the contact point would be reflected as a readily distinguishable movement of .016 inch on a dial, say, of the tip of the pointer or hand. The leverage could be compounded, perhaps as sketch B in Fig. 23 suggests, so that, in this case, an original contact point movement of .001 inch is "blown up" or amplified to .256 inch or, practically, $\frac{1}{4}$ inch on a dial. Or .0001 inch can be amplified to .0256 inch.

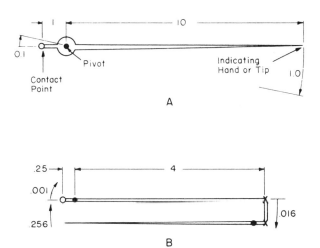

Fig. 23. Schematic diagrams of elementary lever mechanisms used in some commercial test indicators: (A) Simple lever having 10:1 ratio: (B) Compound lever having 256:1 ratio.

Most mechanical measurement devices *amplify* a reading in order to compare or relate the assessed measurement to a scale that is universily understood (a dimension). The amplification of a particular reading or dimension in a measuring instrument is expressed as a ratio in order to understand the relationship:

$$\frac{\text{The movement of the instrument's contact points}}{\text{The movement of the indicator or scale}}$$

Amplification versus Magnification

Incidentally, this idea of one part of a mechanism moving faster than another part, the concept of enlarging a motion or movement, brings up the gaging and inspection jargon of "amplification" and "magnification" and, if instrumentation is to be studied, the definition of one term as distinguished from the other should be made reasonably clear. In the case of the lever–pivot mechanism the measurement is amplified. If, on the other hand, the graduations on an instrument are viewed through a magnifying glass, the measurement is considered magnified. A dial indicator amplifies a measurement through a gear train (as will be seen); the same measurement is then amplified by the pointer or hand and the dial divisions and could be further magnified through a glass. Optical projectors magnify measurements without amplifying them, whereas, as will be seen amplification is accomplished through an electronic system. The reason for mentioning what may seem to be an academic difference is to point out that instrument error may occur from or through the mechanics of amplification or from the visual error of reading a magnified scale, and sometimes it is necessary to analyze the cause of errors. In the shop, the terms are used more or less interchangeably to express the idea of "blowing up" a dimension to readable proportions.

Understanding the Dial Indicator Mechanism

Commercial dial indicators use a gear train to secure amplification. When one gear meshes with another (see Fig. 25), the angular amount the driven gear turns depends not only on the amount the driving gear turns but also on the ratio between the number of driving gear teeth and the number of driven gear teeth. Since the teeth on meshing gears must have the same spacing or pitch, this ratio can also be reduced to terms of pitch diameters of the two gears.

In diagram A of Fig. 24 the driving gear a is considered as having 100 teeth and the driven gear, b, as having 10 teeth. If gear a rotates once, then gear b will be turned through 10 revolutions.

The commercial dial indicator mechanism makes use of this

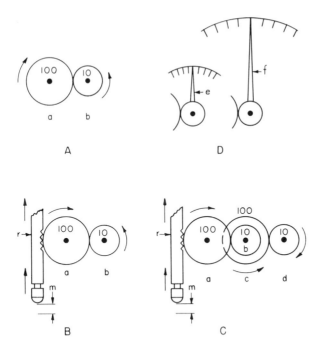

Fig. 24. Schematic diagrams of dial indicator amplifying mechanisms: (A) Simple gear train, 10:1 ratio; (B) simple gear train combined with rack, 10:1 ratio; (C) compound gear train and rack, 100:1 ratio; (D) indicator hand f is twice the length of hand e therefore doubling the amplification.

principle by transferring a linear difference in space to a gear train through a rack gear. Sketch B of Fig. 24 shows the motion m of the rack r. A point on the circumference of gear a consequently moves as far as the linear movement, m, of the rack r and, because of the gear ratio, a point on the circumference of gear b moves through 10 times the angle that a point on gear a moves.

Greater amplification can be obtained by increasing the diameter of gear a, but it is usually secured, for the sake of compactness in a dial indicator, by "compounding" a gear train as suggested in diagram C of Fig. 24. Here, another gear, c, is fastened on the shaft of gear b. If gear c is fastened to the same shaft as gear b, any point on its circumference moves through the same angle as a point on the circumference of gear b or through 10 times the angle of a point on the circumference of gear a. Then, with the ratio of the number of teeth on gear c to the number of teeth on gear d equal to 10 to 1, any point on the circumference of gear d will move through 10 times the angle of a point on the circumference of gear a. Thus, an amplification of 100 times is obtained.

Fig. 25. Cutaway view showing the construction of one type of commercial dial indicator.

In addition to securing amplification of movement through gear ratios, the dial indicator translates the motion of the circumference of gear *d* through a pointer or "hand" swinging over a graduated dial as diagrammed at *e* in sketch D of Fig. 24.

Such a motion can be further amplified by increasing the length of the hand as at *f*. Where the hand, *f*, is twice as long as the hand at *e*, the dial graduations for *f* can be twice as far apart and the reading of a measurement consequently made that much more discernible to the eye.

In other words, dial indicators sense and detail a dimension through a translation of that dimension. These devices relate the actual dimension that is presented for assessment to a translation or mechanical portrayal of the dimension. In order to understand the dimension, the device produces a corresponding motion (mechanical, pneumatic, electrical) or signal that transmits the (perceived) motion into *results* — a measurement or reading on a scale that we can understand. Mechanical contact indicators typically use one of three methods to transmit and amplify the workpiece variation into a dimension or indication. These are: lever amplification, gear train amplification (dial indicators use both of these), and "twisted taut band" magnification, (Fig. 26).

Mechanical indicators (Fig. 27) work by

1. Mechanically contacting the variation on the workpiece;
2. Producing a corresponding motion;
3. Transmitting the mechanical motion through a linking mechanism (gears and levers);
4. Displaying the transmitted motion on a scale (the dial).

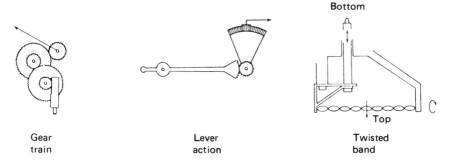

| Gear | Lever | Twisted |
| train | action | band |

Fig. 26. Three types of mechanical amplification.

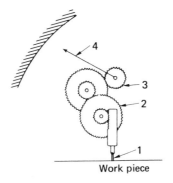

Fig. 27. Operation of mechanical indicators: (1) contact workpiece; (2) produce motion; (3) transmit motion through gears; (4) show display on scale.

Remember, the dimension displayed on a scale can either be the actual dimension of the workpiece (read directly) or a difference between the workpiece and a standard, depending on the set up of the measurement.

Therefore, an indicator is a measurement and display device that translates an assesed dimension through the four steps listed into a scale that is usually expressed as a "digital" or numerical (1, 2, 3, 4, 5, . . .) scale (3'5") or an analog or continuous response scale.

Generally, (mechanical) contact comparators are used in applications that require a relatively high degree of range and accuracy. This type of (mechanical) comparator is one of two general categories of comparators or workpiece contact mechanisms — the other type includes non (mechanical) contact comparators, such as optical or air-driven comparators.

Hysteresis should be reviewed at this point. The concept, as you recall, indicates a situation based on the physical laws of nature that fights repeatability. Friction, gravity, and other tangible overlap of the physical plane ensure that mechanically similar situations are not *exactly* repeatable — there will be some measure of ascertainable differences from one occasion to the next. In the case of dial indicators and mechanical comparators, the presence of hysteresis as a factor affecting inspection results is not to be discounted, depending on the degree of accuracy of measurements that is demanded and expected. Likewise, the parallax error also affects the human interaction with the mechanical measuring system. The floating effect of the needle over the face of the indicator can and does cause a misperception in operators or inspectors not trained to understand and address this issue of inspection.

The principles of amplification and magnification secured through dial indicators have been discussed in some detail in order to emphasize, farther on, certain inaccuracies and shortcomings potential in the dial indicator as a shop measuring tool and the common precautions necessary in using one.

Construction of the Dial Indicator

An indicator consists primarily of a metal case (usually brass) that acts, mechanically, as a rigid frame or foundation for the instrument mechanism, as a protective covering, and also as a means for handling the indicator. Anyone who is a stranger to dial indicators would probably think in terms of clocks or watches the first time he saw one.

This frame, or the case, supports a gear train, the gear rack and its bearings, the graduated dial, and a protective crystal held in a bezel. As in the case of the micrometer, the inspector should have a ready acquaintance with the common names of the various parts and features of an indicator and Fig. 28 has been provided for that purpose.*

A dial indicator is not an independent measuring device in the sense that the steel rule or the micrometer or the gage block is. The dial indicator must be clamped to, mounted on, or fastened into supple-

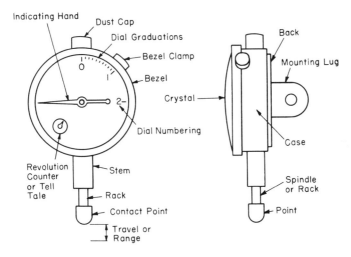

Fig. 28. Nomenclature used in describing dial indicator components.

mentary equipment. In this respect, it might be compared to the vernier height gage which, by deliberate design, is useless in itself as a measuring instrument but must have a surface plate or the equivalent to complement or complete it. In other words, a dial indicator provides a movable or measuring anvil, surface or contact but not the fixed, the reference surface. Thus, it is apparent from previous descriptions and illustrations that dial indicator measurements are secured through the movement or deflection of the indicator point and the travel of the spindle or rack that is, in turn, transmitted and amplified through a gear train and indicating hand which moves around a graduated dial.

The standard (AGD) indicator is assembled so that the hand is in the so-called "9 o'clock" position (see Fig. 28) when the spindle is at

*The inspector should obtain from The Superintendent of Documents, U. S. Government Printing Office, Washington, D.C. a copy of the U. S. Department of Commerce, National Bureau of Standards bulletin CS(E)119–45, the commercial standard on dial indicators for linear measurement. This booklet outlines the agreements reached by a committee on American Gage Design (AGD) in connection with, among other things, standard nomenclature for dial indicator parts and certain standard dimensions and specifications on indicators.

rest. AGD standards also require the standard travel or range of the indicator spindle to be such that the hand will make $2\frac{1}{2}$ revolutions at least from the original 9 o'clock position. Indicators are available, of course, with a range less than $2\frac{1}{2}$ turns, and as many more where the hand turns through 10 revolutions (in some very special designs even more). Long-range indicators, so-called, are usually equipped with small revolution counters or tell-tales, see Figs. 28 and 29.

Courtesy of Brown & Sharpe Mfg. Co.   *Courtesy of Federal Products Corp.*

Fig. 29. Long range dial indicators equipped with revolution counters. Tenths of an inch or ten-thousandths are read on the revolution counter, thousandths of an inch or "tenths" are read directly on the main dial.

Indicator back styles do much these days to contribute to the flexibility of the dial or test indicator families. The backs of these instruments have variable or adjustable connections, to be used in different fashions with the test set. Basically, the range of the indicator can be adjusted by moving the connection on the back of the indicator horizontally or vertically, and resecuring the connection.

The types of connections on these indicators also vary; in addition to the ability to move in stepped increments up and down, various other special connections make unique setups and changes fast and easy. Hole attachments, especially shallow diameter indicator gages (STDs), both special and right angle attachments, and perpendicular indicator attachments, are but a few of these special physical connections available for test and dial indicators.

The versatility provided by these special and loose-joint connections on these indicators allows them to be adapted to use not only surface plates, but also machine tools, or to be used virtually anywhere on an assembly or component. The *X*, *Y*, and *Z* axis relationships, and an understanding of these relationships, is crucial in the proper setup and use of these instruments. The branch of mathematics that deals with these relationships of position and measurement through movement is called kinematics, an interesting and growing field of study.

Balanced Dials Versus Continuous Dials

Indicators come with what is known as balanced dials and continuous dials — with and without revolution counters — as illustrated in Fig. 30. The reasons for using balanced dials versus continuous dials will become more apparent as gages and comparators are studied in the sections farther on, but, as a quick explanation, it might be said that if the tolerances are bilateral, as ± .001, the balanced dial indicator is preferable, and if they are unilateral, as −.000″, + .002″, the continuous dial type would be preferred.

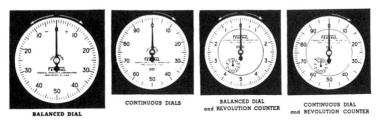

CONTINUOUS DIALS   BALANCED DIAL and REVOLUTION COUNTER   CONTINUOUS DIAL and REVOLUTION COUNTER

BALANCED DIAL

Courtesy of Federal Products Corp.

Fig. 30. Various types of dial faces used on dial indicators.

The use of the revolution counter on the continuous dial indicator is rather obvious. One complete revolution of the big hand on the continuous dial indicator in Fig. 30 registers .100 inch. If, however, the measurement makes the big hand turn rapidly through two or three revolutions, the tell-tale or revolution counter is necessary. The indicator in Fig. 31, for example, registers .379 inch. The revolution counter shows three revolutions of the big hand, or .300 inch, and the big hand has come to rest at 79 main dial divisions, or .079 inch. The sum of the two readings makes .379 inch.

Revolution counters are used on balance dial indicators, see Fig. 30, because, so many times when measurements are taken, the indicator hand revolves more rapidly than the eye can count its complete turns. A balanced dial, for instance, might seem to indicate + .012 inch when actually the hand has made a complete revolution and the actual measurement is .112 inch. In other words, the "tell-tale" informs the user which revolution of the large hand he is reading.

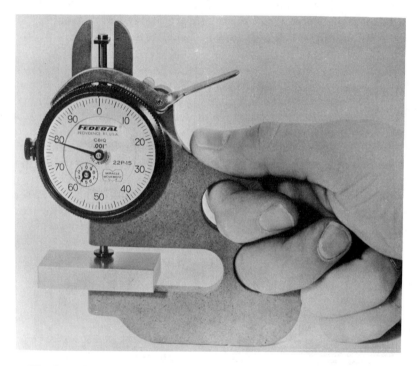

Fig. 31. Dial indicator shows the gage block to be 0.379 inch thick.

Other Factors Affecting Choice of Indicators

Commercial indicators provide a variety of lug arrangements for clamping or attaching them to test sets, gages, and other apparatus and machines. Also a variety of shapes of contact points — spherical, flat, button, taper — are available. Like other measuring apparatus, various attachments such as lifting levers, right-angle, or hole attachments, stem extensions, etc., are made available by manufacturers for use with indicators. The inspector might do well to study manufacturers' catalogs to become conversant with the various useful extra attachments that can be secured.

Assuming that the shop where the inspector works has available or will get a reasonable variety, in size, range, and discrimination, of indicators, the inspector can use a little thought in regard to choosing the indicator best fitted to the job at hand. As for sheer size or diameter itself, AGD indicators vary from about $1\frac{3}{4}$ inches to $3\frac{3}{4}$ inches. Figure 32 shows the relative sizes of the four standard AGD dials available.

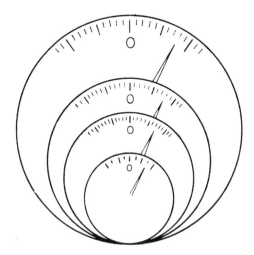

Fig. 32. Comparison of the relative sizes of the four standard AGD dials.

Selecting the Proper Size of Indicator

Several motives might influence an inspector in his selection of an indicator as to size, assuming that his shop's stock of indicators allows a reasonably free choice. Probably the natural tendency would be to use the larger diameter indicator, since its graduations are spaced farther apart — magnified — and it is "easier to read." If, because of physical limitations of the job, the indicator must be some distance from the eye, then this latter contention of easier reading is valid.

However, there are many setups where the largest size indicator is too big and bulky. It gets in the way. Hung on a test set its weight might cause deflection or tipping. Many times the use of a big indicator would compare to strapping the mantle clock on your wrist.

Even though the larger size indicators seem easier to read, it has been found that in many instances more accurate readings are secured and less mistakes are made when a smaller size indicator is deliberately chosen. The mere fact, probably, that you are forced to bring your eyes closer to the indicator, that you lean forward and perhaps squint a little, that basically you concentrate more on reading a small sized indicator tends to produce a more reliable result. The most common commercial indicator used is probably the 2 to $2\frac{1}{4}$-inch size.

The next choice, in connection with indicator selection, ties in with the graduations offered on the dial — the indicator's discrimination in other words. Ordinarily and commercially, indicators come with dial divisions (and internal gear-train amplifications) so that readings can be taken in "tenths" (.0001 inch), "quarter-thousandths" (.00025 inch), "half-thousandths" (.0005 inch) and "thousandths" (.001 inch). Figure 33 shows four of the several dozen available standard dials.*

Coupled closely with the size of indicator to be used for a certain job, its discrimination, and the sort of dial face best suited, is the question of the so-called range of the instrument. Commercially, dial indicators are more commonly supplied with one of the following total ranges or capacities: .010″, .020″, .025″, .050″, .075″, .100″, .125″, and .250″.

By indicator range is meant the nominal working travel of the contact point and spindle or gear rack. If an indicator with .010-inch

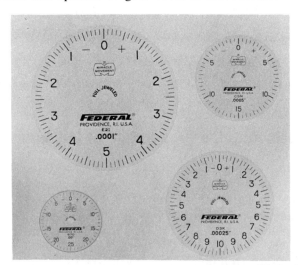

Courtesy of Federal Products Corp.

Fig. 33. Four of the several dozen available dial face graduations.

*Indicators are supplied in standard metric divisions — in millimeters from .001 mm to .010 mm. Also indicators can be obtained with .010 inch dial discrimination or, for that matter, at special and higher prices, with almost any style or discrimination of dial division. Refer to manufacturers' catalogues.

range is selected, its use must be limited to $\pm$ .005 inch tolerances, for instance, while a .125-inch indicator can be used on size variations up to $\frac{1}{8}$ inch, etc. As a quick method for determining an indicator's standard working range (this figure is usually not stamped or printed on an indicator), multiply the number of dial divisions in thousandths by $2\frac{1}{2}$. As an example, the balanced indicator in Fig. 34 is capable of measuring .010 inch in one revolution of the hand. The figures or digits on an indicator dial refer to thousandths of an inch (.001 inch) even though, as in the case of the indicator pictured in Fig. 34, the individual dial divisions register in "tenths" (.0001 inch). You can count on the indicator face in Fig. 34, five one-thousandth divisions down on one side of the dial and five more up the other side, or a total of .010 inch completely around the dial. Multiplying .010 inch by $2\frac{1}{2}$ gives .025 inch as the working range of this indicator, or .0125 inch on either side of the zero position.

Reasons for Indicator Over-travel

Indicators have, of course, over-travel, i.e., the spindle will move several thousandths of an inch more than the standard range before coming to a full stop. At rest, the indicator hand is usually at the so-called nine o'clock position as in Fig. 34. When an indicator is moved into place for measuring, it is so located that the hand moves up to the 12 o'clock position or slightly beyond. (For a continuous dial indicator, this is done by rotating the indicator hand one-quarter turn clockwise; for a balanced dial indicator, one and one-quarter turns clockwise.) *This* position is the start of the working range, *not* the 9 o'clock station. After the $2\frac{1}{2}$ revolutions of standard working range have

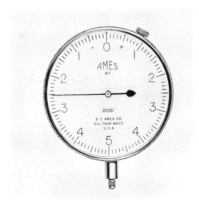

Courtesy of B. C. Ames

Fig. 34. Dial indicator having dial divisions of 0.0001 of an inch.

been used up, with the indicator hand then at the 6 o'clock position, there is usually left at least another quarter-turn of extra travel (back to 9 o'clock again). Measurements are not supposed to be taken within the range of the preliminary quarter turn or in the final quarter turn of over-travel, but always inside the nominal $2\frac{1}{2}$ turns of working range.

The reason for indicator over-travel is obvious since the combination of the setup and oversize in workpieces might well bring readings just a little beyond the end of the standard $2\frac{1}{2}$ revolutions. The normally unused preliminary hand travel from 9 to 12 o'clock, however, is designed for another purpose. It is called starting tension. Look at the ghost view of an indicator in Fig. 25. The motion of the spindle and the internal workings are controlled primarily by a helical pull back spring which appears in Fig. 25 at the left of the spindle or rack gear. The compound gear train turning the hand can be traced out. But then there seems to be a gear left over — a larger gear to the right of the spindle, which has a spiral hairspring attached to it. This gear, with its spring, acts as a balance wheel. Gears will not mesh and operate freely without some clearance, however minute, between the teeth — without some backlash in other words. Any looseness, clearance, or backlash could cause a displacement error between the actual travel of the spindle and the subsequent arc of the hand. The hairspring, however, imposes constant tension in the same direction against the gear train teeth and prevents error from backlash. It is to get the fullest advantage of the backlash correction of the hairspring plus putting some tension on the spindle pull back spring that a quarter turn is usually started on an indicator before actual measurement begins.

Factors Limiting Indicator Range

There are definite reasons why the indicator range is confined to $2\frac{1}{2}$ turns except in special instruments that are particularly designed and specially fabricated and calibrated for longer ranges. It is practically impossible to manufacture a commercial instrument gear that will have perfect teeth around its entire periphery. It is equally difficult to mesh two gears so that no inaccuracy in ratio of rotation will appear throughout complete revolutions of the gears. But it is highly possible at assembly and calibration operations to select sectors of the larger gears to mate with the pinions with sufficient accuracy to ensure the fidelity and reliability of the instrument, provided it is used within the range circumscribed by the mating gear sectors. (Something akin to this procedure appears during the manufacture of micrometers where the spindle screws are individually fitted, by lapping, to each nut.) Look again at the ghost view of a standard indicator mechanism in Fig. 25. It will also be seen that if the hairspring wheel is forced to turn too far, the hairspring itself will get wound up too tight and distort or buckle. Hence the usual limitation of indicator spindle travel to a range of less than $\frac{1}{8}$ inch. In the so-called long-range indicators — $\frac{1}{4}$-, $\frac{1}{2}$- to 1-inch, some of which will be described in use farther on — most of the difficulties suggested are overcome by special-size gears, pinions, and special compound arrangements. However, the longer the range demanded in an indicator, the greater the inaccuracies toward the end of that range. Manufacturers then do not necessarily guarantee $\frac{1}{4}$-dial division repetition and plus-or-minus-one-dial-division calibration accuracy.

Effect of Gear-train Inertia and Friction

A gear train has inertia that must be overcome. The total inertia is increased by the length of the hand. Meshing gears set up friction; there is friction between the spindle and its bearings; the gear pivots turn in bearings (usually sapphire) and additional friction is present. The total effect sometimes makes an indicator appear slow or sluggish, a condition that seems to show up less as the indicator rack is moving in and more as the rack returns to the at-rest position after a reading is made. The general condition has been referred to (erroneously) as hysteresis.

Practically, the inertia lag is no hindrance, since the pointer usually comes to rest by the time the eye has adjusted itself to read the indicator. Practically, also, if an indicator lacked the natural inertia and friction of the gear train, if it lacked this damping effect, the pointer or hand might take so long fluttering to rest for a reading that some deliberate damping device probably would be provided in the instrument.

Mention is made of these indicator peculiarities — peculiarities inherent more or less in any mechanical amplifying system — because inertia, and especially friction, can finally so retard the complete motion of the gear train that the hand either will not quite come up to the final accurate reading or will swing beyond it and fail to return. If an indicator is dropped, if it has been mistreated, or if it becomes filled and clogged with dust, grit, oil, or coolant, it will "tire" and should be turned in for overhauling and recalibration.

In use, the indicator dial's 0 is seldom or ever exactly at the twelve o'clock position. The dial is attached to the bezel and, when the bezel clamp is loosened, bezel and dial may be revolved either clockwise or counterclockwise to a desired zeroing position. However, zeroing is never done until there is pressure against the indicator point and until the indicator has registered about a quarter turn starting tension.

Use, Care, and Calibration of Indicators

As has been said, an indicator is not, of itself, an independent measuring instrument. It is either clamped to a height gage or test set or it is an integral part of an indicating gage or comparator — see farther on. So used, it is to all intents and purposes a gage, because it does the actual measuring or comparing.

The matter of gaging pressure does not need to be considered, because the tension spring system in the indicator automatically exerts the pressure on the workpiece that is optimum for the accuracy of measurement required. The work is introduced or passed under the indicator point (or, as in the case of a moving height gage, the indicator point is moved over the work), a motion that can be made without regard to gaging pressure. Reasonable care should be used, however, not to thrust the workpiece and indicator point together. As in the use of gage blocks and other precision equipment, steady careful motions are the order of the day and the heavy hand is no more desirable here than in dentistry.

Most any dial indicator reading may show the sharp tip of the hand resting somewhere between two dial graduation lines. The regulation recited earlier for the steel rule applies here. Don't try to estimate the reading between dial divisions no matter how strong the tendency to do so. Decide on which dial division you will take as the reading. Or get an indicator with finer discrimination. If an indicator with .001-inch divisions is making you guess, try .0005-inch, .00025-inch, or even .0001-inch graduations.

As with any other type of precision equipment, an indicator cannot be slammed around, dropped, or tossed on a dirty bench and then be expected to register accurately. It should be kept reasonably clean from oil, chips, dirt, or coolant. Ordinarily, an indicator does not have to be wiped thoroughly at each use for fear of subsequent rusting or corrosion. There are practically no ordinary shop conditions where temperature changes from handling, drafts, radiators, or direct sunlight will affect its accuracy enough to be noticed.

If it is mistreated, if it gets dirt-clogged, bent, or broken in some manner, it will tell you something is wrong. It is possible to damage a snap gage, vernier caliper, gage block, or micrometer and fail to realize that it is not measuring correctly. But not so, ordinarily, with a dial indicator. Like a neglected watch, it just doesn't work.

Checking the Indicator for Accuracy

An indicator should be checked regularly for accuracy and repetition. Or, as with any other instrument you are a stranger to, it may be best to check an indicator before putting it to use on measurements you wish to rely on.

Clamp it to a test set over a surface plate or machinist's flat. Wring a .100-inch gage block, say, on the surface under the indicator point, and then regulate the test set clamps and arm so that the indicator point touches the gage block — registers — and turns the hand the prescribed quarter revolution. Then, loosening the bezel clamp, turn the dial till the dial 0 graduation is directly under the tip of the hand. Slide the .100-inch gage block in and out under the indicator point a couple of times. The indicator should regain the 0 reading each time with a repetition error no greater than one-quarter of a dial division. (Be sure in such a test that any apparent error is not due to your manner of manipulating the test gage block. As it is moved in and out under the indicator contact point, it must be kept wrung to the surface under it.)

To calibrate or check the accuracy of the instrument, a .105-inch block, for instance, can be used next. The instrument should read + .005″ with an error no greater than plus or minus a dial division. The repetition test can be repeated with this block, if desired. The accuracy of the remaining range of the indicator can be tested in a similar manner, by using a succession of gage blocks increasing in size by steps several thousandths of an inch apart.

Where an indicator shows test errors exceeding a quarter of a dial division in repetition and plus or minus a full dial division in accuracy, it is best to return it to the manufacturer for repair. (Or a calibration chart can be made and future readings from the indicator in use can be corrected by the amount of actual error showing on the chart.)

When an indicator becomes too sluggish, if it sticks, or is completely broken, it should be returned to the manufacturer. The amateur, though he feels capable of repairing the kitchen clock, should not try his skill on a precision instrument like an indicator. And don't take it to the local watch repair shop. Such artisans may well understand time and clock mechanisms but few of them have the necessary comprehension of precision measurement to properly know the end result of their repair effort even though an indicator does look like a watch. Send it back to the manufacturer.*

Comparators and Dial Gages

The test set, with its indicator, on a surface plate, forms a comparator. As its name implies, a comparator is an instrument that

---

*Some gage repair departments in the larger shops do employ repairmen who have been factory trained by instrument makers. They, in most instances, can supply expert repair service on indicators comparable to instrument factory skill.

compares an unknown dimension with a known size. On a surface plate, of course, the usual procedure is to congregate a gage block stack equal to the blue print dimension being checked and zero the test indicator to it. Then, gage blocks are removed and the workpiece is tried under the test set indicator. Whatever the indicator reads, plus or minus, different from the original indicator 0, shows how much larger or smaller the workpiece is as compared to the master or gage block stack, in this instance. Figure 35 shows such an operation.

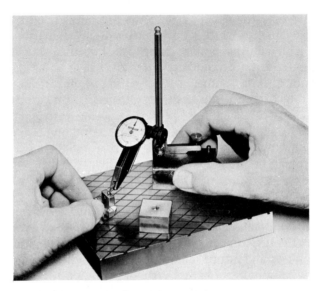

Fig. 35. Dial indicator on a test set has been "zeroed" on a gage-block stack of height equal to the blueprint dimension being checked.

Cylindrical masters are also used with comparators. In fact, the general rule has been established that if the workpieces to be tested are round, the master should be round. Commercial master cylinders are made from hardened steel with even greater precision than standard plug gages. Errors in actual size, ovality, and taper are usually confined within ten millionths of an inch of the specified size of the master.

Test and dial indicators are relatively inexpensive and reliable, keeping their purposes and shortcomings in mind, for purposes of comparing and indicating differences. Their shortcomings are obvious when linear measurements and other tasks requiring fine discrimination and accuracy are demanded. Then the practicalities of the physical world take over — hysteresis, gear problems, friction, etc. — and preclude the finer measurements. The practical limits of these dial indicators is about .0001 inch.

More discriminatory powers have to be derived from the advanced mechanical comparators — reed-type comparators, and mechanical comparators or indicators using modern, friction-reduced mechanisms to obtain greater precision — which are still being improved. The real progress, however, in finer measurement from comparators, is in electronics and optics.

While perhaps the first thought in connection with dial indicators is their use in test sets, actually their most common use is on comparators and dial gages. In Fig. 36 are shown a variety of comparators and dial gages.

In studying comparators, and while using them, the basic geometry of measurement and gaging should be kept in mind. More mistakes are made from ignoring or forgetting the correct geometry, from incorrect manipulation, than from any amount of intrinsic instrument inaccuracy. Usually, however, the instrument is blamed.

Any linear measuring instrument, any gage, and any comparator must have first a reference point or reference anvil. Then it must have a measuring point, a movable or sensitive contact. The measurement must be secured over a true diameter, in the case of a cylindrical workpiece, and across the correct section in the case of other shapes. In addition, a comparator must be correctly "mastered"; that is, the indicator is first set by means of an accurate master to the dimension that is to be measured or, rather, compared.

Elements of the Bench Comparator

The ordinary bench comparator covers the geometrical requirements in the form of a reference anvil, b in Fig. 37, and the sensitive contact, c, which is really the indicator point. The reference anvil is many times referred to as simply the gage's anvil. Sometimes it is called a platen, or the table, and frequently it is dubbed simply the base of the gage.

Figure 37 portrays the typical C-frame idea appearing in almost every type of outside diameter gage from the vernier caliper up. In the commercial comparator the C-frame connection is accomplished by an upright or post, p, in Fig. 37, fastened solidly to the base, b, and by an extending arm, a, which holds the indicator. Usually, on a commercial comparator, the arm can be slid up or down the post and clamped by means of a screw handle, h, or some such device. Another glance at the outline in Fig. 37 brings to mind the familiar conception of the surface plate and height gage, of which the ordinary comparator is a practical adaptation.

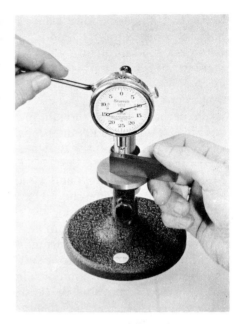
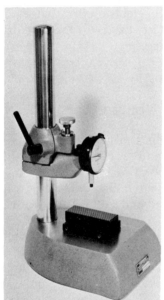

Courtesy of Standard Gage Co.
Courtesy of the L. S. Starrett Co.
Courtesy of Federal Products Corp.

Fig. 36. Various comparators which utilize dial indicators as a measuring
element.

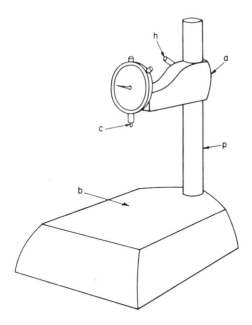

Fig. 37. Illustration of the C-frame construction of most outside diameter gages. Base *b*, column *p* and arm *a* comprise the C-frame.

The base, reference anvil, table, or reference surface of the ordinary comparator is usually machined to the customary accuracy of the commercial surface plate in regard to flatness, warp, and surface finish. On higher-priced and more precise instruments, the comparator's base is more frequently referred to as the anvil or platen, and these surfaces are usually finished to accuracies in millionths of an inch, comparable to the machinist's flat.

Use of Interchangeable Indicators and Contact Points

On many commercial comparators of the general type illustrated in Fig. 37, the indicator is interchangeable. The selection of the proper discrimination of indicator — one with the correct dial graduations — can be used most usually on the rule that the indicator discrimination should be at least one-quarter the tolerance spread being checked, but need not be finer than one-tenth. If the blue print shows ± .005 inch, for example, an indicator with .001-inch graduations has sufficient discrimination in most cases. If, however, the tolerances are ± .001 inch, then certainly the .0005-inch indicator should be chosen (.0005 inch is one-quarter of a tolerance spread of .002 inch). Where working tolerances in ten-thousandths must be checked, then the .0001-inch indicator or even the .00005-inch indicator is used. Fine discriminations in this neighborhood represent the boundary between mechanical indicator ability and the finer discriminations and accuracies possible with air and electronic equipment.

Almost without exception, indicators are equipped with spherical hardened steel contact points. These may be of various shapes as A, B and C in Fig. 38. So-called flat end contact points, as D, E and F in Fig. 38 can also be secured. For the sake of reducing wear, spherical contact points can also be secured with wear surfaces of chrome-plated steel,

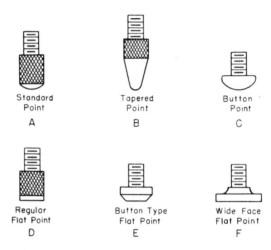

Fig. 38. Various shapes of contact points that are used with indicators.

carbide, Norbide, sapphire, or diamond; flat points (and anvils) come in hardened steel, chrome-plated steel, or carbide, Norbide, or sapphire. (It is said, as a common rule of thumb, that chrome plating increases the wear life of the contact point tenfold and that carbide, Norbide, sapphire, and diamond will wear and last through 100 to 1000 times the gagings the hardened steel point can withstand.)

With the diagrams in Fig. 39 an attempt has been made to illustrate the attributes and shortcomings of flat-end and spherical indicator contact points and the comparison between them.

If a comparator's indicator is zeroed or mastered on a gage block stack as at A, Fig. 39, it is obvious that for ultimate accuracy the gage blocks must not only be wrung onto the anvil but that the flat end contact point must also be wrung, in effect at least, onto the gage blocks. Otherwise, if the workpiece is cylindrical as at B, there can be a discrepancy between the diameter measurement read for the cylinder

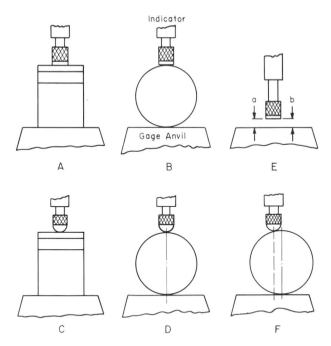

Fig. 39. Various applications for flat and round contact points. (A) Unsuitable application for a flat contact point. (B, C, and D) Proper application of flat and round contact points. (E) Proper position of flat contact point. (F) Result of improper manipulation of round contact point.

and the master reading over blocks that is not due to any actual difference in these two dimensions.

Where the master is cylindrical, as in Fig. 39-B, and the workpiece is also a cylinder, the aforesaid discrepancy is overcome. Nevertheless, the flat end contact point surface should be truly flat and parallel to the anvil as illustrated in diagram E of Fig. 39. In other words, a and b must be equal. Where the comparator has an adjustable arm, where the indicator and the contact point can be moved up and down above the anvil, it is obvious that the parallelism between a flat end contact point and the anvil will be maintained only as a matter of chance.

So, to get around this common difficulty, comparators are most usually supplied with spherical points as at C and D of Fig. 39. Since the contact of a sphere with either a plane or a cylinder is geometrically only a point, the difficulties of parallelism and wringing disappear.

The spherical point presents only one common error, which results from incorrect manipulation, as illustrated in Fig. 39-F; the inspector could happen to measure a chord of a cylinder as shown, rather than the true diameter. The proper technique with comparators, indicating gages, and test sets is always to watch the motion of the indicator hand as the master or workpiece passes under the point, until it reaches the so-called high point — that indicator dial reading where the hand, after advancing in one direction, hesitates and then starts to recede or turn back.

Checking Ovality and Taper

In addition to strict outside diameter applications, the comparator very handily checks ovality and taper. For determining ovality, the workpiece is rotated until the indicator shows a maximum diameter. The workpiece is then rotated 90 degrees and another reading taken. If the piece is oval shaped, this second reading should be lower than the first; in fact, it should be the lowest reading that can be obtained with the workpiece rotated through any angle. The difference between the first and second indicator readings represents the degree of ovality present. Taper, of course, is checked by indicating the workpiece at several points longitudinally along its axis.

On the whole, comparators will stand more abuse, careless handling, and incorrect manipulation than any other type of linear measuring apparatus and still give accurate measurements. This is largely because a master is used and the workpieces are measured, or compared, under exactly the same conditions. A second reason for the comparator's nearly inevitable fidelity arises from the comparatively short range of indicator motion used. Even so, however, the comparator's base or anvil should be checked regularly for flatness and wear and it should be kept free from corrosion or dirt. Its adjustable indicator arm should always clamp up rigidly on the post or upright — the indicator should be tight to the arm. General looseness in any sort of adjustable measuring apparatus is not to be tolerated.

Special Features of Comparator Gages

Figure 40 illustrates some of the accessories that can be obtained for commercial comparator gages which make their use, setting, and operation easier or quicker. In addition to the clamping handle, h, many comparator posts are equipped with gear racks so that by turning a handwheel like g, the indicating "head" can be more readily raised or lowered. Usually, too, this type of movable apparatus has a fine adjustment screw (f in Fig. 40) for the easy final zeroing of the indicator. The comparator in Fig. 40 also displays a set of attachable centers for the ready testing of ovality, curvature, and warp in

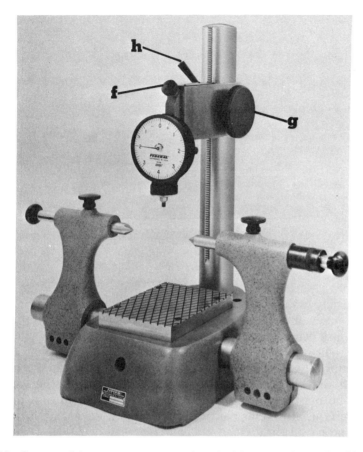

Fig. 40. Commercial comparator gage equipped with accessories to simplify setting and operation.

cylindrical workpieces. So-called "serrated" platens or anvils, as in Fig. 40, can be secured with diagonal grooves that are purposely milled into an otherwise perfectly flat smooth surface so that dirt, dust, or chips will be readily scraped into them, thus keeping the actual anvil measuring surface more or less automatically free from that type of possible measuring error. V-blocks are, of course, used on the platens of comparators, just as they are on surface plates.

The sketch in Fig. 41 illustrates the principle of the so-called "back stop." A back stop, as its name implies, is a fixture attached to a comparator (usually to the post) against which the workpiece butts or stops in the correct position under the indicator point. Its use eliminates the sort of off-center error illustrated in Fig. 39-F. When a back stop is properly adjusted, a cylindrical workpiece will come

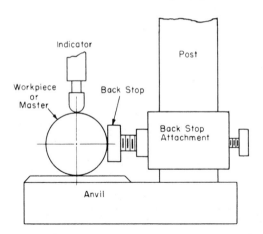

Fig. 41. A comparator may be equipped with a back stop against which the workpiece butts in the correct position under the indicating point, thereby eliminating the error shown in Fig. 39-F.

automatically to the correct measuring position with the indicator point properly over the diameter. The workpiece will occupy the same position as a cylindrical master.

The right way to set a back stop is to loosen its clamping screw or device, move the master or a workpiece under the indicator point until the "high" indicator reading is obtained, and, bringing the back stop against the master or sample workpiece, tighten it in that position. From then on, all similar pieces are in position for gaging if they rest against the back stop.

Portable Dial Indicator Gages

Up to this point on the subject of comparators, the idea of bench gages has been described — the situation where the workpiece is brought *to* the gage. The comparator principle can also be applied to portable gages — to the situation where the *gage* is brought to the work. Two types of portable comparators are shown in Fig. 42.

The comparator principle still applies, however, in that the portable indicating gage possesses an indicator, a sensitive or measuring point, and a reference surface or anvil, that are held in proper position by a C-frame. Portable comparators are almost inevitably equipped with back stops. Portable indicating gages of the comparator type are mastered, in the same manner as their bench prototypes, their indicators are zeroed, and the measurements taken are comparisons with those of the master used.

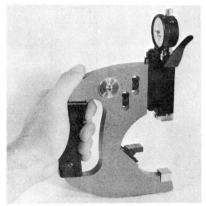

Courtesy of Federal Products Corp.

Fig. 42. Two types of portable comparators

The Indicating Snap Gage

The sketch of an indicating snap gage in Fig. 43 will serve to bring out the usual geometry and mechanics of this type of instrument. Contrary to the bench-type comparator, the reference or fixed anvil on a portable gage (f in Fig. 43) is usually above the sensitive contact. It rests on the workpiece. By designing in this manner, the solid anvil supports the weight of the gage. If it were the other way round, the sensitive contact and the indicator would register some of the weight of the gage and the reading would be a mixture of the comparative diameter of the workpiece and the added contact and indicator deflection caused by the gage weight and the pressure of the hand applying the gage. The fixed contact too, in commercial gages, is equipped with a screw, tongue, or clamping mechanism whereby the reference anvil can be raised or lowered, thereby increasing or decreasing the overall capacity of the instrument by widening or narrowing the span of the jaws.

Usually, in portable gages, the back-stop face (b in Fig. 43) is cut at a 45-degree angle, thus, geometrically, simulating the V-block. The 45-degree back stop assists the inspector in holding the workpiece in the proper position for correct gaging.

The motion of the sensitive contact (many times referred to as the "lower anvil"), shown as p in Fig. 43, is transmitted to the indicator, by various designs of plunger movement, pivot or spring mechanisms, depending on the make of instrument, all of which are contained inside the frame c

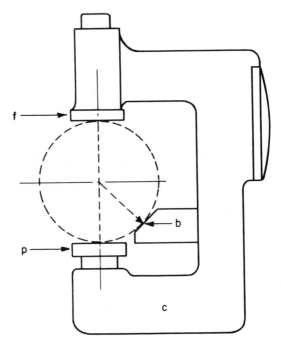

Fig. 43. Schematic diagram of an indicating type snap gage.

Using the Right Gaging Methods

The professional touch to be learned in the use of indicating snap gages is to let the gage do its own work. The matter of suitable gaging pressure is self-regulated in the spring-loaded mechanism of the sensitive contact *p*. No more finger pressure is needed to gage a workpiece than that which will slide the contacts over the workpiece. As soon as the gage has suitably "homed" on the workpiece, release the hand pressure and simply support the gage sufficiently with the finger tips to prevent its slipping away from the workpiece. Examination of Fig. 44 will make this instruction clearer.

The proper manipulation of a portable comparator, then, consists in starting the fixed or reference anvil on to the workpiece, using the fixed anvil as a pivot, and swinging the lower or sensitive contact, *p*, into position, continuing the swinging and pushing motion until the contact of the back stop *b* completes putting the gage in position. Then relax the finger hold on the gage.

Like its larger relative, the bench comparator, the portable indicating gage is rugged — at the same time sensitive and accurate — and will withstand a great deal of abuse. It should be kept clean, and, as

(C)   *Courtesy of the L. S. Starrett Co.*

Fig. 44. (A) Method to be used in applying an indicating snap gage to a workpiece. (B) Checking the measuring surfaces of an indicating snap gage for parallelism and flatness. (C) Using a snap gage on work in process.

much as possible, away from oil, coolant, dirt, and chips. The effect of hand temperature must be watched. Steadily gripping a portable indicating gage can affect its accuracy, sometimes as much as .0005 inch, because of the heat absorbed by the gage's C-frame and the subsequent expansion of the frame. Take a measurement and set the gage down; don't handle it like a shovel or a garden fork.

Checking for Proper Calibration

Since the ordinary portable comparator or indicating snap gage is used with a master, it does not ordinarily need to be calibrated or checked. However, it is a good thing every so often to test the calibration of its indicator and the parallelism of its jaws, if the latter are flat. This is done, usually, by selecting a gage block stack representing the nominal capacity of the gage. The reference anvil is adjusted to the gage blocks until the indicator registers the customary quarter turn. The gage block stack is then slightly wrung into final position between the gage contacts and the indicator is zeroed. The operation is repeated with gage block-stacks .001 inch, .002 inch, or .005 inch longer or shorter, but *without* changing the setting of the indicator. The indicator should faithfully reproduce the increase or decrease of gage block stack length with suitable plus or minus readings.

It is wise also, of course, to give a portable snap gage a repetition test on a cylindrical master every so often. The repetition error should not exceed one-quarter of a dial division on the gage's indicator; calibration, flatness, and parallelism errors should not be greater than a dial division.

With the gage block stack available, it is easy to check the parallelism between the reference surface and the sensitive anvil. Add to the equipment a standard instrument measuring roll or wire plug gage 1/16 inch to 1/8 inch in diameter. The reference contact is adjusted down onto the measuring roll riding on the gage blocks until the indicator registers. Move the measuring roll back and forth and from side to side and watch the indicator hand; any variation in parallelism will show up as a variation in indicator readings. Where the sensitive contact is spherical or cylindrical in shape, the parallelism check is, of course, unnecessary.

The flat anvils of a portable indicator gage should be checked occasionally for flatness. If considerable care is used, the test can be made by sliding a steel ball, Fig. 44, between the anvils registering on it and watching the gage's own indicator. Humps and hollows in the anvils will be transmitted through the ball and the gage mechanism to the indicator. The optical flat test, to be described farther on, is more accurate.

Wear on the gaging contacts of portable gages is reduced, commercially, by the use of chrome-plated steel, carbide, Norbide, or sapphire for contact surfacing in place of hardened steel. However, because the portable comparator-type indicating gage is used always with a master, the effects of wear, uneven anvils, and lack of parallelism are not usually noticed until such errors have reached a comparatively extreme stage.

One more recommendation for this general type of gage should be mentioned. Because a dial indicator and a sensitive contact form its basic measuring principle, workpiece conditions like taper, ovality, crookedness, and lack of parallelism can be readily seen or explored for, conditions which are much more difficult to detect with conventional micrometers and virtually impossible to find with fixed snap gages.

Dial Indicator Gages are Direct Reading

Dial indicator gages embrace a line or selection of direct reading instruments, in contrast to comparators. For this type of gage, the indicator is equipped with a continuous dial, instead of the balanced dial, which is universal on comparators, and a revolution counter or tell-tale hand. The indicators are of special design and are constructed to accommodate ranges and gage capacities of $\frac{1}{4}$ inch, $\frac{1}{2}$ inch, or 1 inch. Such gages are invariably equipped with lifting levers, so-called, or some mechanism by which the indicator contact point can be readily raised and lowered. One such type of instrument is illustrated in Fig. 45. The familiar C-frame caliper principle is apparent, to which the solid, lower, or reference anvil is attached. In contrast to the portable snap gage comparator (see Fig. 43) the corresponding type of direct-reading gage comes with the anvil arrangement reversed because of mechanical difficulties. In using the direct reading gage, care must be used to hold the lower or solid reference anvil against the work and not allow an erroneous indicator reading because of unsuspected pressure on the upper or sensitive contact.

Using the Dial Indicator Gage Properly

Returning to the idea of the professional touch in gaging and inspection, there are one or two techniques in connection with using the sort of direct-reading, dial-indicator gage illustrated in Fig. 45 that the new inspector would do well to remember. If the gage is equipped

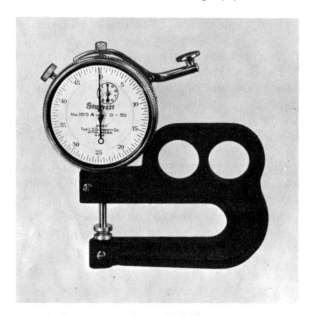

Fig. 45. One type of direct reading dial indicator gage.

with flat-surfaced anvils, they should be checked now and then for parallelism both at the at-rest position shown in Fig. 45 and near the other end of the gage's range where the sensitive contact is raised farthest above the reference anvil. The latter test can be made with a combination of gage block and a small diameter precision measuring roll.

It is also good practice to check the gage's accuracy near the range where it will be ordinarily used. Suppose, for example, the indicating gage has a total range of $\frac{1}{2}$ inch but it is to be used mostly on workpieces whose diameters will vary between about $\frac{1}{8}$ inch and $\frac{3}{8}$ inch. Test the gage's accuracy with a $\frac{1}{4}$-inch gage block or master cylinder. If it is accurate, within plus or minus a dial division, at the $\frac{1}{4}$-inch station, it is fairly safe to presume its accuracy at $\frac{1}{8}$-inch or $\frac{3}{8}$-inch readings. Another similar procedure is to check or master the gage on a gage block stack or master cylinder that equals the nominal blue print specification of the workpieces to be inspected and then, in effect, use the gage as a comparator.

Generally, the discrimination of an indicator used on a direct-reading gage is tied in with the gage capacity. If .0001-inch discrimination is required for dial graduations on the indicator, the range of the gage cannot exceed .250 inch. The 1-inch range instrument is equipped with a .001-inch division dial and the $\frac{1}{2}$-inch gage with a .0005-inch division dial.

Long-range indicators are used on bench gages of the general type illustrated in Fig. 37. Such equipment is especially useful for checking

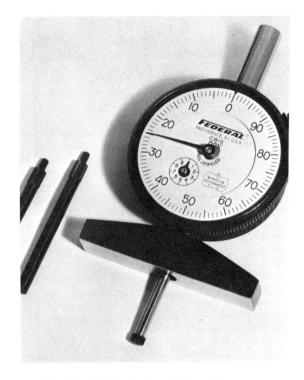

Fig. 46. Dial indicator designed for use as
a depth gage.

thickness, shoulder heights and depths, and other miscellaneous dimensions as well as outside diameters. A special adaptation of dial-gage design,
a depth gage, appears in Fig. 46. The inspector is again urged to study gage
makers' catalogues, bulletins, and advertisements to become up-to-date on
the various types of modern indicating gages and equipment available.

Indicating Micrometers

The subject of longer-range, continuous-measuring, dial-indicator
gages cannot be concluded without mention of indicating micrometers.
While an indicating micrometer, Fig. 47, seems to be essentially the
same as the conventional micrometer described previously, it has,
other than its indicator and indicator mechanism, several different and
valuable characteristics.

In the first place, the reference anvil is spring loaded to impose a
constant pressure of two pounds on the workpiece, as shown in Fig. 48.
As the micrometer spindle is turned down on the workpiece in the

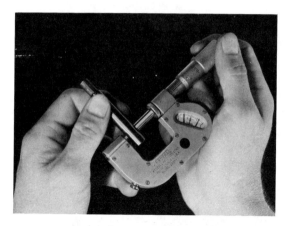

Courtesy of Federal Products Corp.

Fig. 47. Indicating type micrometer

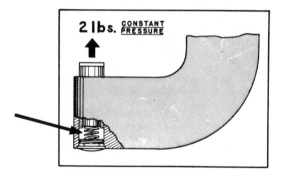

Fig. 48. Cutaway view showing the spring-loaded reference anvil of the micrometer in Fig. 47.

customary fashion, the spring-loaded reference anvil yields as two
pounds micrometer screw pressure is reached and the movement of the
anvil is transmitted through an indicator gear train, enclosed in the
frame of the micrometer, to the indicator dial. When the dial reads 0,
the mechanism is balanced, and the micrometer can then be read in the
customary manner.

How Accurate Readings are Obtained

The practical advantage offered by the spring-loaded reference
anvil is that correct, consistent gaging pressure is always applied to the

workpiece. It is easily possible to obtain an erroneous reading with the conventional micrometer — erroneous from several tenths to several thousandths — by varying the amount of thumb-and-finger pressure on the micrometer thimble, spindle, and screw thread. Let two people try to measure the same diameter and usually different micrometer readings will be reported, because no two workers ordinarily exert the same amount of force on a micrometer thimble. The indicating, spring-loaded micrometer eliminates this universal, potential error. Two or twelve people using it will get the same reading.

Although the indicating micrometer is vernier equipped, it is handier to use the indicator for reading fractions of a thousandth of an inch, following the directions in Fig. 49.

1. Set the Indicating Micrometer anvils on the piece being measured and turn the spindle until the Mikemaster indicator hand swings up to **dial zero.** You now have stable pressure (2 lbs.) on the work.

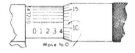

INDICATOR

2. Read barrel and thimble. Thimble graduations may not match barrel 0 reference line.

(Reading: .436″ plus some tenths)

3. Turn thimble up, in same direction until its next graduation matches barrel 0 reference line.

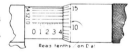

4. Then read tenths (.0007″) directly on the Indicator.

The correct reading is .4367″

Fig. 49. Operating instructions for the indicating micrometer of Fig. 47.

The Direct-Reading Bench Micrometer

The big brother of the portable indicating micrometer is illustrated in Fig. 50, a direct-reading bench instrument known as the Ultra-Mike.* This instrument has the spring-loaded, constant, correct pressure reference anvil. Its indicator comes to the zero position to signal

*Trademarked name Brown & Sharpe Mfg. Co., North Kingstown, R.I.

Fig. 50. Direct-reading bench micrometer.

that the correct pressure has been applied to the workpiece by turning the spindle — in other words, when the Ultra-Mike indicator registers 0, it is the signal for reading the micrometer spindle and thimble. The thimble is enlarged so that it can be read directly to .0001 inch. The capacity of the Ultra-Mike can be increased over the 1-inch range of its spindle because the reference anvil is contained in a tail stock that can be moved back and forth. Initial settings of the tail stock in even inch increments away from the spindle are obtained with the use of 2-inch, 3-inch, and so on, gage blocks. An elevating table makes the introduction of workpieces between the Ultra-Mike anvils easier.

Comparators for Internal Diameters and Indicating Bore Gages

Thus far, the discussion of dial indicator gages and comparators has been based almost entirely on the measurement of outside diameters, thicknesses, or depths. Gaging equipment of similar principle has been devised for measuring inside diameters — holes or bores. Some idea of the commercial types of this apparatus available can be gained from an examination of Fig. 51.

Indicating bore gages also can be used to check the width of slots, to measure between two planes or parallel surfaces, performing a

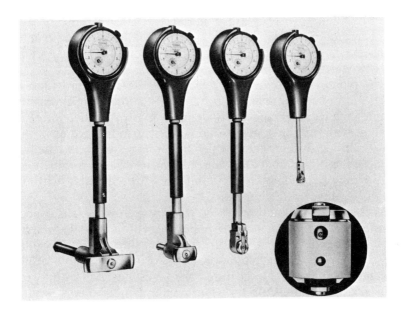

Courtesy of Federal Products Corp.

Fig. 51. Commercial types of indicating bore gages.

function similar to the square plug gage. It goes without saying that extra care in manipulation is to be used in order not to measure unconsciously on some diagonal rather than on the true perpendicular axis between the two planes.

One type of indicating bore gage, A in Fig. 51, is provided with a *pair* of reference contacts or solid anvils and a single sensitive or movable contact all arranged 120 degrees apart. Another variety, B, is furnished with only the single solid contact or reference anvil directly opposite the sensitive contact, which might be compared to the familiar machinist's spring caliper. Still another variety has this same arrangement of opposed single contacts but they are equipped with special centralizing devices, as in C and D.

At this point it would be well to review the discussions concerning the reference point, measured point, repetition and especially the description of the correct use of ordinary inside calipers because the same geometry and principles apply to the proper use of indicating hole gages and comparators. Indicating bore gages must be carefully centralized ("rocked" is the shop term) and manipulated to be sure the true diameter is being measured. While gage makers provide so-called centralizers, they are not necessarily fully automatic and the inspector is not absolved of the blame for getting an incorrect measurement where he trusted a centralizing device too far. An attempt has been made in a series of "right and wrong" diagrams in Fig. 52 to point out possible errors in the manipulation of hole gages.

This type of gage is limited, in any individual use, to the range of the indicator. Depending on the discrimination and precision desired, indicating bore gages have capacities varying between .010 inch and .250 inch. However, the total capacity of the average commercial hole gage may be extended over about an inch range because the reference anvil can be unscrewed from the gage and a longer of shorter solid contact substituted.

Mastering Indicating Bore Gages

Indicating bore gages are, almost without exception, comparators and must be "mastered". And following the rules of the game, which say that if a comparator is to be used to measure cylindrical work, it should be mastered on a cylinder, or on a rectangle if it is to be used on rectangular work, the bore gage is best mastered in a master ring of suitable internal diameter. In other words, the average commercial bore gage is not a complete gage unless it is accompanied by its master ring; the two go together like bread and butter. To all intents and purposes, master rings are standard American Gage Design (AGD)

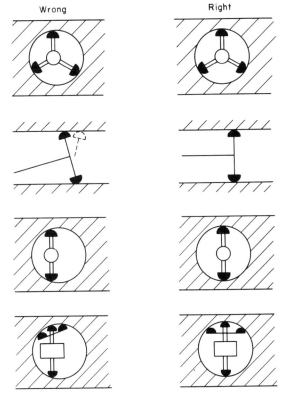

Fig. 52. Diagram showing the results of correct and incorrect manipulation of hole gages.

ring gages and the gage is "set" to the ring. The dial of the gage's indicator is "zeroed" at that location of the indicator hand where, as a result of rocking and centralizing, it reaches its maximum swing and starts to turn back. (Review the identical principle and geometry describing the correct way to "master" a comparator's indicator on a cylindrical master.)

Theoretically, at least, an inside comparator can be mastered in a gage-block caliper as illustrated in Fig. 53. But to manipulate or hold the gage and read the high point of the indicator hand's swing takes the touch of an expert; it is so easy to "master" it incorrectly, on a diagonal, as indicated by the dotted lines in Fig. 54 at A.

Effect of Error in Bore Gage Centralizer

If the gage has a centralizer, that centralizer must have been manufactured, assembled, and set absolutely accurately. The geometry

Courtesy of Dearborn Gage Co.

Fig. 53. Mastering an inside comparator in a gage block caliper.

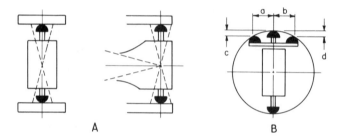

Fig. 54. Dotted lines in (A) indicate incorrect mastering positions for operation shown in Fig. 53. (B) Comparator gage equipped with centralizers to reduce the possibility of incorrectly mastering and using the gage.

of the situation is outlined in sketch B of Fig. 54. Dimensions a and b — the centerline relationship between the centralizer tips and the axis of the sensitive gage contact — must be equal. Similarly, the centralizer cannot be tipped, twisted, or offset; dimensions c and d must be alike. If dimensions a and b or c and d are unequal to the extent of .001 inch, an error in measuring the ID of a hole up to.0002 inch is easily likely where the gage is "mastered" in the rectangle of a gage block caliper and then used in the cylindrical section of a bore.

On the other hand, even though a centralizer is not absolutely true geometrically, it will have little effect on the accuracy of the bore measurement if the gage is "mastered" in a ring and its indicator "zeroed," because the gage is working under the same conditions. This statement holds true with mechanical errors or wear in a centralizer up to about .010 inch. A bore gage's centralizer when twisted, warped, worn, or offset more than .010 inch can produce an instrument error even though the gage is mastered in a ring.

Proper Manipulation of the Indicating Bore Gage

Indicating equipment of a design similar to that of a micrometer plug gage is shown in Fig. 55. In this case, the spreading contacts are moved outward by a spring inside the plug and the indicator registers the amount of that movement in tenths, half-thousandths, or thousandths of an inch. The contacts are retracted by a trigger-type lever.

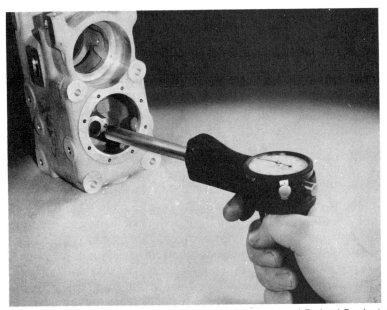

Courtesy of Federal Products Corp.

Fig. 55. Indicating equipment which utilizes spreading contacts similar to those used in a micrometer gage. A trigger-type lever mechanism retracts the contacts.

Centralizing with this sort of gage is accomplished by having the "plug," which is really a metal skirt, made to within ± .001 inch of the minimum inside diameter that is expected to be measured. In other words, if the nose of the plug itself will not enter the hole, the hole has been bored at least .002 inch undersize. Gages of this character usually have a total operating range of .010 inch. They are, to all intents and purposes, single-purpose gages, like equivalent, fixed, conventional plug gages, and almost always a particular plug with its expanding contacts must be inserted in the gage head for each hole size. Since these gages are comparators, they must be mastered, and each plug is usually accompanied by its equivalent master ring gage.

In mastering and using this type of gage, as with its prototype the micrometer plug gage, it must also be rocked a trifle to ensure suitable centralization. If the contacts are retracted by the trigger, if the plug is inserted in a hole, and if the spring-loaded contacts are released to press against the sides of the hole — all this without rocking or "homing" the plug a little — the indicator hand will be seen to move, creep a little, before settling into a final reading. What the indicator is trying to tell you is that the gage is not yet set on the true diameter. When, after a trifle of rocking and "snugging" the gage plug in the hole, the indicator settles down — if you get repetition in other words — then the gage is properly centralized. One of the things to be learned from use and experience with all types of indicating gages and comparators is that the motion of the indicator hand will tell you of errors in the gage or its manipulation.

Up to this point, the discussion has been confined to portable indicating bore gages — to the situation where the gage is brought to the work. Bench-type indicating hole gages, as shown in Fig. 56, are also available for the situation where the workpiece is brought to the gage. And here, one more bit of geometry and professional manipulation should be described.

When the bore gage is brought to the work, the solid reference contact or anvil touches the workpiece first, as shown at *a* in Fig. 57-A. It bears the weight of the gage and the operator's hand and acts as the pivot point about which the gage is rocked, until the sensitive contact at *b* is truly opposite. But when the work comes to the gage, the situation is reversed. As shown in Fig. 57-B, the reference anvil, *a'*, points up, the workpiece rests on the solid contact and is rocked or pivoted about it until the sensitive contact *b'* is diametrically opposite. Thus, our fundamental rules for reference and measuring points are lived up to.

Hole Conditions That Must be Anticipated

It probably seems rather obvious, now, that, with internal indicating gages and comparators, hole conditions such as ovality, taper, hour-glass and barrel shape, and bell-mouth, can be explored for and readily detected. In fact, the great advantage of indicating equipment over the standard plug gage is this ability. Always, when measuring an ID with an indicating bore gage, stop long enough (a) to explore for extraneous conditions and (b), if they are discovered, find the *minimum* diameter, for after all, it is the minimum diameter of a hole that is effective when a shaft or cylinder is assembled in a hole.

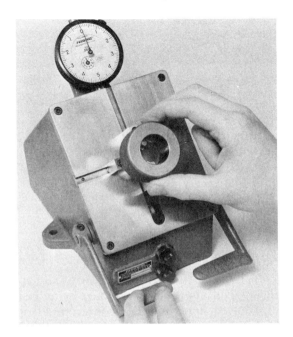

Courtesy of Federal Products Corp.

Fig. 56. Bench types of indicating hole gages. With these types of gages, the workpiece is brought to the gage rather than the gage to the work.

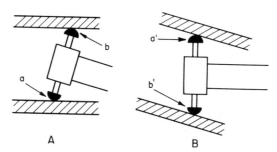

Fig. 57. Methods of manipulating bore gages: (A) For portable bore gages the solid reference contact, *a*, touches the workpiece first. (B) For benchtype gages, the reference contact, *a'*, is on the top and the workpiece is rocked or pivoted until the sensitive contact *b'* is diametrically opposite.

Cement in your mind, by studying the diagrams in Fig. 58, the presence of ovality, barrel shape, hour-glass shape or bell-mouth, and taper, some of which or all of which will undoubtedly appear to some degree in practically every bored hole checked by an inspector. These sketches show in dotted lines how the conventional plug gage may check the minimum diameter, but little else, and, with arrows, how explorations with the contacts of indicating gages disclose actual conditions.

Anyone who knows about machine design realizes that a bearing, for instance, must have so-called bearing area — enough *area* of metal in contact between shaft and bearing to withstand the pressure imposed on it. Pressure is usually thought of in terms of pounds per square inch. The sketches in Fig. 58 point out that the shaft and bearing, under the conditions illustrated, would have practically point or line contact resulting in an uneven distribution of bearing pressures which may result in vibration, accelerated wear, and premature breakdown. Thus, extraneous conditions such as ovality may be more serious than actual digressions from the specifications in diameter tolerances, clearances, and allowances.

Care and Checking of Indicating Gage Equipment

The greatest danger in the use of indicating gages may be expressed by saying that familiarity leads to overconfidence. We pick up a portable indicating gage or go over to a bench comparator and use it without the necessary preliminary precautions to assure ourselves we are going to read the accurate results we require. Where such an instrument has not been used lately, the following routine is worth following:

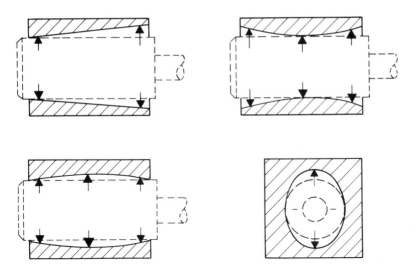

Fig. 58. Various hole conditions which a conventional plug gage will not indicate but which may be explored with indicating-type gages.

1. Assuming the indicator has been checked for repetition, calibration, and free running, look over the way it is clamped to the test set, comparator frame, or gage. Any detectable shake or looseness is not to be tolerated.

2. Checks for looseness or play should be applied to comparator posts, bases, clamping handles, fine adjustment mechanisms, and anvils. It is easy, for instance, to rely on the accuracy of a comparator and find afterward that the reference anvil was not securely clamped down.

3. In the case of portable gages, especially those with adjustable reference anvils (which are supposed to be solid), and also in the case of bore gages with adjustable or changeable extension anvils, test the anvils and extensions to be sure they are secure and can't wiggle loose.

4. If gage back stops are to be used and relied on, make sure they are also clamped tight in the proper location.

5. The sensitive contact points on many portable gages and bench comparators are tipped with wear resisting carboloy, Norbide, sapphire, or diamond inserts. It is well to try such tips to see that they haven't loosened in previous use. Also examine them under a glass. If they are cracked, chipped, or badly scored, their surface conditions may prevent accurate readings or repetition or they may scratch and scar the work.

6. If opposing anvils are supposed to be flat and parallel, give them the wire or ball test.

7. Usually portable indicating gages and bench comparators are as neglected as the back of a small boy's neck. If dirt, dust, grit, chips, grease, scum, and coolant will interfere with the accuracy of gage blocks, they will also offset the precision of comparators. Clean an instrument thoroughly at each use, and, after you are through with it, rust proof exposed iron or steel surfaces.

8. Make as sure of the reliability of master discs and master rings as you would of gage blocks. After all, they are equivalent in precision. Give them the same care you would to gage blocks. Examine them for nicks and scratches and the scars of rough handling. Incidentally, handle masters as carefully as you would gage blocks or machinists' flats.

Pantograph Mechanisms

In the field of gaging, a pantograph is essentially a motion transfer mechanism. It consists usually of two parallel steel blocks or plates supported by four parallel strip springs as indicated by the isometric sketch in Fig. 59. Figure 60 shows pantographs in practical use.

The utility of the pantograph is several fold. In the first place, perhaps, it should be considered as a sort of extension of the indicator spindle and point. Its use enables the sensitive contact of the gage to reach into and bear against the workpiece, many times dodging projections or recesses on the workpiece or overcoming measurement conditions in which it would be awkward to manipulate a standard design indicator, indicating gage or comparator. To put this another way, by using pantograph mechanisms the gage's indicators can be offset or backed off away from the workpiece and the inspector's hands, as he manipulates a pantograph-equipped gage, and thus be more readily readable.

This same attribute makes the pantograph a shock absorber, to a degree, in that the springs of the pantograph help to absorb the blow of lodging a workpiece in the gage, thus protecting the more or less delicate internal mechanism of an indicator.

As has been suggested, the plates of a pantograph, a and b of Fig. 59, are parallel to each other. They are usually manufactured carefully 4-square with flat, parallel surfaces — made, in other words, with about the same care and tolerances used in manufacturing commercial V-blocks. As can be seen from Fig. 60, one plate of the pantograph is usually fastened to the body of the gage and then adjustable contact points, lugs and other accessories are fastened to the other plate.

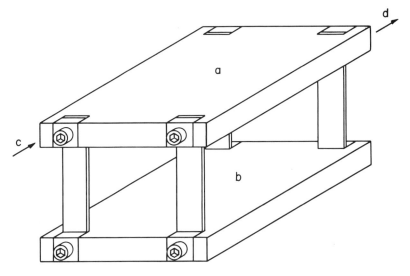

Fig. 59. Basic pantograph mechanism.

Fig. 60. Pantograph mechanism as used with gaging equipment.

If, for instance, the pressure of a workpiece dimension against a pantograph contact point causes a motion of plate *a* as shown at *c*, Fig. 59, that motion will be exactly transmitted at, say, location *d* and if an indicator point is bearing against surface *d*, the indicator, itself, will register as exactly as if it contacted the workpiece directly. In other words, lateral motion applied to any part of plate *a*, Fig. 59, is transmitted equally by any and all other parts of plate *a*.

Application of Pantograph-Equipped Gage

In using a pantograph-equipped gage, it is sensible, of course, to check the attached contact points and any lugs fastened to the pantograph to be sure they are not loose and that contacts, especially, will bear suitably against the workpiece. (The way the fixed plate is fastened to the gage should also be tested for any looseness of course.) At the same time a check should be made, when the pantograph contact bears against a workpiece or master, to be sure the indicator position is adjusted so that its point presses, with suitable quarter turn indicator tension, against the movable pantograph plate or attached lug.

One other advantage of the pantograph design should be cited. There are, in the strict sense, no movable parts in a pantograph — no pivots, bearings, plungers, or gears. All movement is contained in the internal elasticity of the four springs. Hence, there is nothing about a pantograph to get out of order, clogged (within the elastic limits of the strip springs), even though the workpiece is dripping with oil, dust, or dirt, or coolant literally pours all over a pantograph. It is necessary only to be sure that the pantograph contact-point gaging surface, the tiny surface of the workpiece under contact, and the area of the pantograph plate or lug under the indicator point are free from grit, chips, or grease. Similarly, the solid reference points of these types of gages should be cleaned.

Speaking of the elastic limits of strip springs, however, always check a pantograph-type gage at regular intevals to be sure one of the pantograph springs has not perhaps been struck or damaged in such a manner as to put a definite, permanent bend or crease in it.

The Reed Mechanism

Somewhat allied to the pantograph and yet introducing still other basic principles of measurement, amplification and magnification, is the so-called reed mechanism shown in Fig. 61. Figure 62 illustrates a so-called visual gage, in which a reed mechanism is used.

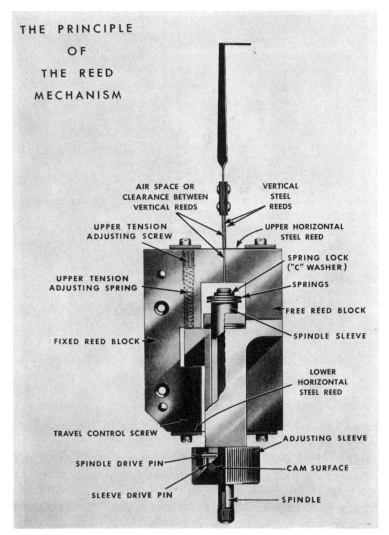

THE PRINCIPLE
OF
THE REED
MECHANISM

AIR SPACE OR
CLEARANCE BETWEEN
VERTICAL REEDS

VERTICAL
STEEL
REEDS

UPPER TENSION
ADJUSTING SCREW

UPPER HORIZONTAL
STEEL REED

SPRING LOCK
("C" WASHER)

UPPER TENSION
ADJUSTING SPRING

SPRINGS

FREE REED BLOCK

FIXED REED BLOCK

SPINDLE SLEEVE

LOWER
HORIZONTAL
STEEL REED

TRAVEL CONTROL SCREW

ADJUSTING SLEEVE

SPINDLE DRIVE PIN

CAM SURFACE

SLEEVE DRIVE PIN

SPINDLE

Courtesy of Sheffield

Fig. 61. Basic components of an amplifying reed mechanism.

The reed mechanism consists, essentially, of two metal blocks, one fixed and one floating, joined by steel strips like a pantograph (the "upper" and "lower" horizontal steel reeds of Fig. 61).

The fixed block is rigidly anchored to the gage head case. The floating block carrying the gaging spindle is connected horizontally to the fixed block by two reeds. A vertical reed is attached at the inside top of each block. There is no contact between them except at their upper ends, which are joined together. Beyond this joint extends a pointer, or target.

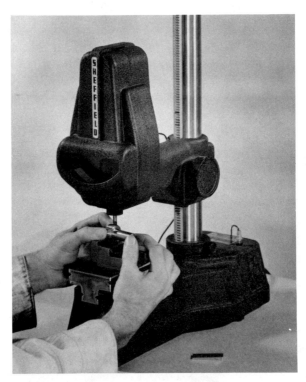

Fig. 62. A visual gage which utilizes the reed mechanism of Fig. 61 to amplify measuring-spindle movements.

The gaging spindle is an integral part of the floating block. When spindle and block are moved upward in the gaging operation, the horizontal reeds deflect slightly but the vertical reed on the floating block tends to slip past its companion. However, as these vertical reeds are joined at their upper ends, instead of slipping, the movement causes both reeds to swing through an arc, and as the target is merely an extension of the vertical reeds, it swings through a much wider arc. The amount of target swing is proportional to the distance the floating block is moved, but, of course, is very much greater.

Through a series of lenses, a light beam projects the shadow of the target on the scale of the gage. Thus, mechanical amplification and optical magnification are combined. Friction is entirely eliminated.

Note that this mechanism incorporates no gears, knife edges, or levers — no rubbing contacts of any kind to wear in service or to wear out of adjustment. It is mechanically positive at all times — no backlash — no lag.

With the incorporation of solid-state support circuitry and gaging, this type of gaging head is capable of impressive levels of precision through high amplification. Systems that incorporate this cooperative venture among mechanical, optical, and electronic measurement and translation are adaptable to many formats, and are represented in stand-alone systems or as support instruments to be used in surface plate setups.

Multiple and Special Design Gages

In general, thus far, the study of indicating gages has been confined to single-purpose gages and comparators. Gages are made, however, that will measure more than one dimension simultaneously. Almost invariably, of course, such comparators must be specially designed and constructed for a particular type, at least, of workpiece. They can be made adjustable so as to accommodate variations in sizes of parts but, generally, the parts must be of the same type.

Figure 63 illustrates a few of almost limitless combinations that can be designed for special and multiple measurement.

The care, checking, mastering, calibrating, and maintenance of special and multiple gages is no different in principle than the supervision of the simplest single-purpose comparator. There are simply "more of the same" as the mother of triplets remarked in regard to the difficulty of bring up a child. Each "station" of a multiple gage is designed to measure a particular dimension or relationship (perhaps such as concentricity) and each station therefore is to receive the same kind of preliminary checks and care the ordinary single-purpose comparator or gage would get. Usually the multi-station master for a multiple gage is made with great care and exactness and the inspector's concern should be extended to make sure that the master is maintained corrosion- and nick-free and that, once the gage has been set, the master is preserved and stored as carefully as any surface plate, for instance, or any other piece of precision measuring equipment

The main purpose of a multi-station gage is to save time. It is usually designed with spring-loaded contacts, centers, V-blocks, platens, and other gage auxiliaries so that the matters of gaging pressures, manipulation, location, and even the error of temperature changes are very considerably eliminated. A little special care must be used in mastering the gage — in setting whatever adjustments the gage calls for and in zeroing indicators. Thereafter, workpieces, of course, should not be literally thrown at the gage. Usually the inspector learns after a few trials the quickest and best way to check an intricate workpiece all in one motion and then, more or less automatically, to check a succession of them.

Courtesy of Mitutoyo Corp.

Fig. 63. A few of the many comparators of special design which measure several
dimensions simultaneously.

One word of warning should be heeded. Where there are several indicators spaced around a special design gage, the inspector should, in theory, get his nose right over each indicator, in turn, in order to read it accurately. In other words, the error of parallax is a predominant one to avoid in multiple-unit gaging.

Relationships in Measurement

Before getting away from comparators and special indicating gages, the subject of relationships should be reviewed because there are certain relationships that are best determined by indicating equipment. By relationships are meant such phenomena as out-of-round, parallelism, taper, and squareness.

The matter of triangle effect or triangular out-of-round, a condition produced on cylindrical work by centerless grinding, has been mentioned before. This relationship is purposely rather grossly exaggerated in Fig. 64. The intent, of course, is to grind the perfect cylinder represented by the dotted circle *a*, but instead the triangle effect represented by the solid lines is obtained.

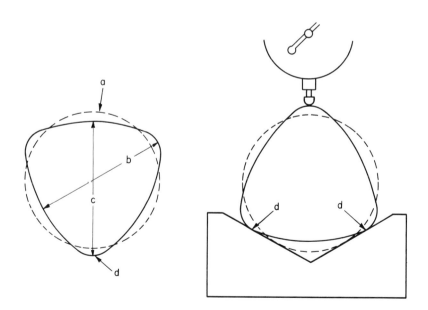

Fig. 64. (Left) Triangle effect produced on cylindrical work by faulty centerless grinding. (Right) Vee-block method of determining the degree of out-of-roundness.

Peculiarly, but true, where triangular out-of-round is present, any diameter measured by simple two-point calipering, as with a micrometer, will read the same no matter where around the periphery of the cylinder the diameter measurements are taken. In other words, measurement *b*, Fig. 64, will be the same as *c* or the same across any other axis and the reading may be the same as the diameter of circle *a* — as if the piece were a true cylinder.

But when the piece is rested in a V-block and revolved, the "high points," like *d* and *d*, Fig. 64, "ride up" in the V and, if an indicator is used as shown, the difference between minimum and maximum radii is detected. Although the dimensions represented by diameters *b* and *c*, and the like, will all be the same when triangle effect is present and while they may well give the same truly cylindrical reading circle *a* would have had, the triangular-shaped shaft will not assemble in a perfectly round hole that is close to the specification diameter of circle *a*, Fig. 64.

While Fig. 64 shows the so-called 60-degree V-block (with a 120-degree included angle), triangle effect can be as readily detected, most usually, in the standard 45-degree V-block. Triangle effect will also appear with 6 or 9 nodes, instead of three, sometimes with 5, 7, or more nodes, but the V-block with an indicator will just about invariably detect it. In a later chapter the analysis of triangle effect with electronic and air gaging devices is mentioned. Since these measuring devices are more sensitive, they are chosen when any minute amount of triangle effect must be detected.

As for other relationships, Fig. 65 points out diagrammatically how conditions of ovality, taper, squareness, and alignment may be analyzed quickly, with practically a single gaging movement, by using combinations of indicators on special design gages.

If two indicators register through special mounting at 90 degrees on a workpiece as shown in sketch A of Fig. 65, the degree of ovality can be seen at once. Taper likewise can be diagnosed in a single pass by sliding the workpiece on a platen under a pair of indicators as diagrammed at B. Squareness is confirmed by two indicators as at C. Finally, two indicators can quickly point out lack of alignment or concentricity as illustrated in sketch D. Here the cup-shaped workpiece must be held and turned on a fairly close-fitting mandrel of some sort.

Special Indicating Gage Applications

A somewhat unusual use of indicating equipment, where there is sufficient production to warrant it, appears in Fig. 66. The relationship being checked is hole center location. The gage consists essentially of

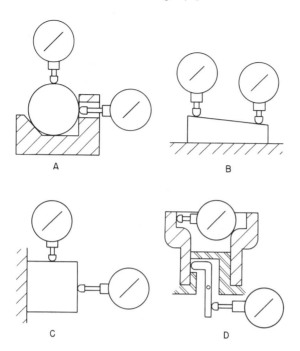

Fig. 65. (A, B, C, D) Conditions of ovality, taper, squareness, and alignment may be analyzed in a single gaging operation by using combinations of indicators in specially designed gages.

two precision plugs, one of which is solid to the frame — the reference plug — and the other of which moves laterally — the measuring or locating plug. The movement of the sensitive member registers on the indicator. The gage is, of course, "mastered" on a piece or master whose holes are correctly centered and the indicator is zeroed. As the gage is tried in each workpiece, the indicator hand position away from 0 shows the amount of error in center distance.

Measurement of Soft Materials

Inspection, of course, is not an operation exclusive to metal cutting plants. Inspections are necessary on plastic materials, rubber, paper, textiles, wood, and other products. While many things we buy are made of metal, their assemblies often include rubber, cloth, or plastic, and the manufacturer must solve problems in purchasing, producing, inspecting, and controlling materials other than steel or brass.

The inspection of articles, components, and sheets of material such as cardboard, felt, and gasket compounds, as well as the materials

Fig. 66. Special indicating gage which measures the error in the center distance of
two holes.

suggested above, many times involve gaging and measurements. But
the gaging of yielding materials — take sponge rubber as an
example — demands techniques somewhat different from those already
discussed.

For this purpose, gage manufacturers can supply instruments
equipped after the fashion of the gage illustrated in Fig. 67. Such gages,
as Fig. 67 shows, have their gaging pressure supplied by means of a
dead weight riding on the spindle or rack rather than by means of inner
tension springs. The size of this weight can be specified to the fraction
of an ounce or gram. This type of gage is usually equipped with a lifting
lever so that the spindle and the sensitive contact can be raised away
from the solid or reference anvil. Both anvils are usually flat and lapped
parallel to each other. The standard comparator's spherical point will
not do on soft materials because the ball shape so readily deforms the
material being tested.

Another factor to be considered in the case of a gage for measuring
soft materials is the area of the anvils, particularly the area of the upper
or sensitive contact. The effect of the gage on a compressible substance

Fig. 67. Indicating gage equipped with a weight to provide uniform gaging pressure on soft, easily deformed material.

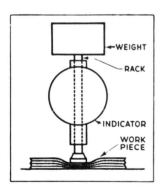

Fig. 68. Effect of gaging pressures on the thickness dimension of a soft workpiece.

sketched in Fig. 68. The amount of compression varies not only with the amount of weight bearing down on the spindle but also with the area or cross section of the movable anvil. In many instances the thickness of the material is determined by measuring several layers of it at a time — several thicknesses counteracting the gage pressure more than a single ply — and then dividing the reading by the number of layers to secure an average figure for the thickness of the single sheet.

Specifications for this type of gage design and the technique to be employed are usually worked out by a factory's engineering department. Where definite specifications are lacking, and the inspector is faced with the necessity for making a measurement that will indicate the true dimensions of a compressible material, he can be guided by the following suggestions in contriving some suitable apparatus from available equipment.

Gaging Soft Materials Correctly

See that the gaging pressure is as light as possible, but still positive. Provide an upper anvil with an area between one-half square inch and one square inch, the upper anvil to be parallel to the platen.

In using such apparatus, the ability to get repeat readings is the objective. Usually, when soft material is gaged, the upper anvil will sink into it a little and the indicator will show some reading. If the gage is left untouched for a few minutes and another reading taken, the second reading will frequently differ from the first one. The upper anvil, in other words, has been imperceptibly sinking into or compressing the material further. If the gage is watched, there will come a time, after a given compression of the material, when the gage reading no longer changes. This is the reading which should be taken, especially if it can be repeated in another similar trial.

This stabilized reading can be obtained more quickly, perhaps, by reducing the amount of weight on the gage spindle or by increasing the anvil area, or both. If repeat readings cannot be obtained, there is probably insufficient gaging pressure. The thickness reading obtained may not be the dimensionally true thickness of the sheet but such readings may be sufficiently accurate, especially when used as comparisons between different sheets, to satisfy the requirements. Another method is to master the gage with a sheet of the material that has been judged to be acceptable by some other test (say at assembly or in the laboratory) and then to compare the sheets to be inspected with this sample or "master."

One other warning. Don't persist in trying to measure or compare what simply can't be measured with the means at hand. If repetition cannot be secured or if the true measurement means nothing when it is secured (because the material becomes too distorted or compressed under gaging pressure), turn to some other technique. Too many times, inspectors solemnly go through measuring operations which are fruitless as far as measurement or control are concerned.

The ideal, of course, in connection with compressible materials, is some method of measurement that secures results without actually imposing gaging pressure or even contacting the work. Many times optical means are devised for this purpose. Noncontact gaging systems, so-called, are being developed in order to meet the needs of gaging many soft materials more precisely and accurately. Laser-based systems and other optical-dimensioning systems are available for particular applications. Additionally, many new electronic measuring systems, some of which do employ sophisticated contact-type gaging heads, can offer predictable and accurate measurements of soft and varying materials through the use of special probe materials and shapes.

Continuous Measurement With Indicating Gages

There are many occasions in inspection work where the gaging and inspection must be performed right at the processing station. Figure 69 shows the type of gage used at a wire insulating machine. The type of measurement illustrated is known as continuous measurement. The production process in many industries — wire, paper, rubber, sheet metal, textile mills, and sheet plastics operations — is continuous right around the clock. The product is shipped in coils, rolls, rods, or sheets — the rolling, the piling of sheets, or the bundling of rods also being accomplished very rapidly by automatic machinery. It is commonly difficult for the inspector to get at individual units of such products for visual inspections and gage checks. Certainly it is heresy in modern American mill production to ask for the machinery to be stopped momentarily so that an inspection of the product can be made, hence the type of on-the-job, continuous measuring gages of the general principle illustrated in Fig. 69.

Wherever metal, plastic, rubber, and the like are passed through rolls or extruding dies in such continuous processes, there are certain principles in connection with gaging the conformance of the product to specifications that the inspector should have firmly in mind.

The material rarely leaves the machine with absolutely flat and parallel surfaces as intimated in diagram A of Fig. 70, instead, it contains so-called local bumps, humps, hollows, pits, and waves, as shown somewhat exaggerated at B in Fig. 70. Many manufacturers who purchase raw sheet or strip stock to use in their own products find themselves forced to reroll, flatten under hydraulic pressure, grind off, or, in some manner, redress the surfaces so the stock is more like the ideal of Fig. 70-A before using it.

If the measurement of the stock is made with roller gages like those shown in Fig. 69, the gage's indicators will, of course, vibrate back and forth, sometimes so rapidly as to be practically unreadable unless the indicators are equipped with electrical or mechanical damping devices. In such cases, the measurement obtained is approximately the average

Courtesy of Federal Products Corp.

Fig. 69. Indicating type of gage which gives continuous measurements of wire diameter while the wire is in motion.

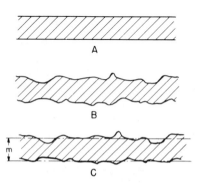

Fig. 70. Two types of surface which the measuring contacts of a continuous indicator must measure across, (A) smooth surface, (B) rough surface for which the indicated measurement will be the average value *m* as shown in (C).

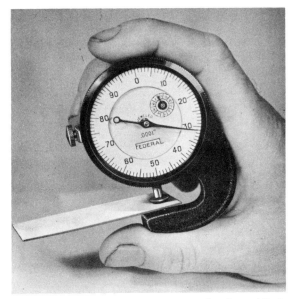

Courtesy of Federal Products Corp.

Fig. 71. One type of indicating gage used for measuring the thickness of short lengths of strip stock.

Courtesy of Federal Products Corp.

Fig. 72. Another type of indicating gage for measuring the thickness of sheet material.

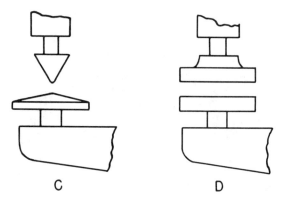

Fig. 73. Gaging contacts of the type shown in (A) will measure the minimum thickness of flat stock. Contacts of the type shown in (B) will measure the maximum thickness of the same stock.

thickness of the moving sheet as suggested by dimension m in Fig. 70-C.

Where units — sheets, rods, or short lengths of strip — happen to be available for measurement and inspection, a gage like that shown in Fig. 71 is useful or the more precise instrument of Fig. 72. But the same condition or averaging over high spots, low spots, and waves prevails.

If it is necessary to determine by measurement the thinnest section, another principle in gage equipment must be adopted. In the case of insulation coating on electric wire, for example, it may be essential to know how thin the thin spots of insulation are so as to calculate its minimum dielectric or insulating strength if current at high voltage is to be transmitted. There are also situations, especially in assemblies (the thickness of gasket stock for instance), where the maximum thickness must be determined.

The gaging solution is fairly simple. For thin sections the combination of cone point and a fixed anvil with a small area of flat as in Fig. 73-A will dig down into the hollows. If it is the maximum thickness of stock that is required, the gage anvils should be flat, parallel and wide enough, as in Fig. 73-B, to ride on top of the bumps and the waviness cycle of the stock.

CHAPTER 10

Electrical Gaging Equipment

The development and growth of electrical and electronic gaging apparatus, circuits, and systems has been so rapid that the inspector unacquainted with the attributes, uses, and applications of electronic gaging is as much behind the times as if he still drove a horse and buggy.

Today, many manufacturing processes and operations have gone "automatic." Automation is a "household" world. Machines that are electrically operated and electronically controlled are in all factories. Nevertheless, inspection and measurement are still required functions although they, too, are being upgraded and made automatic with the aid of electronic equipment.

Consequently, the modern inspector will want to be as conversant with electric and electronic gages — their use, setting, checking, and care — as he has been with verniers, micrometers, and mechanical indicating gages. Here again the field is so wide and varied and is expanding so rapidly that only its basic principles and conceptions can be touched on in this text. More than ever the inspector will find it necessary to study gage manufacturers' catalogs and instruction books for exact details concerning the proper operation of the particular make of electric gaging equipment he is dealing with.

Today's (and tomorrow's) inspector is liable to find himself using, setting, adjusting, or maintaining the simpler electric and electronic comparators. He may be responsible for the operation of automatic sorting gages, which vary in bulk from a unit about as large as a typewriter to a piece of machinery which would fill a large room.

Again, he may be setting and checking machine-control gages on automatic machine operations. These vary in size from a relatively small, automatically operated caliper gage to a complex system of automatic control gages, measuring ID's and OD's, and dimensionally controlling automatic machining operations that stretch several hundred yards along a shop floor.

In the following pages the reader's attention is called to the rather careful distinctions made between *electric* and *electronic* gaging.

Most electrical and electronic gages work by translating the movement of a mechanically contacting gage head or contact lead into an amplified electrical signal. This amplified signal is then read or displayed from an indicator of some type that illustrates the proper scale and reading on that scale. As with mechanical comparators, these electronic devices typically translate a direct contact with the gaging head into an amplified electronic signal that is displayed on a measurement scale, because the ratio of the (gaging head) movement to the electronic signal is understood. Based on the resolution of the gaging setup, the actual workpiece variation is usually displayed on an incremental or digital scale.

This assessment of "measurement" through variations in an electrical field or displacement has made possible, over the years, the measurement and discrimination of increasingly finer increments. These improvements manifest themselves in more and greater amplification of the measuring devices — thus greater sensitivity or resolution, and potential for discrimination, when they are used properly and maintained and calibrated regularly.

Additionally, electronic measurement systems are now commonly integrated into control devices through computerized assessments of the variation that is being detected or through programmed parameters for a machine or production tool. In other words, electronic gaging is not only less time consuming and increases efficiency, precision, and discrimination in measuring, but it can also be "plugged" directly into those pieces of hardware that maintain control over variation in the manufacturing environment. Reductions in error and feedback time ultimately results in greater cost savings, because the inspection routine provides detailed dimensional data and, therefore, control immediately.

Solid-state devices have all but replaced the early electrical translations of the measured gaging movement; now, the power of the vacuum tube is completely encapsulated by the refinement and miniaturization of transistors and integrated circuits. Still, all electrical and electronic measuring or gaging systems contain the same basic constituent elements.

Basic Units of Electric Gaging Equipment

One of the simpler forms of basic electric equipment is illustrated in Fig. 1. It is essentially a dial indicator that can be mounted in the usual fashion on a comparator or gage base. Adjustable microswitches have been assembled in the little chamber appearing just above the indicator base, switches that are actuated by the motion of the indicator spindle.

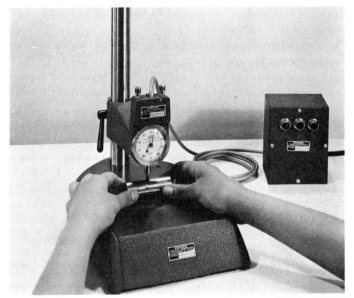

Courtesy of Federal Products, Inc.

Fig. 1. Single-purpose type electric comparator with signal light limit indicator.

As the microswitches are closed or opened by the indicator spindle travel, they will light the warning bull's-eye lights in the "power box" shown (Fig. 1) electrically connected to the indicator.

The microswitch contacts can be adjusted by the thumbscrews above the indicator in such a manner that when the indicator hand reaches prescribed tolerance limits the preset switches close. A yellow light usually signals undersize work and a red light oversize, while, frequently, a green light continues shining as long as the work being measured is within specifications. The electric current can be made to ring a bell or buzz a dial tone in a headset instead of or in addition to lighting a bulb. Blind people have become highly successful inspectors and sorters using such sound equipment where different tones designate go/no go; undersize or oversize work.

The simplest of electrical gages, which are still used in many areas of industry and inspection, directly activates a switch by mechanical contact. The accuracy of switches used in this type of electric gaging is within about .00025 inch and the lever ratio of the switches provides a magnification of only about 2 to 1.

These inexpensive electrical gaging devices represent go/no go classifications by activiating sets of single pole, double-throw switches that indicate a measurement that is too large or too small depending on which way the switch is thrown; no contact means the piece is within the specification cutoffs.

A typical electrical gage in use for some time adds the dimension of variables or real-number measurements to the distinction of some go/no go classifications. This gage will display a red light/green light for small or undersized pieces, and can also delineate the true size of a part on a dial or digital readout. In fact, smaller, faster, and less expensive electronic versions of these gaging setups are quite popular today with the renewed interest in statistical process control by line operators. But the older electrical versions of this are extremely limited in their discriminatory capabilities and range, and have, for the most part, been effectively superseded by newer electronic versions of devices for similar purposes, with increased discrimination through greater amplification.

Electric gages of the type of apparatus shown in Fig. 1 are also found on multiple checking gages, on automatic sorting gages, and on machine control equipment. Once a microswitch is closed, the current impulse can be amplified to perform many services the gage designer desires.

The same design principles are also used in multipurpose gages of the kind shown in Fig. 2 where several dimensions on a workpiece are checked simultaneously. This type of equipment involves a group of measuring contacts complete with microswitches and warning lights. Since, as the illustration shows, the lights are often "patterned" to an outline of the workpiece painted on the light panel, this sort of gage is sometimes referred to as a "picture-panel gage."

The types of gages just described are solely go/no go in operation; the lights only tell when an upper or lower limit has been reached. They have, however, the advantage in use of being speedier than normal indicator gaging. The eye perceives the lighted bull's-eye much more quickly than it can follow an indicator hand to a certain dial graduation; the human reaction to accepting or rejecting a workpiece is much more rapid under the stimulus of the flash of light.

In setting up such gages for use, first follow the manufacturer's instructions as to adjusting microswitch screws and other suggestions. Other than this, the customary setup precautions previously described for single and multiple mechanical indicator gages are to be followed. The microswitches and all the electric apparatus in this type of gage are

Courtesy of Sheffield

Fig. 2. Air-electric, multiple-dimension gage with signal light limit indicator. Thirteen different dimensions are being gaged simultaneously. No signal light indicates dimension is within tolerance; a green light indicates dimension is undersize; and red, oversize.

designed and made to allow an error in measurement not exceeding 5 millionths of an inch. Hence, if such a gage is not accurate in action, look to mechanical causes.

If, however, the warning lights or signals fail to function and if the mechanical action of the gage has been thoroughly checked, the services of an electrician may be required. However, embarrassment may be saved by being sure, first, that the electric gage has been plugged into a live factory circuit. Peculiarly, many service calls on electric gages end up with the report that the gage was receiving no electricity! Electric gages are designed so that only a milliampere or so of current goes through the switch contacts (or sparking is dampened with condensers) and most of them will operate at least a thousand hours before a contact corrodes or a component wears out. On the other hand, dust, dirt, coolant, oil, and moisture can creep inside, and switch points and connections may have to be cleaned free of these enemies of electric conductivity.

The Electronic Comparator

The history of precision measurement displays many types of instruments and devices with a variety of discriminations or accuracies. Back in colonial days, Vernier added his invention to the steel rule, capable of measuring as close perhaps as 1/64 inch or 1/100 inch, and gave artisans and mechanics an instrument that would detect a size difference of 1/1000 inch. Something like a century later the micrometer was invented which brought us measuring precision in terms of 1/10,000 inch.

Within another half-century Johansson put his label on the "Jo" block which, while it gave us a measuring standard varying only two or three millionths of an inch from the actual, could not be handled consistently without a probable error in manipulation ranging as high as 50 millionths. Even mechanical dial indicators, developed soon after Johansson's invention, cannot ordinarily be trusted to accurately detect size differences of less than 50 millionths (.000050 inch). In the meantime, too, progress had been made with interferometry and the optical flat (described in Chapter 12) with its resulting discrimination of .000012 inch.

With the advent of the electronic comparator — so-called because of its accompanying electronic circuits and devices — we are able to read a size difference of one millionth of an inch (.000001 inch) on a dial or meter scale.

In these more modern gages, electricity is successfully employed to transform the motion of a sensitive or measuring contact to a readable, accurate figure. Some electronic gages make use of the increase or decrease in reactance when a plunger actuated by the measuring contact moves in and out of an enlarged solenoid. In another type, the spindle motion changes the positions of high-frequency coils in relation to each other. Others make use of a change in capacitance and still another is based on unbalancing a Wheatstone Bridge circuit. In any case, the electrical change caused by any movement, minute as it may be, of the measuring contact is picked up and amplified by a circuit somewhat resembling those used in radios and then converted to visible readings on a dial.

The first and foremost component in these systems consists of the contact or probe point, typically a transducer, which transmits direct or proportional information about a measurement into an electrical signal. These contact points can be made from many different types of material, and their specific configuration will also vary tremendously with the application, the expected degree of discrimination and accuracy, and the material that is measured. The technology that drives changes in these probe points is advancing daily.

Second, the electronic comparator contains a device or series of circuits that amplify the contacted or discerned variation — a modification of the electrical signal that is sent from the transducer. These circuits and their attending "connections" actually see no limits electronically — their potential for amplification and expression of workpiece variation is tremendous. This part of the system is, however, limited in scope of amplification and resolution by the physical interface between the contact point and the workpiece.

Finally, a display instrument, sometimes similar to those used in a mechanical indicator or comparator, is essential to the human interpolation and understanding of the variation that is being assessed by the electronic device or machine. Digital incremental display of variation is of course very popular. This display instrument also be bypassed in favor or increased automation and direct recording or machine control through electronic means.

Advantages of Electric and Electronic Comparators

It has been suggested, previously, that indicating comparators of the mechanical type have a discrimination in measurement down to .0001 inch, special designs being obtainable that will measure to .00005 inch. Electric comparators have comparable discrimination. It is the measurement niche in which tolerances are below .0001 inch that the electronic comparator particularly fills, although it is a perfectly satisfactory instrument for use on work in which tolerances exceed a tenth of a thousandth. No other instrument so satisfactorily, quickly, and easily handles measurements where the specifications are for .0001 inch tolerance or less. Accuracies to .00001 inch or to microinches, .000005 inch (and even .000001 inch), can be obtained.

In addition to superior accuracy, the electronic comparator possesses two other favorable characteristics. It is faster than the usual mechanical indicating gage; the electric meter flashes the correct reading faster than the hand can adequately or correctly manipulate the workpiece and faster than the eye can comprehend the signal. The other property is the lack of wear. True, there are moving parts in connection with the sensitive contact or spindle, but the manufacturer of electronic gages is able to so design the mechanical moving parts and the electric relationships in the head of the instrument that no appreciable error is detected from what might be called the use or wear of the instrument. Gear trains, bearings, bushings and the like in mechanical type indicators do wear, of course, and an equivalent instrument error is then noticed.

In addition to the types of electronic comparators appearing in Fig. 3, the gaging industry also furnishes extra sensitive, extra accurate

Courtesy of Federal Products, Inc.   *Courtesy of Brown & Sharpe Mfg. Co.*

Fig. 3. (Left) Commercial type of electronic comparator being used to measure external diameters. (Right) Another electronic comparator with a special adaptor for gaging internal diameters.

"gaging machines," so-called, for outside diameter, length, and, particularly, superaccurate inside diameter measurement. Two of these are shown in Fig. 4.

The results of measuring and translating electronically were significantly enhanced by the (Wheatstone) inductance-bridge circuit gages. Variations of this type of gaging head enjoy extensive use today. These devices use a mechanical contact, which is a reed-floated armature between a pair of coils. As the armature physically reacts to dimensional changes on the workpiece, the reactance of the circuit can be read directly on a meter. This translation of a circuit imbalance can be read directly on an indicator down to .000025 inch with no additional amplification of the signal. The addition of signal amplification makes this method a mainstay of many electronic measuring machines.

Electronic gaging systems have the advantages of supplying a continuous analog output, which, when amplified, can be presented as a dynamic, real-time output or display that minutely describes all

Courtesy of Colt Industries,　　　　　Courtesy of Federal Products Corp.
Pratt & Whitney Machine Tool Div.

Fig. 4. (Left) An extra precise electronic gage or measuring machine. (Right) Close-up view showing zero comparator reading being checked with a gage block set-up.

changes or variation on the piece being measured, or along a continuous production system. This analog display can of course be converted to a digital scale based on a specific resolution level.

Switching on electronic measuring machines is electronic, not mechanical, and solid-state circuitry greatly increases the reliability of the units, making the mechanical/electrical interface much more troublefree. Solid-state circuitry includes devices, such as transistors, resistors, capacitors, inductors, transformers, and crystals, that are interconnected on simple and complex integrated circuits.

The circuits that amplify these electronic translations of mechanical change or measurement tend to cost less, work better and longer, are smaller, and require less time for set up and actual measurement and display or recording as improvements are made.

In order to better understand the principles of these measuring systems, which are here to stay, and constantly changing, an investigation of the common *types* of electronic gage heads would be appropriate.

The lever-type of gage head, one of the simplest, consists of a dial test indicator support fixture and base, a spindle tip, and a suspension system that allows the tip of the lever to pivot. In this setup, electronics simply assumes the role of some of the more purely mechanical aspects of the same configuration that would use a dial indicator. The relationship between mechanics and electronics is one of facilitation and cooperation.

The second type of gage head is a reed-spring type — a contact block connected with frictionless reed suspension; this time the device operates with axial displacement of the contact head. This type of head offers a great variety of applications, and is similar in many ways to the third type of electronic gage head — the cartridge or plunger type.

The cartridge type of gage head also offers in-line, or axial, displacement of the contact mechanism. In the case of this cartridge-style mechanism, though, the suspension of the plunger is usually provided by spring steel discs or a strictly controlled helical spring.

No matter what type of contact or displacement method is used in electronic gage heads, they must have some mechanism that translates that (physical) displacement into a coherent and regular (analog) electronic signal. In many cases, this method of translation is with a transducer — a device that converts physical motion into an output voltage.

A common type of transducer design is the linear-variable-detector transformer (LVDT). An LVDT produces an output voltage from a supplied alternating current that is proportional in amplitude and phase to the movement of the core, plunger, or lever. Primary and secondary coils are located around the movable core. When the magnetic core moves, changes in the voltage through the coils is induced in a very precise fashion — so that these changes in voltage can be related to dimensional variations or movement of the core.

Other electronic gaging systems use a variation of this principle, but the notion remains precise and impeccable — extremely small physical changes or motions produce a discernible change in voltage that can be amplified and properly scaled to announce the dimension represented by the movement and corresponding voltage change.

The flow of electricity through the coils of these gaging heads is assessed from a so-called "balance point" or "null position," wherein the output voltage through coils, based on the respective position of the contact point to the coils, is zero. When the plunger or contact point is moved away from this null position in either direction, the associated measurement may be described in terms of positive (+) movement or negative (−) movement, equal in opposite directions.

With this feature or characteristic, electronic comparators have the advantage of fast "zeroing," without having to manipulate the setup physically — it can be done electronically. Likewise, adjustments or changes in the setup of the configuration, or multiple measurements can be accomplished quickly. Adjustments to different scales of measurement, or tolerance limits, can be entered and rapidly measured because of the flexibility of the "electronic" points of reference, or position, on the machine.

Electronic comparators and gaging systems, once again, are used mostly in much the same fashion as a mechanical gaging setup is constructed. The gaging head functions just as a dial test indicator or other mechanical gage would function. But the measurement or reading is taken from the amplifier — the unit that is connected to the gage head which displays the electronic translation. This translation can be changed quickly from metrics to inches, in the digital mode, or displayed on an appropriate analog scale.

Needless to say, apparatus of this kind is costly. Although such gaging machines are designed and built to detect a size difference of .000001 inch, the inspector, unless he is specially instructed and practiced, may find it difficult to achieve an accurate reading precisely to a millionth of an inch — a microinch. An inspector assigned to using a measuring machine should, if at all possible, be instructed by the manufacturer of the equipment in its exact use.

However, the newest in electronic gaging is also designed to be easy to use, and reliable and consistent in terms of producing accurate measurements — in spite of the operator interface. The trend to in-process and on-line measurement *and* control has assisted in this recent trend. It is now necessary for fairly painless and automatic readings, with greater discrimination, to be taken by manufacturing employees who are often less skilled than professional inspectors.

Not only must the measurements be accurate and reliable, they must be readily available and transferable into usable data. So, more often than not, the act of measuring also includes the act of *taking* data, or recording it, and actually processing these data, so they can be used for control purposes. This whole sequence of events needs to be as foolproof as possible, and most modern gaging is designed to contribute to these enhancements. Now that the world of discrimination, accuracy, and precision has been conquered, gaging instrument manufacturers are turning their improvement efforts to making the machines more foolproof and easier to use, and able to deliver more information based on the measurements that have been taken.

Precautions in Using Electronic Comparators

Because electronic gages are capable of measuring to .000005 inch, the inspector must display extreme care in manipulating masters and workpieces on the solid reference anvil and under the sensitive contact. It takes mighty little "cramping" of the part being measured, only a minuscule of mishandling, to produce a .000005-inch error in measurement. In fact, the electronic instrument so readily amplifies and magnifies minute dimensional variations that even an otherwise experienced inspector using it for the first time will frequently claim the instrument is inaccurate. Many an inspector and mechanic, when starting to use electronic gages, has come to realize for the first time his habitual carelessness and ignorance in the manner of holding and manipulating workpieces and gages.

More than usual precautions must be observed in making ready to measure with an electric comparator. Make sure the instrument is not exposed to direct sunlight, steam radiators, open doors or windows, or other forms of draft, nor to excess vibration. The reference anvil should be frequently checked for flatness, scratches, nicks, and corrosion. The solid reference anvil and the clamping device for the adjustable head must be tight. The real meaning of instrument looseness, deflection, and the effect of uncontrolled temperature changes becomes evident when sensitive electronic equipment is used.

The standards of cleanliness applied to the reference anvil and essential parts of an electric comparator are the same as that applying to gage blocks.

Most electric and electronic gages are designed and built to measure accurately at some standard voltage, usually 110 volts. If, after an electric or electronic gage has been mastered, after its zero and tolerance limits for the particular work at hand have been established, the supply line voltage should drop more than 10% (1 or 2 volts), or surge up a volt or two, the gage should be remastered. An electric or electronic gage which has been set at 110 volts, may show an appreciable error if the voltage drops to 105.

Figure 62 in Chapter 9 shows the reed-type instrument, which employs an electric bulb — with mirrors or prisms to flash the reed movement onto a scale — and is frequently referred to as an electric gage. It must not be confused with the electric gage, or electronic gage, previously defined, however, which employs electricity directly, and not merely an electric light beam, for amplification or magnification.

Automatic Sorting Gages: Economics

In certain operations, especially on small parts in screw machines and presses, it is sometimes found less costly overall to maintain fairly good control over quality, but to allow the quantity of rejects to run as high as 5 or 10% and then to screen out the defectives with an automatic sorting gage. Such a plan, under certain conditions, will make the final cost of the machined components lower than their cost would be if an attempt is made to hold the rejects and scrap at each machine down to 1% or less. In the latter case, machine down time and the extra inspection force required to check or screen the work sometimes make the piece cost higher.

Hence, plain manufacturing costs and economics often recommend the introduction of automatic sorting gages in place of line or patrol inspections, and, consequently, the inspector finds himself charged with the responsibility of setting and checking an automatic sorting gage and overseeing its operation.

Another factor of plain economics brings about the use of automatic gaging. The production of the familiar automobile engine wrist pin offers a concrete example. One way is to set and constantly police a grinder, and to have an inspector constantly checking its output, so that each wrist pin is within a tolerance of .0002 inch — practical duplicates of each other. This method requires an operator on each grinder and considerable machine "down time," plus several inspectors checking a battery of grinders.

The more modern way is to widen each grinder tolerance to, say, .002 inch and then have the output from the line of grinders put through an automatic sorting gage of the type pictured in Fig. 5. Here the pins are automatically sorted into size classifications .0001 inch apart and marked. This method reduces down time at each grinding machine and usually allows an operator to take care of several grinders instead of one — not to mention a reduction of inspection time and expense.

The holes in the corresponding pistons are bored under similar wider tolerance conditions and are similarly size-classified, as shown in Fig. 6, and marked. At assembly, a certain size pin is assembled into piston holes of the same size classification. This results in even closer

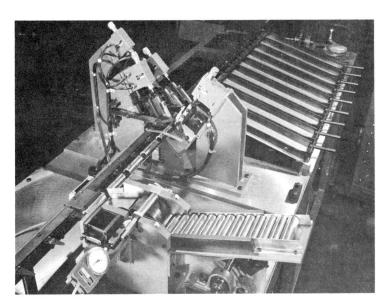

Fig. 5. Automatic electronic sorting gage being used to classify wrist pins by .0001-inch increments.

Courtesy of Sheffield

Fig. 6. An automatic electronic gage capable of sorting and marking pistons classified as to wrist-pin holes sizes.

fits than under the older method, greater engine life and a quieter automobile. The method described is generally known as selective assembly. Substantial cost savings are effected not only at machining operations but also at assembly. Sorting and selective assembly is practically a must in the ball bearing industry, for instance, because of the extremely close assembly tolerances required.

The third type of economy acquired through the use of automatic electric sorting gages comes from those situations where, for a number of good reasons, parts or components must be gaged and inspected 100% and where it has been the practice to accomplish the screening through squads of inspectors doing the work manually. Sometimes a visual inspection is required along with the piece-by-piece manual gaging.

Under such conditions, substantial savings in time and effort are made by introducing automatic sorting gages. Depending on the nature of the product or component to be screened by automatic gaging, the gage may replace the effort of a number of 100% inspectors, freeing them for other and more valuable work in the plant. An automatic gage can do the work, in many circumstances, of a dozen or more inspectors and get the job done more quickly and more accurately.

Figure 7 pictures an automatic sorting gage performing, in a sense, a dual operation. Here the single operator is performing a 100% visual inspection for burrs and defects on typewriter bars and at the same time feeding them into the automatic gage which makes the essential measurements, sorts the bars into specified classified sizes and rejects those which are oversize and undersize.

Courtesy of Federal Products, Inc.

Fig. 7. An inspection operation that combines manual, visual inspection with automatic size gaging and sorting of typewriter parts.

Automatic Sorting Gages: Installation, Setting, Care, and Checking

Pieces of equipment of the size, intricacy, and cost of automatic electronic sorting gages are often, if not usually, selected, bought, and installed without the Inspection Department being consulted or notified. Such proceedings are generally assigned to Engineering. But, just as often and in fact, usually, the routine daily setting, checking, and care of such equipment, after it is installed, becomes in a short while, by a normal process of industrial delegation, the responsibility of Inspection.

To repeat a previous suggestion or warning, the inspector should first make himself thoroughly conversant with the gage manufacturer's instruction books and catalog material, pertaining to the particular automatic sorting gage, for detailed information concerning how to set, adjust, check, and maintain the gaging machine.

The inspector can and should back up the gage manufacturer's instructions with his own fundamental knowledge of gaging principles. After all, the primary function of the electric–electronic automatic sorting gage is correct measurement. Setting, checking, verifying the accuracy of this function on an automatic gage is essentially no different than for any hand or bench gage.

For example, Fig. 8 offers a close-up of the measuring contacts of the wrist pin gage spoken of just previously and pictured in Fig. 5. The gaging machine's conveying mechanism automatically feeds successive wrist pins to the measuring contacts and carries them on by. One pair of measuring contacts checks diameter and out-of-round. (A special device revolves the wrist pins under the measuring contacts at this point.) The next pair of measuring contacts checks taper.

Thus, it might be said that the gaging machine substitutes for a pair of hands in that it "feeds" the pieces in under the contacts and revolves them, holds them down to the reference surface or platen and prevents the workpieces from tipping or producing any manipulation errors, in

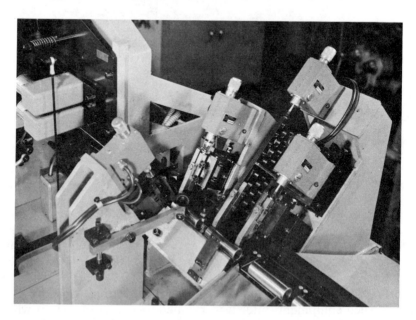

Fig. 8. Close-up view of wrist-pin measuring contacts for sorting machine shown in
Fig. 5.

much the same fashion an inspector would work were he manually checking such work under a bench comparator. Hence, in setting, mastering, checking, and maintaining an automatic sorting gage, the inspector treats each measuring unit much as if he were dealing with individual bench comparators.

Periodically or at every reasonable opportunity the inspector should check the platens for roughness, scratches, wear, and particularly, dirt. Similarly, the movable or sensitive contacts would be examined with a close watch being kept for flat spots or undue distortion of the contact profiles from wear. The gaging pressure a movable contact exerts on a workpiece can be adjusted — be sure it is correct for the purpose. Be sure adjusting screws and other parts are tight; looseness may spoil the gage's accuracy and repetition. Follow closely any other instructions the gage manufacturer may offer.

With the gage in operation observe the way the pieces are "feeding" through the measuring stations to be sure the conveying mechanisms are not in any manner cramping or misaligning the workpieces or otherwise creating inaccurate readings. The gaging machine should be free from undue vibration, thumping, or jarring (such as may result from nearby machinery), which could interfere with its precise measuring function. The workpieces should be clean — so free from grit, oil, or coolant that such foreign material does not interfere with getting a correct measurement nor wear the gage contacts.

In most cases where an automatic sorting gage is in use, the inspector would provide himself with a separate hand gage or bench comparator whose accuracy and discrimination are equal to or better than the rated discrimination of the automatic gage. With such equipment he can periodically check the output of the automatic gage and either assure himself of its continued accurate operation or get a warning that something may be amiss in the automatic gage's measuring function. For instance, in checking the wrist pin gage, Fig. 5, the inspector would make independent checks on sample workpieces from each size classification chute, in this way not only verifying the measurement accuracy of the gage but also making sure the electronic classifier equipment is functioning properly.

Machine-Control Gages

One of the forerunners of the present-day automatic machine-control gage is the type of indicating caliper gage shown in Fig. 9. The gage is swung down on to the workpiece, after the latter is centered or chucked and the machining action started, and the gage's caliper is

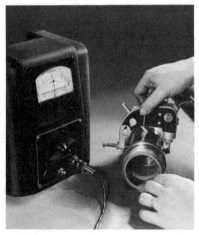 

Courtesy of Brown & Sharpe Mfg. Co.   *Courtesy of Federal Products, Inc.*

Fig. 9. (Left) An electronic caliper gage for hand operation. (Right) Electronic caliper gage which measures as the work piece is being machined.

engaged on the OD. The gage (except in the case of plunge grinding) traverses the work with the cutting action and its indicator shows the amount of metal being removed at each traverse. The machine operator watches the indicator, which tells him when the work has reached specified size.

Where such gages are in use, the amount of inspection required on the work put out is reduced if not practically eliminated. However, it is wise to have an inspector check the work at the start of a run and occasionally thereafter to be sure that the machine operator is reading the indicator correctly and also that the operator, through neglecting the finish size, is not getting oversize work from stopping the machining too soon or putting out undersize work from not acting quickly enough.

A gage, and its indicator, like that pictured in Fig. 9 can be originally set or "mastered" to correct finish work size with a disc master or the equivalent in the same manner any indicating caliper type comparator gage is mastered. The general practice, however, is to machine the first workpiece, checking its diameter from time to time with a separate hand gage, until it is at specified finish size. With this workpiece still revolving, but without any cutting action taking place, the machine caliper gage's indicator is set to finish size. Thereafter, each successive workpiece is machined until the indicator signals its finish size. Finished workpieces are checked periodically and independently with a separate hand gage, as previously suggested, not only to verify work size but also to determine whether the caliper gage contacts are wearing back unduly.

Electric Attachments for Machine Control Gages

Another step in "automation" progress, in so far as the contribution of gaging is concerned, has been that of adding the microswitches to the indicating caliper gage previously described. An example of this sort of equipment is shown in Fig. 10. The size-limit electrical impulses from this equipment are used to light warning lights or, through amplification, to stop the machine at the proper time or control the

Fig. 10. This electric caliper gage not only indicates but also controls the machine operation through the attached microswitch.

machine feed mechanisms. Figure 11 pictures another step in machine control gaging where the caliper on the revolving work transmits size signals through air–electric apparatus to control the operation of the machine and the finish size of the work.

Oftentimes the gage system is an integral part of the machine, designed specially for it and built into it. In Fig. 12 you see a practically completely automatic machine that loads itself, chucks and unchucks the work automatically, and machines it. The gage, just at the left of the chuck, takes over the finish-size control function.

Although a great deal of ingenuity has been displayed in the creation of such machining automatons, a certain amount of independent inspection is still valuable. By periodically checking samples of the

Fig. 11. A combination air–electric gage that operates to control the machine operation and the finish size of the work.

work produced, including a piece or two as each job is set up and starts off, the accuracy of the setup and the continuation of that accuracy can be verified. Only under the impossible combination of perfect men using perfect machines on perfect material would consistently uniform and precise work be produced hour after hour.

Additional Precautions to Observe with Automatic Gages

On the more or less fully automatic machine, represented by the equipment pictured in Fig. 12, where the workpieces are loaded, machined, and unloaded automatically, the caliper gage is moved out of the way, also automatically by solenoid or air cylinder mechanisms, during the unloading cycle. After the new workpiece has been automatically loaded and machining starts, the same solenoid or air cylinder mechanisms set the caliper on the work, automatically, in measuring positions. When the work is at specified size, the gage signals the retract–unload–load cycle to start.

Like anything subject to friction, pressure, dirt, and wear, the gage actuating mechanism can fail, until corrected, to place the caliper on

Fig. 12. A fully automatic machine with automatic loading and machining size controlled by electronic gaging.

and over the workpiece in proper alignment. If the machine is failing to produce work to specified size, in other words, the inspector should look to the gage handling apparatus, as a possible source of trouble, in addition to the customary search for looseness, binding, or wear in the caliper itself or for error in setup or mastering. Sometimes in a similar manner, the machine's automatic loading mechanism, due to chips, dirt, friction, or wear, may fail to "set" the workpiece correctly for machining and create a condition where it is impossible for the machine to bring the work to size.

Since the inspector is often charged with the responsibility not only for detecting off-size work but also for analyzing the causes, in connection with the use of electric automatic sorting gages and machine control gages, two suggestions in review are appropriate here.

Between the gage manufacturer's directions and his own knowledge of gaging fundamentals, the inspector should be able to search out any inaccuracies caused by the gage parts themselves — looseness, wear, binding, misalignment, etc., in the calipering contacts and immediate connecting mechanisms. But, as has been intimated, the trouble may come from the material or gage *handling* apparatus, which can "spoil" a precise measurement just as readily as an awkward novice trying to measure with a precision bench comparator.

Occasionally the trouble is electric or electronic. The gage mechanism itself senses the measurement accurately but the electric signal or impulse necessary for correct automatic sorting or machine control may be transmitted incorrectly or not at all. The trouble may be from fouling of switch contacts, from a burnt-out tube, resistor, relay, or condenser, or from a loose connection. Or the electrical system may not have been correctly "hooked up" in the first place in accordance with the proper circuitry.

Another source of difficulty, at first seemingly obscure, may arise in the particular case of automatic machining, where the machine itself may not be able to respond fast enough or accurately enough to the control signal the gage supplies. Or the trouble may be as simple as the need for sharpening the cutting tools or dressing the wheel.

When an automatic sorter or an automatic machine fails to produce work to tolerances, the tendency is first to blame the gaging mechanism when the difficulty may actually have its sources in any number of other everyday mechanical or electrical causes.

Postprocess Gaging

The general type of calipering gages used in automatic machining, like those illustrated in the preceding few paragraphs, are sometimes described as "in-process" gages because they control, or at least measure, while the work is being machined. Other types, which are just as fundamentally in-process gages, are continuous measuring gages, like the examples pictured in Fig. 13, which measure and control were diameter, insulation, strip or sheet thickness, and the like, while the process is going on. One of the limitations of in-process gaging is that it controls random-size deviation only and cannot compensate for cumulative deviation or machine drift. Even a machine in good condition will drift; that is, it will produce work gradually changing in size as it gathers the heat generated by machining as well as the heat of its own operation. To correct this, "postprocess" gaging is employed.

The basic conception in postprocess gaging is to have an automatic gage that will measure each piece immediately *after* it is fully produced. The effect is similar to the arrangement where an inspector or the operator checks each piece produced with a hand gage or bench comparator. It has the advantage over in-process gaging in that it can be performed under more favorable conditions, free from coolant splash, heat, and machine vibration. A postprocess gage can be designed to flash simply a warning light, when off-size pieces appear or,

Courtesy of Federal Products, Inc.

Fig. 13. Two types of electronic gaging for continuous manufacturing operations. (Left) The gage here is continuously measuring insulation thickness and controlling its application. (Right) Plastic sheet thickness is being gaged continuously and the rolling process automatically controlled.

like other automatic gages, it can have control equipment to stop the machine or change the tool or wheel setting automatically.

Figure 14 pictures one such installation where each finished part slides down a chute directly from the machining area to a postprocess gage. The gage is essentially a condensed version of an automatic sorting gage (see Fig. 5 and accompanying description) that checks the part and then directs it into the proper disposal chute for those which are undersize, oversize, or acceptable. At the same time the gage transmits a signal to the machine if the workpiece is out-of-tolerance.

Another development in the field of automatic size control on fully automatic machining operations has been the installation of both in-process and postprocess gaging equipment on a machine. The in-process gage exercises close control over random deviations and the postprocess gage controls machine drift. When correction for drift is necessary, the postprocess gage does not control the machine directly but signals its partner, the in-process gage, and the latter in turn adjusts itself or automatically compensates in its own setting to eliminate the size deviation and signals the machine also what to do to compensate.

Several advantages accrue from this partnership. One is a final output of at least 99.9% perfect, in-tolerance work. Another advantage is the ability of the partnership to control an older, worn machine with minimum machine down time.

If an automatic machine is not sensitive enough, if it cannot respond quickly to an in-process control gage signal; the delayed action

Courtesy of Federal Products, Inc.

Fig. 14. One type of postprocess gaging setup used for automatic size control and screening.

will permit the production of off-size work. And if the gage control is the type that shuts a machine down entirely until an operator can make the tooling adjustments, the machine may be stopped too high a percentage of productive time. The postprocess gage partner offsets such disadvantages because, in addition to signaling the appearance of off-size work, it automatically screens out the few scrap or oversize pieces which may appear while the machine is either automatically correcting itself or is being corrected by an operator.

Often, also, a postprocess gage is equipped with an electric time delay feature. The time delay can be manually set or regulated to delay (or hasten) the shutdown or control signal to the machine. Usually the time setting is based on deliberately allowing a small percentage of off-size pieces to be made. In many circumstances, permitting a postprocess gage to screen out a prescribed amount of scrap is less costly, overall, than to have the machine's production continually shut down or delayed, for size adjustment.

For the inspector, the checking, setting, mastering, and maintenance of a postprocess gage involves the same fundamental activities that have been described previously for other automatic electric gages.

CHAPTER 11

Air Gaging Equipment

While the ideas and principles of fluid and pneumatic methods of making measurements were known and applied more than a century ago, their outgrowth, now known as air gaging, failed to get much specific development until after World War II. Now the use of compressed air as a basic means of sensing and securing precision measurements is commonly accepted and widespread in industry. Along with the adoption of air gaging, however, too few have taken the trouble to study some of the underlying principles of air gaging and especially to appreciate the advantages, as well as the few shortcomings, of air gage applications and equipment.

Air gaging, in many applications, is faster, requires less manual precision, and is, in the long run, less costly than many other inspection methods. But as with all other inspection and measurement methods, the technique or application must fit the need.

Air gages are used to measure outside diameters, being equal to other types of portable and bench comparators in many respects. Air gage equipment also is being used to check relationships like concentricity, squareness, and flatness. But the air gage's great popularity is undoubtedly due to its unique aptitude for measuring hole and bore conditions. In many applications an air gage is superior to other types of internal gages. Examples of the use of air gaging will be brought out farther on.

Basic Elements of the Air Gage

The principles on which the air gage is based will be gone into in some detail, because if these are clearly understood, the inspector can then get the most use and the greatest accuracy out of an air gage and will realize more readily when or why it is not performing as it should.

All air gages require a reasonably steady source of compressed air. It may be taken, usually through a reducing valve, from a factory air line or it may come from a separate, special air compressor. Air gages require from about 30 to 80 pounds per square inch pressure and the air supply should be filtered or cleared in some manner so that it is free of dirt, grit, oil, and excess moisture. (A section farther on describes the maintenance of air gages and offers a few suggestions for getting clean, steady air to them.)

Air gaging supplies a great deal of amplification in the measurements being taken. This high-sensitivity aspect of pneumatic gaging provides accuracy in measurements to as small as .00010 inch. However, the range is limited by the physical realities of the air gaging concept.

The direct nozzle air plug — the most common and versatile type of air gaging head — has no moving parts, so friction is not a consideration in the combination of the gage with the workpiece. Parts can be measured without direct contact with the gage head, and, in many cases, multidimensions may be assessed simultaneously.

The gage head can be attached to the end of an air source, providing great flexibility in the gage setups. The flow of air from the head helps to purge the workpiece of contaminants, an advantage in some applications.

In addition to a constant air supply, air or pneumatic gaging requires an "indicating means," and a metering means, or nozzle.

To understand the principle of air gage action, refer to the schematic diagram in Fig. 1. The compressed air first goes through a so-called master jet, which is essentially an orifice or hole with a

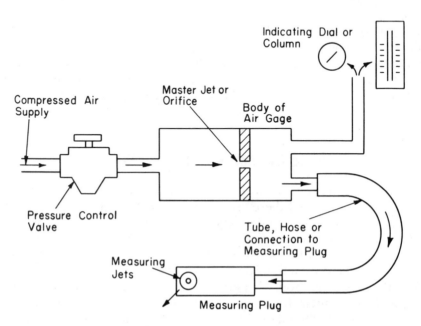

Fig. 1. Schematic diagram showing the basic elements of an air gage.

precisely controlled inside diameter, placed in the line of flow. The master jet imposes a decided restriction on the free flow of the air and builds up a steady base pressure, which, depending on the make of air gage, may or may not show on the indicating device.

The air then travels along through tubes, if the measuring plug is attached to the main frame of the air gage, or through plastic hose to the measuring plug, if the latter is to be more or less portable.

Measuring plugs for air gages are made of hardened steel. Frequently they are chrome plated; in many cases they are equipped with carbide wear liners. In general, they closely resemble conventional inside diameter plug gages. Before the plug is hardened, however, the air passages are drilled in it and the hardened "calipering jets" are inserted (Fig. 2).

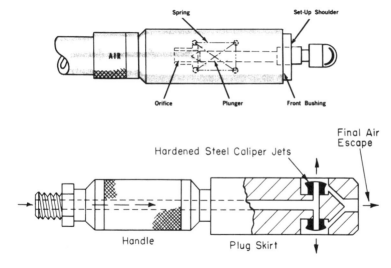

Fig. 2. (Top) Indirect reading air gage head. Fig. 3. (Bottom) Direct reading air gage head.

If an air plug is not directly fastened to the air gage body, it is usually equipped with a handle, as shown in Fig. 3, and a pipe nipple for connecting it to an air hose. The plug itself is also known as the skirt. The calipering or measuring jets, usually separate hardened steel tips, are recessed into the plug or skirt. The jets are connected to the gage's air supply by holes or tubes drilled in the solid plug or skirt as shown in Fig. 3.

Gage geads are of two designs — direct and indirect. The direct air gaging method uses open jets on the measuring device or gage that do not contact the work surface, or piece to be measured. By contrast, the indirect gage method uses an intermediate physical surface or mechanical device to contact the surface of the part to be measured. This intermediate contact device can be a cartridge or plunger mechanism not unlike the gage head on electronic comparators.

From another point of view, these two types of gage plugs are commonly of the plug, ring, and snap (gage) variety, depending on the application. Additionally, the number of jets or orifices in these gage heads can vary according to the application.

The plug or skirt is essentially a guide, as well as a holder, for the calipering jets. Its main purpose is to put the calipering jets into position for measuring. The plug is usually anywhere from .0005 inch to .003 inch under the size of the holes to be measured.

If the bore sizes to be measured, for example, are, per blue print, .785″ ± .001″, the air plug would be made probably at about .783 inch or .7835 inch, a half-thousandth of an inch smaller in diameter than the smallest size hole it would be expected to measure. Where the workpiece bore happens to come smaller than .783 inch, say, the air plug will not enter the hole and it acts then in the manner of a conventional "Go" plug by rejecting, mechanically, an undersize bore. The air gage plug and calipering jets make an air gage a single-purpose measuring instrument. Of course, various diameters and sizes of plugs can be interchanged with the gage body and indicator but the plugs themselves are single purpose.

The range of an air gage plug is .003 inch total. Where the holes to be measured come .003 inch larger than the plug skirt diameter, the air gage indicator will not register the true workpiece diameter correctly, if at all.

The principle on which the air gage operation is based is that back pressure will be created when air in motion is deflected from its normal direction of flow. Referring to Fig. 4 the flow of regulated, filtered, compressed air is restricted at the master jet m and the indicating device shows what is known as "static pressure." The air continues on through the tubes and the plug passages and issues freely from the pair of calipering jets. The friction and further restriction imposed on the flow of air by the hose, the tubes, and the small diameters of the calipering jet holes of course builds up the static or base pressure of the system a little more.

How the Air Gage Functions

Now the plug is introduced into the workpiece hole, w, Fig. 4. The air, hissing freely away from the jets, is deflected. Most of it bounces off the walls of the workpiece and escapes through the final escape

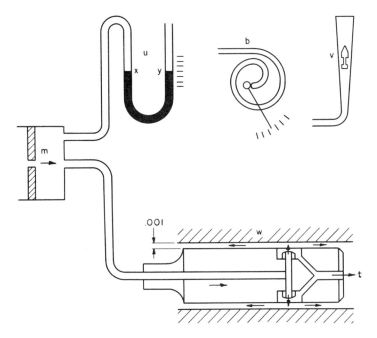

Fig. 4. Schematic diagram showing manometer tube (*u*), Bourdon tube (*b*), and Venturi (*v*) types of air gages and the path followed by the air flowing in an air gage setup.

outlet, *t*. A little of it may leak away, after it leaves the jets, along the workpiece and plug walls as the direction arrows indicate.

Since the direction of air flow has been changed, back pressure immediately builds up and increases the pressure at *m*, in front of the master jet, causing the indicator to register greater pressure because the escape of air at the jets has been deflected and restricted at the calipering jet orifices.

The smaller the inside diameter of the workpiece, the closer its walls come to the jets, the more abrupt the deflection of the air (there is an effect of a harder bounce off the workpiece wall) and the greater the general restriction of its escape. Consequently, the back pressure is greater and the indicator registers the decrease in diameter literally by a higher pressure reading. If the inside diameter of the workpiece is greater, the whole effect is reversed, and if the inside diameter of the workpiece exceeds the outside diameter of the plug by too much, the back pressure created is too small to register on the indicator.

The sketch of Fig. 4 shows three methods of registering air pressure. There is the U-tube or manometer type, *u*, where pressure unbalances a mercury or water column, the difference in height between *x* and *y* registering pressure. The manometer is the most sensitive of the several methods in registering air pressure changes and, on the whole, gives the most accurate reading. However, mercury is expensive and the water column requires air pressure reductions to about two pounds. Both are inclined to be unwieldy as shop equipment. Hence, manometer gages are usually found only in gage laboratories.

Diagram *b*, Fig. 4 illustrates the so-called Bourdon tube design (characteristic of steam pressure gages for instance) where the air pressure tends to uncoil the spiral coiled tube. This movement is amplified, as in the mechanical indicator with gears and a pointer or hand.

Sketch *v* shows the Venturi tube method where the air escapes through a tapered tube. Inside the tube is a light weight bobbin, somewhat like a small cork, which floats in the stream of air. The greater the back pressure, the greater the velocity of air flowing through the tapered tube. It carries the bobbin or float up with it to a point where the increasing taper of the tube offsets the "corking" effect of the bobbin.

A later development brought out the balanced circuit type of air gage. In this equipment the pressure indicating mechanism is cut in *across* the two air paths and a built-in gage zeroing valve is provided. Such a "balanced circuit" is shown schematically in Fig. 5.

This design uses, in effect, two master jets, *m* and *n*, one in each branch or circuit, to effect the basic restriction in the flow of air. From the master orifice *m* the air flows to the air plug, as Fig. 5 indicates, where it meets the further restriction of the workpiece or the master setting ring. This latter restriction builds up back pressure inside the bellows device, *B*, causing the bellows to expand or elongate and move the indicator, *I*, above.

At the same time the other half of the air is flowing through the other master jet, *n*, to the setting screw, *S*, Fig. 5. By closing or opening the valve, *S*, the pressure in the sealed chamber, *C*, surrounding the bellows, *B*, can be altered. When the back pressure in *C* is regulated to match the back pressure from the measuring plug, inside the bellows *B*, the bellows does not expand and the indicator registers 0. An advantage of this system becomes apparent after studying the "linear scale" principle described below.

Linear Scale Principle

The air gage takes advantage of the fact that minute differences (down to .000005 inch) in the internal diameters of workpieces will

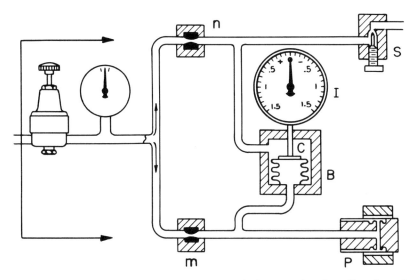

Fig. 5. Elements of a balanced-circuit type of air gage showing direction of air flow.

affect the back pressure sufficiently so that the pressure change can be amplified on the indicating mechanism — a difference in .0001 inch in workpiece ID showing up as about $\frac{1}{4}$ inch movement on the indicator dial or column. Such an observation will be accurate, however, only when it is a linear scale reading.

The linear scale thesis is best explained, perhaps, by means of a curve like that in Fig. 6. If with an air gage you seal off the air plug caliper jets (see Fig. 3), say with finger and thumb, so that no air escapes, the gage indicator will register maximum back pressure. Such a condition is shown at a on the curve in Fig. 6, where the inside diameter is 0 and the indicator reading at maximum.

Suppose then you release finger and thumb a trifle and let a little air escape (an amount that might be equivalent, for instance, to the minimum internal diameter of a workpiece), the indicator would register a little less pressure — an amount such as shown at e in Fig. 6, which would be equivalent to an inside diameter like s for example. A greater inside diameter, as at t, would register still lower on the indicator as at f.

If you kept on using a series of workpieces of increasingly larger inside diameters until finally you caused practically no deflection of the air escaping from the calipering jets with an inside diameter as at u, you would get a condition of practically no indicator reading as at position g on the curve in Fig. 6.

The correct interpretation of the curve in Fig. 6 shows that any indicator readings between a and e, or between f and g, do not register the

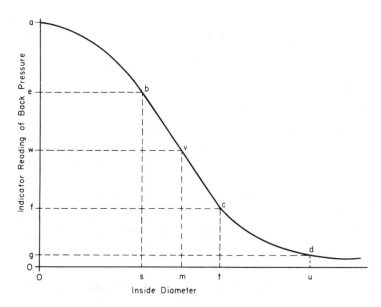

Fig. 6. Air gage pressure–diameter curve illustrating linear and nonlinear scale readings.

proportionate or correct inside diameters (because the graduations on an indicator or a column are equally spaced) but that the indicator readings between *e* and *f* are directly proportional* to the changes in inside diameter between *b* and *c* (or *s* and *t*). The section of the curve between *b* and *c* is known as the "straight line" section in air gage parlance, and the trick is to design and match air gage elements so that any measurements taken by the air gage will be within the dimensional range of *s* to *t*.

With most air gages and under the majority of practical conditions, the dimensional range of *s* to *t* is usually about .003 inch. In other words, an air gage will register with high precision the dimensions within this range but its readings are subject to error outside of it. An air plug made and marked to check 1.433- to 1.436-inch holes, for example, would be inaccurate for 1.437- to 1.440-inch holes.

Basically an air gage is a comparator; it cannot be used as a direct-reading instrument except within the .003 inch range suggested above. With each air gage plug comes either one master ring or a pair of them, depending on the make and type of air gage. Such master ring gages are usually made to tolerances of .00001 inch. If the air gage requires only the single setting master, it is usually made to the mean

---

*Strictly speaking, there is a slight variation from direct proportionality since the "straight line" portion is really a slight curve, but the effect of such variation is negligible as compared to the value of one scale division.

dimension, halfway between the upper and lower tolerance limits. Where the pair of rings is prescribed, the dial or column markers are usually set to upper and lower tolerance limits. The air gage circuit is adjusted, in other words, so that it will register internal diameters between *b* and *c* on the curve of Fig. 6. All air gages are provided with pressure regulating or setting knobs. The operation of "zeroing" one type of air gage, using a single ring is pictured in Fig. 7. Also shown is another type of gage being checked for the lower limit with one of two ring gages, the other being used for the upper limit.

Courtesy of Federal Products Corp.   *Courtesy of Sheffield*

Fig. 7. Two types of air comparators. (Left) A single ring gage is being used to check the zero reading. (Right) Here, the lower limit reading is being checked with one ring gage; the other will be used for the upper limit.

Effect of Crusted Dirt, Grease, Coolant, and Surface Roughness

On the whole it is safer to clean workpiece surfaces that are to be air gaged. While the air from the jets will probably blow off most of the oil, dirt, or coolant, it may not happen to do a thorough job. If the cutting oil is heavy, if the workpieces have been greased or rust-proofed or if they have become encrusted with dirt, the air may not clean it all off and the internal diameter measured will then include twice the thickness of a foreign film which may be .0001 inch to .0003 inch thick. Some care must also be used, if the workpieces are not cleansed, to make sure that air jets do not splatter oil or coolant onto the clothes or into the eyes.

The matter of surface finish is important in air gaging. The air gage really measures what might be called the pitch line of surface roughness. See the exaggeration in Fig. 8. The air gage measurement indicated would be m and not m_2, which might be called the root depth or major diameter of the hole. Nor would the air gage measure m_1 or what might be called the minor diameter of the hole.*

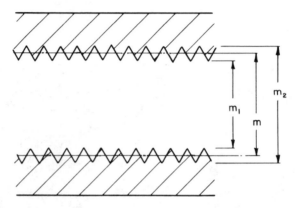

Fig. 8. An air-gage measurement indicates the mean diameter m when the surface is rough.

Confusion sometimes arises where the air gage registers dimension m, while a conventional plug gage, or a mechanical indicating bore gage of the types shown in Figures 55 and 56, Chapter 9 will show the hole size to be m_1. (If the tool marks are quite coarse and if the mechanical indicating gage reference and sensitive points have small radii, it is possible for the mechanical gage to indicate diameter m_2.)

The rule of thumb commonly applied is that air gages will show a troublesome error where the surface roughness of the bore equals or exceeds 100 microinches and for this reason air gaging should then be discarded in favor of mechanical-type gages. However, since the range of the air gage does not exceed .003 inch (the equivalent of tolerance limits of ± .0015 inch), the surface roughness of the bores to be measured by this type of gage preferably should not exceed 60 microinches.

This surface roughness limitation can be overcome by using masters with a surface roughness that is about the same as that of the hole to be measured. Remember the air gage is strictly a comparator and that it is usually mastered with rings. Ring gages are customarily lapped to surface finishes of about 5 microinches. Thus an error would result if

*The terms used here — pitch diameter, major and minor diameters — are borrowed from the common industrial terminology for the elements of screw threads.

the air gage were zeroed against a 5-microinch surface and then used to compare the diameter of a bore with a 200-microinch surface roughness. A good rule of thumb is to be sure the surface conditions of masters and workpieces are within 50 microinches of each other.

Proper Use of the Air Gage Plug

Examination of an air gage plug will show that the hardened spherical jet tips are recessed inside the main plug diameter, usually about .001 inch. The plug proper, or skirt, is really only a guide sleeve, which also prevents the jet tips from being worn away from constant application to workpieces. The plug should be repaired, replated, or replaced when its outside diameter has worn down to the jet tip diameter.

The plug acts as a centralizer or guide. Ordinarily, it can be introduced into a workpiece with a perfectly natural plugging motion. However, where precisions of .0001 inch or greater are desired, one or two precautions must be observed. The air gage plug must not be used in bores whose internal diameters are .003 inch greater than the plug outside diameter. If the air jet tips are in a vertical line as in sketches A and B of Fig. 9 at the time the gage is mastered, the jets should be kept

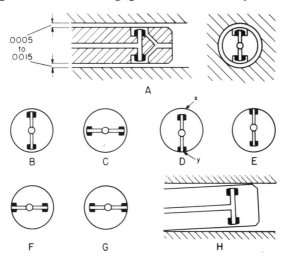

Fig. 9. Air-jet tips should be kept in the same position in measuring the workpiece as when the gage is being mastered. (A and B) Optimum vertical position for mastering and measuring; (C) optimum horizontal position; (D, E, F, and G) various extreme positions in which unequal spacings of the jets from the surface being measured may lead to errors; (H) a tilted measuring element usually will not produce a noticeable error.

in the same position when measuring the workpiece. If the jets are horizontal for mastering — as at C, Fig. 9 — they should be horizontal for measuring. Don't mix the conditions.

While the theoretically ideal measuring position of the plug is shown at A, Fig. 9, with equal clearance between each jet and the bore wall and with the jets perpendicular to the axis of the bore, practically no appreciable error occurs when the clearances are a little unequal. Sketches D, E, F, and G show various positions in which the unequal spacings are extreme. These may lead to some errors since the air is deflected more sharply at, say, y in sketch D and less back pressure is built up at the opposite jet at x. But remember the two jets are connected to a common tube. The air gage registers a back pressure that is a mixture, a balance, of the individual pressures set up at x and y, which tends to reduce the indicated error.

Following the same general reasoning, the air gage plug can be tilted a little, as suggested in sketch H of Fig. 9, without noticeable error. All the slight variations in position are acceptable provided, in general, the plug outside diameter is within .003 inch of the bore. internal diameter.

As ordinarily set up and used, air gaging is the speediest method for checking internal diameters since the operator simply inserts the plug and reads the indicator from piece to piece about as fast as he can move.

Application of Air Gages

Ovality can be explored for by turning the air gage plug; barrel and hour-glass shape, bell mouth, and taper can be explored for by moving the air gage plug back and forth. The manipulation is simpler than with mechanical indicator equipment because the elements of rocking and centralizing can be ignored. Air gages can be used on yielding materials and on fine finish bores, where the mechanical indicating gage and even the conventional plug gage fails, because the air gage exerts but nominal pressure against the walls of the bore.

The type of gaging plug previously described has certain limitations. It cannot be used for gaging porous materials, and any side holes, oil grooves, recesses, or keyways in bores must be watched. Special consideration must be given to projections into the bore. Some of these conditions are illustrated in Fig. 10. A diameter like x, for instance, will not be accurately determined with this type of gaging plug. As a minimum width or "land" for which this type of gaging plug is adaptable for accurate measurement, dimension w, Fig. 10, must exceed .100 inch ordinarily, if diameter y is to be accurately gaged.

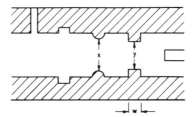

Fig. 10. Diameter *x* is difficult to determine by air gaging. Correct air gaging at diameter *y* ordinarily requires that land *w* be at least 0.100 inch.

To overcome some of the difficulties mentioned, a gaging element is used in which a mechanical component is interposed between the gaging nozzle and the workpiece to provide direct contact with the geometrical feature to be inspected. This mechanical component may be a ball, lever, plunger, or blade. To escape, the air from the gaging nozzle must pass between the component and the work surface or, if the material is porous, through the workpiece, itself. The closeness of contact between this component and the work surface governs the back pressure on the air gage, which provides the reading. This type of gaging element can be used to measure porous metals, and for narrow lands or the mouths of holes.

Maintaining Air Gage Accuracy

In the manufacture and calibration of an air gage and plug great care is used in controlling the three diameters, *a*, *b*, and *c*, diagrammed in Fig. 11. Unless certain ratios are maintained, the gage will be inaccurate. If the master jet orifice (diameter *a*, Fig. 11) becomes fouled with moisture, oil, or dirt, the gage will not function accurately. Naturally, the measuring jets (diameters *b* and *b*) must be kept clear. Diameter *c* must not be allowed to wear down. Hence, the general requirement that the air supply be clean and moisture free and that dirt, oil, grease, or coolant on workpieces shall not be so excessive as to clog the measuring jets.

Should there be reason to suspect air gage inaccuracy, first check for air leaks in all fittings, tubes, and connections. Leakage can usually be adequately indicated by corking the measuring jets with the finger tips and watching the indicator. If there is line leakage, the indicator will not remain motionless after the jets are sealed off. Repetition is another test. Try the gage several times in its master to be sure you get repeat readings. If not, suspect, among other things, fouling of the master orifice or the measuring jets. If you are sure they are clean and

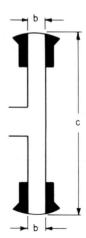

Fig. 11. Three diameters, *a*, *b*, and *c*, which must be closely controlled in the
manufacture of air gages. In use, diameters *a* and *b* must be kept clean.

you still fail to get repetition, suspect irregular air pressure from the
supply line. If the workpiece is dirty, clean it and test again for
repetition. Check for unusual surface roughness, rifling, pits, or holes
in the workpiece.

One of the major causes of air gage difficulty is the quality of
compressed air supplied it. Commercial air gages come with filters
adequate for ordinary air line dirt, water, and oil conditions but not for
the dollops and slugs of water and oil often also supplied in some shop
air lines along with the air. Commercial air gages also come with
pressure regulators satisfactory for most conditions but they are not
always adequate for the sharp drops and peaks in pressure present in
some unregulated shop lines. Such extreme conditions must be overcome of course before any sort of satisfactory service can be expected
of any air gage.

Much harder to detect, however, is an air supply condition that
forces an air–oil–water mixture or *mist* into the air gage passages.
Sometimes this mist condition will cause more trouble by fouling
master jets and plug jets, with resulting inaccuracies in readings and
repetition, than an internal stream of clear, condensed water in an air
line.

The cure for all of the trouble implied above is adequate and
proper trapping of air lines that feed air gage setups. Suitable air
cleansing and trapping equipment is available on the market and a
shop's maintenance crew has little excuse for supplying badly contaminated compressed air to air gage service.

As temporary expedients, to secure accurate results from air gages over short periods in spite of air supply lines loaded with moisture or oil, the air gage's filter can be drained off every quarter hour or so. Another trick is to open a petcock or valve in the airline, which will allow the line to bleed a little steadily.

The diagrams in Fig. 12 are offered to illustrate the importance of

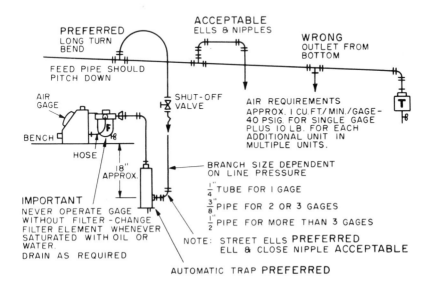

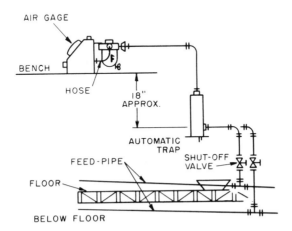

Fig. 12. Some suggestions for correct installation of air supply piping. (Above) Layout with air line on ceiling. (Below) Layout where branch comes from floor level or ceiling below.

clean air and to give a few more suggestions for the installation and maintenance of air gage supply lines.

Again, setting masters is very important in air gaging. They are used to establish maximum and minimum tolerance limits on an air gage or comparator. The shape of the gage head dictates that of the master — air plugs, of course, are usually set with rings, while ring and most snap gages are calibrated through the use of discs or plugs. Two are used, to represent the + and − limits of the gage.

Indirect gage heads, using a plunger mechanism, are set by varying the tension on the cartridges or plunger itself.

Air Gaging Applications

Within the scope of a book like this it is not feasible to try to describe the manifold uses of air gaging in industry. Indeed, little more than a hint can be offered. Gage manufacturers advertise air gaging equipment widely and will supply catalogs and descriptive pamphlets. An inspector, to keep adequately abreast of his job, should arrange for access to such literature on modern air gaging and keep informed on this important phase of today's technology.

As has been suggested, the air gage is one of the faster and more precise methods of measuring hole sizes and checking hole conditions. the photographs making up Fig. 13 will give an idea of single and multiple uses of such air gaging equipment.

As mentioned previously, air gaging equipment is also used for checking taper, ovality, triangular out-of-round, flatness, parallelism, squareness, hole location, as well as outside diameter, width, and thickness. The sketches in Fig. 14 visualize three of such relationship checks. Air gage equipment can be also made up, for example, to check the outside diameter (of a shaft) and the inside diameter of a bearing and, with a third meter, show exactly the clearance (or interference) between the two. Equipment of this nature is used for the rapid and accurate selection of mating parts in a selective assembly operation.

There are several reasons why air gaging equipment is popular, useful, and, in fact, profitable to use. In the first place it is accurate. It will easily and accurately determine a size difference of ten millionths of an inch (.00001 inch). In the second place it is easy to use. It is direct and easy reading as compared, for instance, to reading a vernier or even a micrometer scale. Trained experts are not necessary. No special manipulation is necessary, again as in the case of vernier calipers or inside mikes. Gaging pressure does not enter the picture as it does where conventional fixed plug gages are used. All these advantages also add up to time and labor saving. Where it takes an accomplished

Courtesy of Federal Products Corp.   Courtesy of Sheffield

Fig. 13. (Upper left) Using an air gage to check a lapped hole surface. (Upper right) Using an air comparator and height gage stand to check the external diameter of a cylindrical part. (Lower left) Simultaneous inspection of hole diameters and center distances with an air gage. (Lower right) Using an air comparator and height gage stand to check surface flatness.

inspector 43 seconds to measure the inside diameter of a bore with vernier calipers (and then not come closer than .001 inch to the true size), an inexperienced operator can check the same hole with an air plug and gage in 4 seconds and be accurate to .00001 inch.

    Another development has added to the extent and flexibility of air gage use. This is a specially made attachment which looks and acts

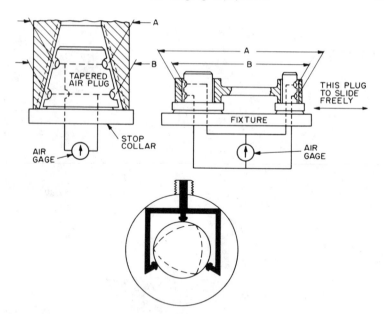

Fig. 14. (Upper left) Using special type of air plug to check accuracy of the taper of a hole. (Upper right) Using air gage and two air plugs to check parallelism between two straight holes. (Below) Construction of an air gage ring for outside diameter measurement. This type of gage is useful for checking taper, ovality, and triangular effect, in addition to checking the outside diameter.

(from the outside) like an indicator stem and spindle. It is in effect an air valve.* Figure 15 pictures one of the fountain-pen-size attachments and shows one of them in use in a simple comparator setup. One of the great advantages of the Airprobe type of attachment is that it can be so readily used in a multiple gage setup like that also illustrated in Fig. 14. These "valves" act externally like an indicator stem and spindle. Internally the spindle motion tends to open and close an air jet, restricting the air flow, in about the same way a ring restricts the air flow at the calipering jets of the regular air plug, and thus transferring the measuring motion of the spindle to the air gage indicator or column.

Designs have also been developed whereby the movement of the air gage indicator or meter also simultaneously actuates electric micro-switch, resistance, or reactance mechanisms. Thus the measurement or reading can be transferred to electric impulses, which, in turn, can direct machine control or automatic sorting functions. Figure 16 displays several of these various applications.

*Common trade names are Airprobe, Federal Products Corporation, and Plunjet, Sheffield Corporation.

Courtesy of Federal Products Corp.

Fig. 15. Left) Direct-contact type of air gage pick-up cartridge with moving spindle. (Right) This type of gaging element can be used in a height gage stand as shown.

In his part of setting, mastering, checking, and maintaining multiple and air–electric and special design air gaging equipment, the inspector treats each individual unit as he would a single gage, applying the same fundamentals of correct measuring. In each case he should be guided by careful reading of the gage equipment manufacturer's instructions. Usually, too, he will be provided with separate manual gaging equipment, of equal discrimination and accuracy, so that he can verify any gaging equipment accuracy and calibration by checking sample workpieces.

Flexibility of Air Gage Systems

Many plants have discovered that once air gaging systems — compressed air supply, traps, filters, valves, and piping, plus air gage meters or columns — have been installed, they can be used for a wider variety of fast, accurate, precise measurement purposes than originally planned for.

As a simple example, consider a machine setup for drilling and reaming different size holes. An air gage with a valve manifold and a group of air plugs, as suggested in Fig. 17 can be installed at the machine for operator use, or process inspection, as at an inspection batch checking center. One air gage and meter setup can thus serve to get multiple measurements quickly. The air plugs can be readily changed for different size holes to accommodate diversities of production.

Courtesy of Federal Products Corp. *Courtesy of Sheffield*

Fig. 16. (Upper left) Use of air gaging units of the type shown in Fig. 15 with multiple indicator panel for checking various dimensions on a shaft. (Upper right) Eight cylinder bores in an engine block are being automatically checked simultaneously at four places for diameter and taper with this air gage equipment. (Lower left) Machine control and automatic sorting units with air–electric indicators used in postprocess machine control. (Lower right) Air–electric gage control unit designed to combine, in a single unit, multiple automatic gaging and feedback control of a series of machines.

Air gage manufacturers are also supplying adjustable bore gage heads, which attach to an air gage, and by means of which a large variety of hole sizes, from $\frac{1}{2}$ inch to 8 inches, can be precision checked. This arrangement is shown in Fig. 18. The adjustable air bore gage often solves the problem of precision measurement of large diameter holes and is especially adaptable to the small lot sizes of job shop work.

While air gaging is one of the faster and better ways of measuring holes, its use is not confined to inside diameters. It is being brought in increasingly to check outside diameters. Gage manufacturers supply air snap gages of the type pictured in Fig. 19, which have the advantage over fixed snap gages and indicating caliper gages of being able to detect size differences readily and accurately in terms of 20 and 10 millionths. Air rings like that pictured in Fig. 20 are used for shaft measurement and,

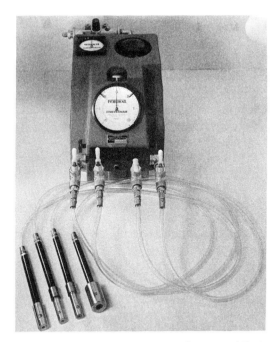

Courtesy of Federal Products Corp.

Fig. 17. By connecting a valve manifold to the air gage outlet, the use of a single air gage can be rapidly adapted to a diversity of plug sizes or other accessories.

because they offer three-point air jet contact, are especially useful on centerless ground parts.

The addition of air cartridge devices to air gages has given the latter almost unlimited measurement facility. There are many situations where the use of the air cartridge is to be preferred over the everyday gear train indicator. In fact, the air cartridge has solved measurement problems that the regular air plug has been unable to crack. Their small size enables the gage designer to place them in spaces where neither indicators nor air plugs could be used. Figure 21 offers three examples of many such quandaries resolved by air cartridges.

The air gage has been found useful for checking internal tapers. It satisfactorily detects surface waviness or lack of flatness, and in addition, out-of-round, eccentricity, hole location, width, and thickness. It is also especially adaptable to multiple gaging equipment. Figure 22 illustrates an air gaging set arranged for the measurement of 17 different characteristics. The inspector would do well to get acquainted with the variety of modern air gaging applications available, information which air gage manufacturers freely offer in a multiplicity of catalogs, brochures, and bulletins.

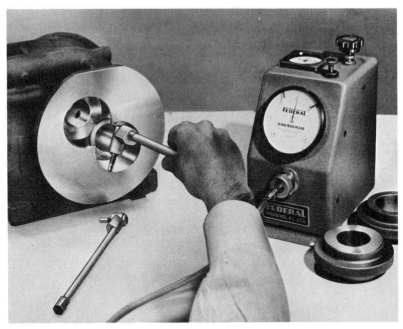

Courtesy of Federal Products Corp.

Fig. 18. Hole gage heads, similar in construction to dial bore gage equipment, are available for connection to air gages. These feature adjustment mechanisms permitting their use in an extended range of hole sizes.

Courtesy of Federal Products Corp.

Fig. 19. Featuring an air gage attachment built into an indicating snap gage design.

Courtesy of Federal Products Corp.

Fig. 20. Shaft and cylinder ODs are measured by an air gage through an air ring.

Selective Assembly by Size

Air gaging is unique among all systems in its ability to match cylinders and holes — shafts and bearings — to any desired fit, and in this era of close tolerances this property of an air gage circuit is of immense help. To produce shafts in any sort of quantity consistently within a tolerance of .0001-inch or less is difficult, expensive, and sometimes practically impossible. Likewise, or even worse, is the job of boring, reaming, grinding, and honing a series of holes to close tolerances. The frequent result is fits between shaft and bearing that are too much on the loose side with assemblies that are noisy, that vibrate, or that wear rapidly or in reverse, fits that bind and heat up.

Air gaging equipment can be made up into a "matching gage" with a manifold connector and a special internal circuit that ties simultaneous ID and OD measurements together with the meter. Figure 23 illustrates one such piece of equipment. The diameter of the hole is sensed on one unit — usually a plug — while simultaneously the OD of a shaft is being sensed in, usually, an air ring. The two measurements are so interconnected through special internal air circuitry that the meter registers only the *difference* in the two sizes. Another way of looking at it is to consider the shaft, say, as the master and that the hole size is being compared with it, or vice versa.

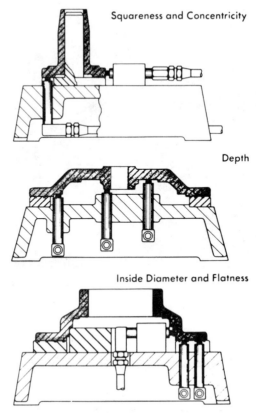

Squareness and Concentricity

Depth

Inside Diameter and Flatness

Courtesy of Federal Products Corp.

Fig. 21. Illustrating gaging fixtures embodying air cartridges for unusual and difficult measurements — top, a check for squareness and concentricity; middle, for depth; and bottom, for ID and flatness.

One way to use a matching gage is to mark a shaft (or use a master cylinder) and put it in its side of the gage. Then succeeding hole pieces are tried on the other side. Those holes showing clearances within tolerance are marked. On other occasions, shafts are mated against a master ring. A third way is to gage in the manner described above until a pair — shaft and hole — is found that suitably matches. The two are marked and the routine continued for the next pair.

Depending on the magnification and resolution of the air gage, pieces can be matched within a few millionths clearance. Taper, out-of-round, lobing, and similar troubles can be detected and their effects as obstacles to a good fit considered. A zero or negative reading on this type of air gage spells interference. It must be remembered that actual diameters of holes or shafts are never registered on mating gages, only the differences in sizes.

Fig. 22. With this air gage setup, measurements of 17 different characteristics can be taken simultaneously or individually.

High-Magnification Air Gaging

Another recent trend has been the introduction and use of air gages for measurements closer than 50 millionths with corresponding meter dial or column scale divisions of .000020 inch, .000010 inch, or .000005 inch. However, as measurements with air are undertaken that are finer than the more or less standard resolution of 50 millionths — half a "tenth" — difficulties and inaccuracies emerge. A quick study of one principle of air gaging may supply clues as to why an air gage is not always reliable in that last millionth or two of range which look so authentic on the dial or column.

The round column or shaft of air issuing from the plug orifice or jet and impinging against the workpiece is often the key to the situation. Its true shape is important to ultimate accuracy. If the air stream spreads, "mushrooms," too much, if its shape is distorted, if part of it tends to stream away from the air column, there may be variations in back pressure and, consequent, fluttering and inaccuracy at the metering device.

Sketch A in Fig. 24 gives an idea of the "air curtain" as it is sometimes called and its cylindrical shape (which is perhaps the ideal), while sketch B pictures the air going astray. The design of the jet

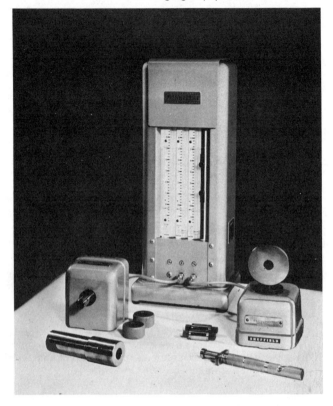

Courtesy of Sheffield

Fig. 23. Special air gage equipment assembled as a "matching gage" unit.

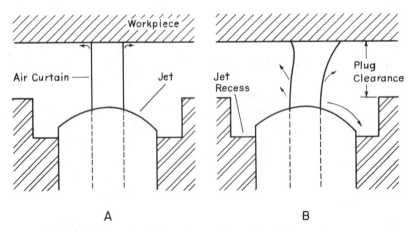

Fig. 24. Diagram of the air curtain and its deformation.

outlet, and the workmanship applied in making it, affect the air curtain favorably or unfavorably. Damaging the mouth of the jet, or letting dirt and oil collect and harden there, lowers effective accuracy. The outer shape and dimensions of the jet are also important if they are not properly designed in the first place or are not kept clean and unscarred, accuracy sapping air turbulence can occur.

The amount of clearance between the air plug jet and the work-piece surface (see sketch B, Fig. 24) becomes more important and more exacting as the air equipment is required to gage size differences in fewer and fewer millionths. The closer the jet outlet comes to the workpiece, the shorter the air curtain and presumably the more accurate the back pressure. Such clearances can be as little as .0002 inch, their amounts being determined by the gage manufacturer. Hence, master ring wear now readily affects measurement accuracy and rings need to be checked and replaced more frequently.

Workpiece and master surface conditions have an immediate effect on the millionth measurement accuracy of extra-high-magnification air gaging. If there is no measurable surface roughness on either master or workpiece, surface finish error can be forgotten. If the surface roughness of master and workpiece are alike or differ by something less than a microinch, the measurement error is considerably reduced if not nullified. Type, direction, and lay of surface condition also has its effect on the impinging air curtain and cause unpredictable discrepancies

Make a point of checking possible manipulation errors by delibe-rately wiggling, turning, twisting, or canting the master ring or work-piece on the plug of a high-magnification air gage, or a shaft inside an air ring, to see whether the position of either the master or workpiece could cause any discrepancies in repetitive readings. If plug or air ring are used steadily, they should be checked regularly for wear. Wear down must not increase the permitted "clearance" shown in Fig. 24 beyond tolerances specified by the gage manufacturer.

The design of each make of air gage with its plugs, air rings, and accessories might be called a closed corporation. This is to say that each manufacturer has adopted his own combinations of jet diameters and clearances, of internal master jet IDs and other features of his circuitry. Any size plug made by manufacturer X will work on X's gage, for example, but not on manufacturer Y's gage.

From time to time "replacement" jet plugs for all makes of air gages appear on the market, plugs and air rings that are not made by any air gage manufacturer but as a product of some machine parts company. Usually the advantage mentioned is that of lower price or faster delivery. Probably this type of part had better be carefully calibrated especially before use on a high magnification gage.

The inspector should guard against another source of confusion where a few high-magnification air gages may be mixed in among a number of regular air gages on the factory floor. Each air gage manufacturer marks his plugs, air rings, air cartridges, and other accessories for the particular magnification of his make of air gage they are to be used with because he knows high-magnification plugs or accessories will not work on regular gages and vice versa. A little care and attention may save embarrassment in this respect.

Fork Gaging

Another technique in air gaging has been labeled "fork gaging." Its use is limited to internal grinding. It continuously registers the change in an inside diameter as the grinding proceeds, acting in most respects in a hole, in the same fashion as the type of machine control caliper gage described in Fig. 9 of Chapter 10 does on shaft, cylinder, and OD grinding.

A fork gage is yoke or U shape, so formed and dimensioned that it can occupy the crescent-shaped area between the work and the wheel. The view in Fig. 25 and the diagrammatic sketch in Fig. 26 offer a pretty clear idea of how it works. Air is fed from the air gage (which can be mounted anywhere convenient on or near the internal grinder) through air passages in the yoke to a pair of sensing jets, one at each tip of the fork as indicated in Fig. 26. The air fork is usually clamped in some sort of special design hinge device (one of which appears in the photo of Fig. 25) so that the air fork can readily be flipped into gaging position after setup or readily retracted. Such a jig locates the fork on the hole centerline, without the fork touching the sides of the hole; thus ensuring accurate measurement and no burnishing. The jig also permits longitudinal adjustment of the gage.

A fork gage can be mastered and the air gage zeroed by substituting carefully centered master rings in the internal grinder in place of a workpiece. It should receive the same careful cleaning and maintenance as any air plug and be checked for OD and jet surface wear and damage. It is necessary, too, to check the two arms of the U or yoke to be sure they are reasonably parallel and not warped, twisted, or bent out of line.

Production people sometimes resent air forks because, at best, they do occupy some of the space within the hole and hence they are unable to start internal grinding operations with maximum wheel diameters. For similar reasons there is a minimum ID (about 1.25 inch) below which a fork gage cannot be used. Information of this sort can be sought from the air gage or internal grinder manufacturer. Also

Courtesy of The Heald Machine Co.

Fig. 25. Air fork gage in working position on an internal grinder.

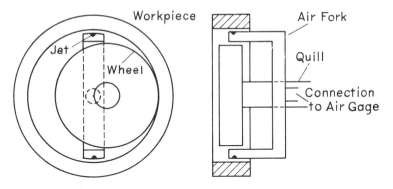

Fig. 26. Diagram of the fork gage principle.

information can be obtained concerning air plug mechanisms that reciprocate in unison with the grinding wheel to register accurate size control of smaller holes.

The air fork can be used in a "manual" sense in that the grinder operator can continuously read his air gage meter and stop grinding when it shows finished size. More frequently in modern machining, however, the fork gage is connected to an air–electric system whose switch or transducer elements automatically signal the grinder what to do and when.

CHAPTER 12

Optical Measuring and Inspection Equipment

Visual inspection is magnifying a workpiece or a part of it and using cross hairs, graduations, and scales, and other methods of measuring to assess the workpiece. This assessment can be a dimensional assessment. One such visual apparatus is the toolmaker's or measuring microscope, equipment like that illustrated in Fig. 1.

Courtesy of Brown & Sharpe Mfg. Co.   *Courtesy of George Scherr Co.*

Fig. 1. Two commercial types of toolmakers' microscopes.

This instrument consists of a microscope mounted so it can be readily adjusted and focused over the work, and a table or stage with adjustable lighting units to illuminate the work. The table is equipped with clips, clamps, a vise, and/or centers for holding various sizes or shapes of workpieces within the field of the microscope.

To facilitate measurement, cross hairs appear also in the view seen through the microscope. Depending on the equipment furnished or specified, the cross-hair arrangements may appear as in any of the sketches in Fig. 2 — as a pair of intersecting center lines, sketch A,

432

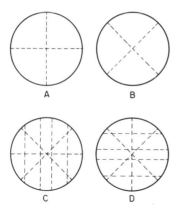

Fig. 2. Two cross-hair arrangements (A, B and C, D) used with toolmakers' microscopes.

which can also be turned as at B, or as two sets of parallel lines, C, which can be turned as at D.

The table moves on cross slides and is equipped with micrometer screws for fine adjustment. Usually the micrometer thimbles, see Fig. 1, read directly in .0001 inch-units. The table can be moved laterally or longitudinally across the field of the microscope and is usually equipped with clamps and stop devices so that the exact amount of this movement, beyond the 1-inch range of the micrometer screws, can be controlled and measured by inserting gage block stacks.

How the Toolmaker's Microscope is Used

The toolmaker's "mike," as it is sometimes called, has its greatest utility in measuring odd profiles, hole locations, and the locations of odd profiles, angles, etc., especially on thin flat stock where conventional methods of measurement are difficult. It is useful, too, in die-sinking problems and for checking jigs. Most measuring microscopes are equipped with a transparent protractor attachment, the view of which can be included in the microscope field for measuring angles. To assist further in odd measurements, the table or stage can be revolved under the microscope or it can be tilted.

As a simple example, the outline of a reasonably intricate workpiece is offered in Fig. 3 with the relevant, desired dimensions shown by letters. When the piece is clipped on the table of the micrometer and the glass focused, only a small part of the object or workpiece will be seen through the eyepiece. This magnified area is known as the field of the microscope.

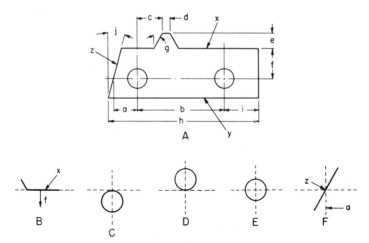

Fig. 3. (A) Workpiece and relevant dimensions that are to be checked using a toolmaker's microscope. (B through F) Steps in obtaining the various dimensions shown in (A).

Perhaps dimension f would be checked first. The workpiece is moved around on the table under the microscope until the edge x appears parallel to the horizontal cross hair. The workpiece is clamped down and the whole table is slid up until workpiece edge y appears under the cross hair, just as a check to see that the workpiece is clamped 4-square to the table.

The whole table would be moved again until edge x is registered squarely under the horizontal cross hair, as shown at B in Fig. 3. The reading of the micrometer spindle regulating the up and down motion of the table is read and recorded at this point. The workpiece is then moved up by turning the up and down micrometer screw and over by turning the right-to-left micrometer screw. The combined motion will bring one of the holes into the microscope field. The hole is centered on the vertical cross hair and its upper edge is located tangent to the horizontal cross hair as shown in sketch C. The reading from the up and down micrometer screw is then recorded.

Again the workpiece is moved until the horizontal cross hair is tangent to the lower edge of the hole, as at D, and the micrometer reading is recorded. The difference between reading C and reading D is, of course, the diameter of the hole. Half of this figure is now added to reading C and the micrometer thimble is turned to this figure to bring the horizontal cross hair into the center of the hole as at E. The difference between the reading at E and the previous recorded reading showing in sketch B gives dimension f of the workpiece.

Now the table is moved along horizontally until intersection z, see sketch A in Fig. 3, appears in the field and the cross hairs are centered as shown in sketch F. The right-to-left micrometer reading is then taken. The workpiece is then moved over until the edge of the left-hand hole appears. Applying the technique just described for getting dimension *f*, dimension *a* is found.

The rather long and detailed instruction just given, in connection with Fig. 3, will be better understood, of course, after the inspector has had a chance to use a toolmaker's microscope. The same geometrical and locating principles apply to securing all the other dimensions of the workpiece shown in Fig. 3. For angles like *g* and *j*, the protractor attachment is brought into use, the protractor hairline being registered on the slanting edges and the angle being read where the hairline intersects the protractor scale.

Microscopes are used for a variety of applications in industry and inspection. This method of direct magnification can vary from a simple 7× glass and light to sophisticated microscopes set up for one or several very unique operations. They are often equipped with many extras and options, not the least of which involves cameras — both still and high-resolution video systems.

Microscopes can add the dimension of measurement to visual inspection through their magnification systems. Most microscopes have a very limited depth of field, and require precise lighting and focusing coordination to extract a "flat" view of the workpiece. But microscopes offer a high level of detail and ability to compare and measure workpieces. The main types of microscope — fixed-scale, filar, traveling, and draw tube — differ mostly in their measurement techniques. Some have moving reticles with a readout mechanism, and others incorporate a fixed reference scale. Toolmaker's mikes move about the work, as opposed to a traveling-stage microscope. These units are usually uniquely suited to their work and involve special setups; their discriminatory abilities vary tremendously.

Visual inspection of component surfaces — especially in the field of electronics, is important to industry and the satisfaction of functional criteria. From contamination to pass/fail assessments of form, fit, and function, assessment of a visual configuration is essential to calling the component good or bad. And this attribute or quality of a component can be as critical to the performance of the workpiece or component as direct linear measurements, in some applications.

Visual inspection in the workpiece can often be carried out without the aid of magnification, or with relatively low magnification, if proper lighting is used. Many different types of inspection lighting are on the market — from normal bulb to fluorescent to fiber optic illumination systems, which provide an intense illumination to a critical inspection area, making it possible to visually detect defects or contaminants on a component or workpiece that otherwise would not have been visible.

Adjusting the Microscope Lighting

One or two warnings in connection with the accurate use of a toolmaker's microscope should be issued. The light from the lighting attachments and the general illumination should be so directed and modified that a clear, clean image of any workpiece edge appears in the eyepiece after the microscope is sharply focused. If the light is too intense, too close, or at the wrong angle, a reflection or "halation effect" appears under which it is difficult to exactly register a hairline on the true edge of the workpiece. If the workpiece is thick, if it has depth, and this is especially true in trying to pick out the true edge of a hole, a degree of fuzziness known as aberration confuses the eye. This is because the eye sees down into the hole. For greatest accuracy change the lighting and the focus until halation and aberration are at a minimum.

The use of fiber optics and advanced optical technology has introduced an inspection tool that has proved to be amazingly adaptable to many inspection and measurement uses, from the aircraft industry to nuclear applications to the field of medicine. This inspection tool is the borescope, and fiber-optic scope, which makes use of unique light-transmitting materials to make it possible to see into and view areas of a machine or hidden space that are impossible to view with ordinary methods (Fig. 4).

These visual inspection systems are comprised of an eyepiece and lens system connected by a shaft of light-conductive material that is either flexible or rigid, depending on the application. The inspector looks through one end of the system, into the article to be inspected, which could be very close, or quite far away — as much as several yards. The lens end of the system has the capability of fitting through a very small opening (1 mm) to view internal configurations without having to disassemble the workpiece or item that is being inspected.

The field of vision offered by these devices can be quite surprising — from a direct, small aperture view to 180, 240, and even 360 degee view of the area to be inspected. Rotary mirror systems, magnification capabilities, extendors, lights, grapplers that can physically manipulate objects within the viewing area can be added to the basic set ups. The uses for these units are quite varied in today's technological world; they are also available with camera and video connection options.

The Measuring Machine

Related distantly to the toolmaker's microscope and also to the Supermicrometer is the measuring machine illustrated in Fig. 5. In

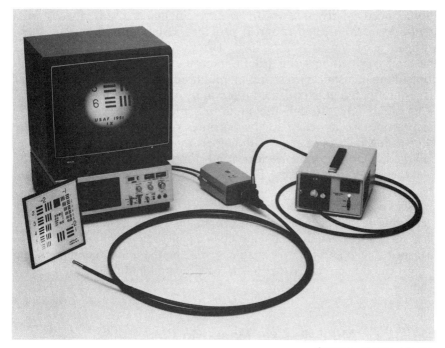

Courtesy of Diaguide Inc.

Fig. 4. Borescope.

Courtesy of Colt Industries, Pratt & Whitney Machine Tool Div.

Fig. 5. Commercial measuring machine that measures directly to 0.00001 inch.

those shops where one of these instruments is used, always in a temperature-controlled room, it provides a standard of measurement with an accuracy to .00001 inch. Briefly it consists essentially of a master bar, dividing screw (like a micrometer screw), and a means of controlling measuring pressure, all mounted on a rigid bed. The master bar is graduated at each 1-inch interval by extremely fine hairlines that are visible only through the mounted microscope. Hairlines in the microscope are matched to the hairlines on the measuring bar. The intervening inch of space is subdivided by the micrometer screw and enlarged thimble, which, also read through a glass, enables the .00001-inch discrimination. The tail stock contains the spring loaded ($1\frac{1}{2}$-lb) reference anvil.

The machine utilizes a mounting bed that holds a tailstock and a measuring head. The measuring head contains a transducer to translate the physical changes that are encountered. The microscope is used to align the measuring head and take the readings, making the units something like a sophisticated micrometer that has to be prepared and read with a microscope.

Multiple-Axis Measuring Machines

Multiple-axis measuring machines are a common extension of the single-axis measuring machines, and are a logical progression to the search for increasingly accurate and reliable discrete measurements. The blend among optical, mechanical, and electrical or electronic systems is quite prominent in these measurement units, and represents an ultimate progression in direct-contact mechanical measurement devices.

The expansion and variation of multiple-axis measuring machines has made this type of inspection device one of the most versatile and sought-after of all measuring units. Multiple-axis measuring machines use many different methods and techniques for comparing dimensions and extracting measurements. But like all other categories of measuring devices, there is a measure of commonality shared by all units in this category.

First, multiple-axis machines possess capabilities that are reflected by their name, which suggests the highly mechanized corollary to the surface plate systems. While the single-axis measuring machine deals with a linear measurement, these units are capable of compounding the linear expression with the measurement and characterization of other geometric relationships of form and position. Three-dimensional definition is derived from these units.

The spindle or measurement contact point is generally mounted on an axis coincidental with the Z axis. Think of this device as a pencil that is perpendicular to the horizon or work surface or table. From this point, which is defined by the end of the spindle or the point of the pencil, the X and the Y axis are extended and defined.

The spindle, on most direct-contact multiple-axis measuring machines, rotates with a high degree of accuracy and defines an absolute zero point. The physical movement of the measurement or contact spindle is poised over the work surface and situated mechanically in order to precisely define linearity, all manner of angles and angle definitions, and roundness, as well as arcs and all the various constituents of geometric relationships.

The gage head is capable of defining circles, as well as all possible combinations of three-dimensional definition. So the units function well as comparators, referencing the true positions of a workpiece to the predictable travel of the gage head, and can display the measured variation to an increasing level of accuracy and precision.

The measurements and deviations or comparisons that are derived from the use of these machines can be comprehended in one of many fashions by the inspector. The amplified readings of the gage head can be displayed on a dial, digital readout, or some type of recording device. At the same time, these systems may employ a microscopic system to view the visual detail of the measurement or comparison, and many units hold a video system for recording and documenting the physical detail. Strip charts and other types of recording mechanisms are used for expressing contour or surface variation over a preprogrammed routine of measurement.

Proper setup and programming of these machines allow the full range of their applications to be revealed. Detailed progressive and continual measurements can be set up for the entire automatic measurement of a fairly large part. Whatever contour is expected from a workpiece and its attending specifications can be fed into one of these machines, and a nonstop and fully automatic, repeatable evaluation of the entire workpiece can be accomplished.

Inspection work like this changes the nature of an inspector's job. The uniqueness of these machines, and the speed at which their technology is changing and progressing, necessitates up-to-the-minute training of the operation of the unit. And operation usually means just that — operation, not particularly "inspection," as it has always been known. The modern measurement machine operator is a programmer, an expert in mechanical measurement like his predecessors, but more in terms of setting up the inspection, not performing direct readings from an operator-dependent measuring device.

In other words, inspection by this method can still be unreliable if the inspector or machine operator does not follow a prescribed method of work, and is careful to attend to a high level of technical detail. But that detail is applied slightly differently — this time in the direction of preparing the *machine* to do the delicate work of direct measurement, in order to serve the demands of speed, increased accuracy, precision, and higher inspection reliability.

Needless to say, an ultra-precision instrument such as a measuring machine is not made generally available. Usually, one man is assigned to use it and to take care of it. Any and all of the rules pertaining to instrument care recited thus far apply to this machine. Even the heat transfer from the observer's body may affect its accuracy, to say nothing of a fleck of dust, a minuscule of grease or sweat, or the outrage of the minutest scratch or nick. It is checked and calibrated only with master gage blocks.

The Optical Flat

Essentially, an optical flat, as Fig. 6 illustrates, is a highly polished piece of transparent material such as plate glass, optical glass, pyrex, or fused quartz, the latter being the best though the most expensive material. Optical flats are cylinders varying anywhere from about $\frac{3}{8}$ to $\frac{3}{4}$ inch in thickness and from about 2 to 4 inches in diameter. At least one circular surface is polished so perfectly flat that surface waviness, warp or irregularity is virtually immeasurable. Optical flats can be obtained with both circular surfaces guaranteed flat and perfectly parallel to each other. However, the single specially flat surface is the more customary and is less expensive; it is differentiated by an arrow pointing to it as shown in Fig. 7. For the best use of optical flats a source of monochromatic light usually is supplied. Such a light source is shown at the right in Fig. 8.

Optical flats provide a simple and rapid means of checking the flatness of surfaces that have been made very accurately. In testing such surfaces, the flat is placed on the work, after removal of all dust or dirt, and a monochromatic light is directed onto the work. If the work surface is now viewed through the optical flat, a series of alternative light and dark bands will appear as in Figs. 6 and 7. The dark bands are called interference bands.

How the Optical Flat Works

When a series of straight interference bands is seen, as in Fig. 7, it indicates that there is a very slight wedge of air between the work surface and the bottom surface of the optical flat. The bands take a

Courtesy of The Van Keuren Co.

Fig. 6. Typical optical flat.

Courtesy of The Van Keuren Co.

Fig. 7. The working surface of an optical flat is indicated by an arrow pointing to it. Some optical flats have two working surfaces flat and parallel to each other, hence are marked with a double arrow as shown at the left.

direction at right angles to the slope or direction of the wedge. The number of bands per inch indicates the steepness of the wedge, which increases in thickness from the point or side of contact at the rate of one-half wavelength (0.0000116 inch for commercial monochromatic light sources) per dark band. A pronounced light spot or line indicates the point or line of contact.

Fig. 8. (Left) Set of optical flats. (Right) Source of monochromatic light.

The dark interference bands show the points or spaces where the light waves, reflected from the work surface, interfere with the waves reflected from the under side of the optical flat. The light spaces show the points or spaces of reinforcement. It is the dark or interference bands which indicate the highly exact and useful measuring unit of 0.0000116 inch. Straight, parallel, and evenly spaced bands, as in Fig. 7, indicate a flat surface, while curved irregular bands, as in Fig. 6 indicate a curved or irregular surface, as will be explained.

When the optical flat surface is in perfect contact with a perfectly flat workpiece surface, no interference occurs and no bands appear. However, workpiece surfaces almost inevitably possess irregularities or the contact between the flat and the workpiece is seldom perfect and effects like those exaggerated in Fig. 9 are registered by interference bands.

For successful use of an optical flat the workpiece surface must be smooth, clean, and bright enough to reflect light. An optical flat can be

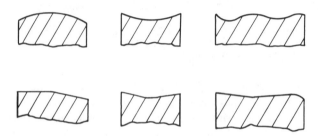

Fig. 9. Some of the surface conditions (exaggerated) that may be detected with the aid of an optical flat.

used with daylight or ordinary electric light, but the interference fringes are vague and ragged if they are discernible at all. Daylight and most electric light is a mixture of some part or all of the spectrum and each color light has a different wavelength. A monochromatic light, or light having predominantly a wavelength of 0.0000232 inch, is used in industrial work to secure sharp interference bands.

Correct Way to Use the Optical Flat

Professionally, the optical flat is used in the following manner. First be sure that the flat and the workpiece are clean and free from grit, dust, oil, or fingerprints. This is usually accomplished by swabbing them with grain alcohol and polishing the surface of each with clean chamois or paper.

Rest the optical flat carefully on the workpiece. Never wring an optical flat; it scratches too readily and the scratches are an abomination. The flat will ordinarily contact the workpiece at one edge or area, but a so-called air wedge will be built up under the rest of the flat. The flat is carefully but firmly, and evenly, pressed down on the workpiece with two fingers until interference bands become discernible. The closer the flat is pressed to the workpiece, the thinner the air wedge or, in other words, the wider apart will be the interference bands.

Lift the flat from the workpiece; never slide it off. If it appears that the contact with the workpiece is not good — if the light bands are not satisfactory — don't slide or wring the flat and workpiece together. No, lift the flat and set it down again, applying vertical finger pressure at several locations on the upper surface of the flat until satisfactory bands appears. In other words, the flat may be rocked and pressed but it never should slide, creep, or wring. For best interpretation, the adjustment of the flat should produce interference bands between $\frac{1}{8}$ and $\frac{1}{4}$ inch apart and, depending a little on the area of the workpiece, three fringes at least should appear.

Even though the operator follows the above procedure carefully, interference bands may not appear. Such circumstances may be caused by dirt or dust between the work and flat, a burr on the work, by grease, or by lack of sufficient polish on the work to reflect the light.

Whether the bands are far apart or close together, each one counted from the point or line of contact of work surface and optical flat indicates a separation of 0.0000116 inch between them. The diagram in Fig. 10 illustrates this in exaggerated form. At a distance of seven bands from the line of contact c-c, the optical flat is separated from the work surface by a distance of 7 × 0.0000116 inch or .0000812 inch. Thus, the thicker the air wedge between flat and work surface, the more fringes closer together or, vice versa, the thinner the air wedge, the lesser count of fringes farther

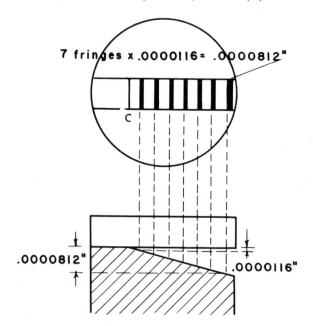

Fig. 10. Diagram showing the geometric relation between the half-wave lengths of the monochromatic light and the thickness of the air wedge under an optical flat.

apart. Actual measurement is taken by counting the fringes and multiplying that count by .0000116 inch.

For practical inspection purposes the optical flat is used mostly to check surface flatness such as the condition of gage anvils, gage blocks, surface plate areas, and other precision surfaces. Optical flats are also used to calibrate the thickness of a gage block in comparison with a master block.

Interpreting the Light Band Patterns

Interpretation of the pattern of light bands is, of course, also important. If a surface is concave in the form of a conical crater, the band pattern will look something like Fig. 11-A; if it is a convex cone, like 11-B. For a circular crater or hump, the bands would be uneven — narrowest and spaced closest together where the sides of the crater or hump were steepest and widest and spaced farthest apart where the slope was the least. To test whether a crater or hump is present, press down in the middle of the flat. The rings will move inward if it is a crater; outward, if a hump. In other words the rings will move toward the thickest part of the air wedge.

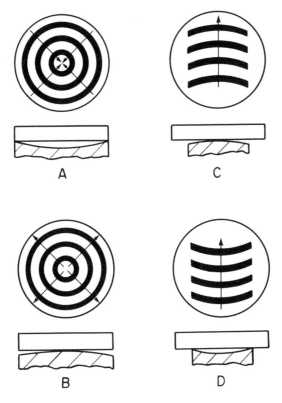

Fig. 11. Various conditions of a workpiece surface and the corresponding interference band patterns when viewed through an optical flat.

The effect of a cylindrical surface on a rectangular workpiece is
illustrated in Figs. 11-C and 11-D. To determine whether the surface is
cylindrically concave or convex, press down on the middle of the flat. If
the rings move as if they were bows shooting an arrow, the surface is
convex; if they move in the opposite direction, the surface is concave.

All sorts of effects may be seen through an optical flat. A few of
them are interpreted in Fig. 12. Needless to say, if the interference
bands are straight and parallel to each other, the workpiece surface can
be considered flat at least within a few millionths of an inch.

Where the inspector's work calls for a great deal of exacting work
and checking with optical flats on a variety of equipment, he would do
well to supplement the description just completed with further study of
catalogues, instruction books, and other literature issued by the
manufacturers of optical flats.

Edges worn round.

Edges worn with
hollow in center.

Partly flat — changing
to hump or hollow.

Partly flat — falling off to
or sloping to hump.

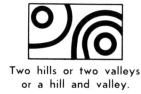

Two hills or two valleys
or a hill and valley.

Fig. 12. Some of the patterns formed by interference bands and their interpretation.

Optical Comparators

Optical comparators are commonly called optical projectors or contour projectors, and are used for special measurement applications that would be difficult to assess with other means. The units have special adaptabilities because they make two-dimensional comparisons or projections of the workpiece to be measured. Out of these comparisons come the capability for indirect and direct measurements. They are often used in special configurations with other mechanical measurement and electronic amplification and control systems.

An optical comparator, so-called, is an apparatus for projecting the enlarged shadow of the profile of an object or workpiece on a ground glass screen. The image seen is usually a silhouette, but may be a detailed image. On the whole, an optical projector is a bulky piece of apparatus because space or distance is needed in order to enlarge or "blow up" the object optically; however, bench models for enlarging or magnifying and silhouetting small parts are available. Figure 13 shows a few of the commercially available models, including one which shows a detailed image of the workpiece.

Whatever its overall size, the optical comparator uses many of the principles and attachments found on the toolmaker's microscope. It is usually equipped with a "table" that can be moved from side to side or laterally, and from front to back. It can also be readily elevated or lowered as well as revolved like a turntable and tilted at an angle, all within certain limits, of course. With such facility, the object or workpiece can be moved into position so that its silhouette will take the

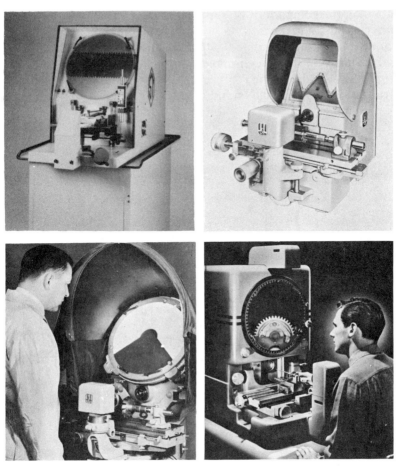

Courtesy of S-T Industries
Courtesy of Bausch & Lomb Optical Co.
Courtesy of Jones & Lamson Machine Co.
Courtesy of Eastman Kodak Co.

Fig. 13. Some of the commercially available optical comparators.

desired position on the screen. By one means or another — micrometer screw and thimble, indexed elevating screw scale, protractor scale, precision gage blocks, or an indicator — the movement of the workpiece clamped to the table can be measured or registered.

The comparator screen is frequently provided with cross hairs so that, as on the toolmaker's micrometer, the movement of the workpiece from, say, one edge to another can be registered. Separate translucent plates with templates and outlines etched on them can be clamped on the comparator screen so that the outline of the workpiece profile can be compared with a preestablished model.

Some models are equipped with a selection of magnifying lenses so that the image on the screen may be projected as 10, 20, 50, or more times the size of the original object. A study of the optical projection and reflection "circuit" shown in Fig. 14 will help in understanding the optical comparator.

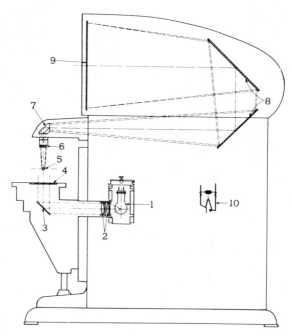

Fig. 14. Schematic diagram of the optical projection and reflection "circuit" of an optical comparator.

These units display proportionally correct "blow-ups" or images of the unit to be measured. In order to accomplish this, these contour projectors hold a light source, a lens system, a staging table, a viewing screen, and an array of special fixtures and measuring devices to fit the particular application. These fixtures and accessories aim to hold, align, and move the workpiece in relation to the lighting and amplification system for fast and ready comparing and measuring.

Light is projected past the part to be measured, and the image of that part is shown in a silhouette on the screen. Depending on whether the workpiece is being measured or gaged, the screen will hold a chart gage, with an appropriate reference scale, or a scaled contour of the workpiece, for purposes of comparison with the workpiece. The units have the capabilities of measuring in both methods, and setups are often devised to combine elements of each.

Lighting for these measuring machines is delineated as a horizontal, vertical, or surface (source) arrangement. Horizontal and vertical systems produce an image on the screen after it has passed the workpiece, and differ in the direction from which the light emanates, and the configuration around the workpiece and staging area. Surface lighting is different because it produces a reflected image of the surface of a part. Combinations of the three layouts are available on some machines.

Many types of light sources are used on these units, depending on the intensity of the light that is needed, and the coherence of the light, or the ability of the light (and lens system) to produce parallel paths for the light. Very-high-intensity light sources and illumination are needed for the surface-reflected configurations.

The work table or surface area, which holds the workpiece, is critical to effective inspection and measurement. The work table generally provides motion or mobility to the part being measured in three planes or dimensions — up, down, and into the focus plane for sharpness of detail or clarity. Additionally, the table will be equipped with special fixturing, holding, and comparing devices that make possible the comparison or measurement from any given position.

Projection systems are varied within these units, again, designed according to the intended use or application. These systems vary the amount of magnification used — from low to high power, according to the contrast and types of projection sought. The systems use different combinations of mirrors and lenses to accomplish their purpose.

After the light and image have passed through the projection system, they are directed to the viewing screen, composed of glass with one surface ground to an opaque viewing surface.

There are three basic measurement methods or techniques for measuring with optical comparators. These are measurement by comparison, by movement, and by translation. Measurement by comparison is the most common use for these units. The workpiece is compared in a go/no go fashion to a master. A special grid, moving table, and other measurement devices make possible measurement in a direct fashion. And, finally, measurement by translation involves the use of special tracer accessories.

In performing measurements by comparison, the setup is very important to accuracy and timeliness of inspection. The positioning mechanism holds the workpiece in the focal plane of the lens system for projection, and is then compared with a chart that outlines high and low end tolerance condition of the piece being measured.

Measurement by movement is even more dependent on the setup and gradient system of the unit. The projected image is related to a reference point that is the optical axis of the comparator. This system works best for single piece inspection.

Finally, measurement by translation means using a probe or pantograph stylus mechanism to trace contours of the part that cannot be directly projected onto the screen. One arm of the pantograph traces the part, and the other arm traces a projected image onto the screen.

These tracer units are usually used with coordinate slides to provide the reference to a dimension or measurement. The types of tracer followers or the projected images vary with the type of measurement application, the size, and the shape of the workpiece. Probe-type, dot-type, and reticle-type tracer followers are all used on these units.

Final detailed instructions in the use of an optical comparator can be best gained from reference to manufacturers' instruction books or from experienced users. While the several available commercial makes are much alike, basically, each has its mechanical and adjustment peculiarities. Manufacturers' pamphlets also offer many suggestions for practical uses of optical projectors, prominent among which are, of course, checking of screw thread and gear tooth form and angle, thread lead, gear pitch, gear, hob and cutter form, tool form, hole centers, and a great variety of special profiles.

There are some profiles on workpieces whose conformance to specification is practically impossible to measure by any other system than to have a draftsman carefully lay out the greatly enlarged facsimile on a translucent screen and then project an image of the workpiece, which has been enlarged to the same size, on the screen for comparison. Other shapes are measurable by other mechanical means, but the process may be infinitely slow in comparison to the speed and facility with which the optical comparator will do the job.

Optical Gaging Charts for High Accuracy

Considerable progress has been made in the design and production of charts for accurate gaging with optical contour projectors. Charts are now produced by a scribing process for which accuracy is claimed to be within 0.0002 inch over the entire chart area. This would mean that for an image or shadow that has been magnified by 10, the error in terms of actual size of the part would be 0.00002 inch while at 50 magnification, it would be only 0.000004 inch. Thus, chart errors need no longer be a significant source of inaccuracy in optical gaging.

For very close tolerances, a unique method of arranging chart tolerance limits, known as "the optical bridge," enables an operator to detect as little as 0.0001 inch variation in a part even at a viewing distance of several feet from the projector screen. Selective grading is readily accomplished by using a special arrangement of chart lines that designate the amount by which the part is under or over the prescribed tolerances.

The basic feature of the optical bridge is the ease with which a band or "sliver" of light can be detected between a rather wide black gaging line and the edge of the projected shadow profile. This is illustrated by Fig. 15 in which two wide black bands and two narrow lines are shown with the same width of space between them. A gaging line, known as the Micro-Gage* Bridge Line, is shown at A in Fig. 16. If even a sliver of light is seen within the bridge, as at B, it denotes that the part is within tolerance limits. If no light is seen within the bridge, as at C, the part is oversize. If light is seen beneath the footings of the bridge, as at D the part is undersize. Correct size parts will show a pattern of lighted rectangles as in Fig. 17.

Courtesy of Optical Gaging Products, Inc.

Fig. 15. Diagram shows ease with which space between two wide black bands can be discerned as compared with same width of space between two narrow lines.

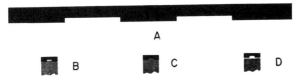

Courtesy of Optical Gaging Products, Inc.

Fig. 16. (Above) Special design of gaging line for optical comparator use. (Below, left) If enclosed light is seen within bridge, part is within tolerance limits. (Below, center) If no enclosed light is seen within the bridge, part is oversize. (Below, right) If light is seen beneath footings of bridge, part is undersize.

A row of accurately positioned alternate rectangles, as shown in Fig. 18, comprises a Micro-Gage centerline. The top edges of the lower rectangles and the bottom edges of the upper rectangles are exactly collinear. Any overriding of the shadow above or below the gaging edges of the rectangles is readily discernible and permits repeat readings to be made within 0.0001 inch.

*Trade name, Optical Gaging Products, Inc, Rochester, NY.

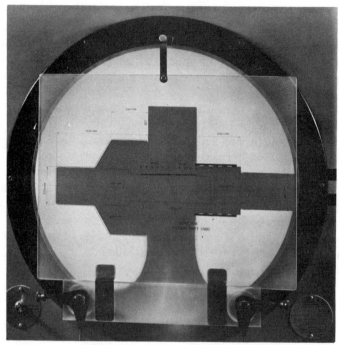

Courtesy of Optical Gaging Products, Inc.

Fig. 17. Correct size parts will show a pattern of lighted rectangles as along right hand edges of this shadow profile.

Courtesy of Optical Gaging Products, Inc.

Fig. 18. This row of accurately positioned alternate rectangles provides a very accurate centerline. Top edges of lower rectangles and bottom edges of upper rectangles are exactly colinear. Any overriding of shadow profile above or below gaging edges of these rectangles is readily discernible.

A somewhat similar arrangement in which two broadened lines without the "bridge" openings have their gaging edges precisely located at the maximum and minimum angles is used for accurate gaging of small angular tolerances.

Locating Points in Space

Inspection difficulties frequently arise from the necessity of checking dimensions from points and lines not located on the surface of the workpiece, i.e., the centers of holes, the intersections of two straight lines that are the extensions of rounded corners, the radius center of a fillet, etc. The use of an optical contour projector greatly simplifies the task of inspection when these problems are present.

In Fig. 19 a standard radius chart is shown being used to locate the radius center of a fillet. In Fig. 20 a standard centerline screen is being employed to find the point of intersection D of sides CD and ED where a corner has been rounded. This same type of screen can be used to find the intersection of two lines joined by a rounded corner when they are not at right angles to each other. Full details concerning this procedure and also for finding the centers of holes and other similar problems can be found in the Eastman Kodak Pamphlet U-3 "Points in Space."

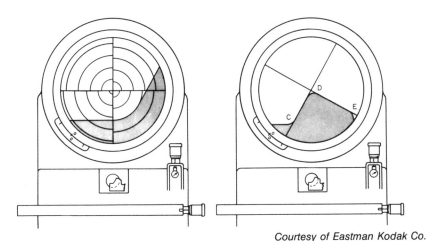

Courtesy of Eastman Kodak Co.

Fig. 19. (Left) Standard radius chart being used to locate radius center of a fillet.

Fig. 20. (Right) Standard centerline screen being used to find intersection D of sides CD and ED where corner is rounded.

Checking Profiles that Cannot be Projected

Another problem now solvable with the aid of an optical contour projector is that of checking internal or hidden contours which are difficult or impossible to project on the contour screen. This is accomplished, as shown in Fig. 21 (left), by means of a special tracer unit. With this unit, a ball-tipped tracing stylus is moved over the part contour while an identical ball moves through an identical path and is projected on the screen. These two balls are connected by a pantograph arrangement that causes them to move in identically similar paths. Thus, a chart gaging profile can be placed on the screen and the path of the projected ball with relation to this gaging contour can be observed as the stylus ball passes along the contour of the part. An interposer fixture that makes use of an interposing bar or lever, similar to that just described, permits profiles of large workpieces that cannot be brought into the field of view to be gaged accurately. Other types of stylii, such as the disc-shaped one shown in Fig. 21 (right), are used depending upon the kind of internal contour that is to be measured.

Courtesy of Optical Gaging Products, Inc.

Fig. 21. (Left) Ball-tipped tracing stylus used to trace part contour where it is impossible to project it on the contour screen. (Right) Similar type of tracer unit using a disc-shaped stylus.

Another adaptation of this idea, which permits any desired magnification to be obtained on the screen, is to project a fixed reference circle that is drawn to the desired magnification of the tracing stylus ball. The chart gaging profile, known as a reticle, which is also enlarged to the same magnification, is then moved along the fixed reference circle as the stylus moves over the contour being inspected. The essential differences between these two methods are shown schematically in Fig. 22. Note that in the second method (illustrated by the right-hand diagram in the figure) the moving chart profile is necessarily inverted.

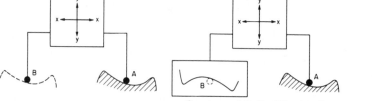

Courtesy of Optical Gaging Products, Inc.

Fig. 22. (Left) Setup where projected ball B moves over drawing of part contour for optical comparator gaging operation. (Right) Setup where projection of accurately drawn part profile is moved over a fixed reference circle, as at B. Note that profile here is inverted as compared with that in the left-hand diagram at B.

The use of multiple-position fixtures that permit a part under examination to be moved in distinct steps of accurately known amounts is another way of handling the optical projection gaging of large work pieces. In Fig. 23 is shown a specially designed broach-locating fixture which permits the broach to be indexed across the path of the optical system in accordance with accurately positioned indexing notches.

Courtesy of Optical Gaging Products, Inc.

Fig. 23. Specially designed broach-locating fixture, which permits broach to be indexed across path of optical system for checking tooth form, spacing, and wear.

A Sharp Image Is Required

The major difficulty in optical projection is to secure the true edge of the silhouette, to obtain a sharp image free from halation, aberration, fuzziness, or double image. It is easy for the shadow on the screen to be a sort of composite of silhouettes. An attempt has been made in Fig. 24 to diagram in exaggerated fashion what is meant by the possibility of double shadow in connection with a hole location problem. The silhouette on the screen may show rim a of a hole, for instance, or rim a', or a composite of both and the operator may wonder as to which shadow edge to locate against the measuring hair line on the screen.

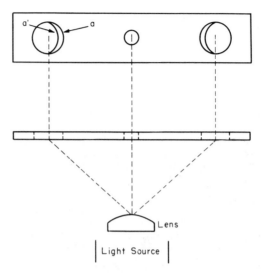

Fig. 24. The appearance of a double shadow in an optical projection hinders accurate measurements.

In any event, every effort should be made to so locate the workpiece on the table in the path of the projected light beam and to focus light and lens so carefully that double shadow, false silhouette edges and other aberrations are avoided as far as possible. If there is real doubt over the measurement secured on the optical comparator, if the tolerances are close, and if other suitable mechanical gaging methods are available or feasible, perhaps these should be employed as a check on or substitute for measurement by optical projection.

For both the toolmaker's microscope and the optical projector, good housekeeping is a prime requisite. Dust, moisture, oil, hair, fuzz, chips, and particles readily collect on such apparatus. Too often they are left uncovered for long periods. More frequently they are as neglected as an attic room so far as keeping them clean is concerned. rust and corrosion, scratches and digs, are as much taboo on the table, centers, vises and holding mechanisms as they are on surface plate equipment. These instruments contain graduated scales — vernier, protractor, micrometer — and should receive the same care as any precision measuring instrument.

Alignment Telescopes and Autocollimators

The highly successful tooling use of alignment telescopes and autocollimators in place of wires, plumb bobs, transits, and levels for accurate positioning of aircraft assembly jigs and accessories has led to their adoption for a number of other manufacturing purposes including checking and inspection.

The autocollimator is an instrument that is a combination collimator and telescope that projects a bundle of parallel light rays through a lens system, and receives the light in a reflected image. The device uses a mirrored target that references a measurement scale and provides a means of dimensional assessment.

The alignment telescope has an internal focusing optical system built into a tubular metal case, the external circumference of which is ground to be precisely concentric with the axis of the optical system. When this telescope is used in conjunction with a sighting target, it is possible to measure lateral displacements from an established line of sight with an error of the order of .003 inch at 120 feet or .001 inch at 40 feet. Furthermore, this highly accurate optical reference line is much more easily maintained and reestablished than a physical reference surface or a reference wire.

Figure 25 shows an alignment telescope with vertical and horizontal optical micrometer adjustments. By means of these adjustments the cross hairs in the telescope can be aligned with the cross hairs or other pattern in the target. Any vertical or horizontal displacement of the target from the line of sight can be read directly on the micrometer scales that are clearly visible in the eyepiece.

When the alignment telescope is used with a collimator, small angular displacements from an established line of sight are readily detected and measured. The collimator is a ground steel tube of the same diameter as the optical telescope and contains a glass reticle or target illuminated from behind and a lens system that causes the rays of light which pass through the reticle to leave the collimator in parallel paths. If the alignment telescope is aligned with the collimator so that the target pattern on the reticle is in view, any displacement of this

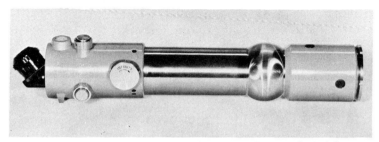

Courtesy of Farrand Optical Company, Inc.

Fig. 25. Alignment telescope with vertical and horizontal optical micrometer adjustments.

target pattern as it is viewed against the cross hairs of the telescope will indicate an angular displacement of the collimator from the line of sight of the telescope. A scale may be provided on the reticle of the collimator to permit readings of the angular deviations.

An autocollimator is shown diagrammatically in Fig. 26. In this instrument the illuminated reticle is within the instrument itself, hence only a mirror is needed as a target. An optical flat is used as the mirror. As shown in this diagram, when rays of light pass from the lamp through reticle No. 2, they are reflected by the beam splitter so that they pass through the objective lens and are projected as parallel rays to the mirror target. These rays are reflected back through the objective lens of the telescope and pass through the beam splitter to be viewed in the eyepiece. If the mirror target is not exactly at right angles

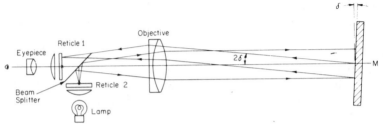

Courtesy of Farrand Optical Company, Inc.

Fig. 26. Diagram of autocollimator showing that observed angle 2δ is twice the actual angle of displacement δ of the target.

to the optical line of sight of the autocollimator, the reflected image of reticle No. 2 will be displaced from that which appears on reticle 1. As can be seen in Fig. 26, the apparent angle of displacement (2δ) is twice that of the actual angular deviation (δ) of the mirror target from the perpendicular, a factor which improves the accuracy of measurement. By using an autocollimator, targets can be aligned to the line of sight within an error of the order of one-half second.

Figure 27 shows a checking fixture on which seven autocollimators are mounted for close range checking of a rocket missile component where extremely small angular tolerances are called for.

Modern Optical Inspection and Measurement

The newest innovations in measurement machines involve the use of new technologically sophisticated gaging systems that deploy non-contact heads, offering no physical interface with the workpiece. These

Fig. 27. Checking fixture for rocket missile component in which seven autocollimators are used.

expensive and flexible units have high-resolution optical sensors that transmit the perceived variation on the workpiece to a measurement scale through electronic application.

Obviously, these units represent the last word in measurement, and, as state-of-the-art gaging devices, are expensive and still rapidly changing. But for the inspection operation with special needs — high resolution and accuracy, and unique demands of workpiece sensitivity or contamination control — these units offer infinite promise.

Visual inspection units are also being introduced that make use of sophisticated video and computer link-ups to assess more than just direct measurements, in the traditional sense of a comparator or gaging system. Printed circuit boards, for example, can be automatically "inspected" at a high rate of speed for the presence or absence of components, for proper orientation, and a high degree of detail, through a set of programmed expectations for the video or optical system.

Indeed, the means and times for measuring 1/10th of one millionth of an inch are upon us. The use of optical interferometers to assess the smallest of physical variation is, in many areas of industry, an everyday or standard requirement. Interferometers are available that work from the measurements and interactions of many different light sources. The setups and configurations of these measuring devices vary greatly — from units that look and act very much like traditional CMMs, to machine-top setups that are movable, with incredible range and flexibility, to very specialized setups intended for a single purpose. In many cases, these units utilize the unique properties of laser light in the measuring scheme — light that, unlike most other types of optical illumination, is very coherent and predictable.

Laser-measuring devices typically measure to microinch accuracy, and, like other highly modern and sophisticated measurement devices, rely heavily on electronics and computerization for their effectiveness. Previously long set up and preparation time that involved measurement and compensation for temperature, humidity, and so forth, are now automatically calculated and compensated for. Specially designed machines measure tolerances below a millionth of an inch. Measurement systems are available that accurately use light-wave interpretation techniques to measure in angstroms (10^{-8} m; 10^{-10} cm), and approaching the atomic level. Laser-measuring units and laser-based interferometers can, like many new measuring tools, perform simultaneous dimensional assessments, process and display the variation, and provide a video representation of the form.

CHAPTER 13

Gaging and Inspection of Screw Threads

The range of screw thread measurement is quite extensive: inspectors involved in the direct manufacture of these fastening units must of course be familiar with the intricacies of the mechanical processes and methods used in the production arena. But others in the inspection and measurement fields must also understand nuts, bolts, screws, studs, and tapped holes. Design and dimensional considerations of these simple devices, as well as taps, dies and chasers, and turned, rolled, and ground threads are important in many areas of inspection, manufacturing, assembly, and measurement.

Few inspectors can avoid contact in some fashion with screw threads in a plant that performs machining operations on metal or plastic parts and any extra hours he spends studying threading techniques and screw thread elements will be profitable. Information and data on screw threads can be found in *Machinery's Handbook*, published by Industrial Press, and treatises on the subject are available in technical libraries. Inspectors could review profitably the contents of the Screw Thread Standards, Handbook H28, U. S. Department of Commerce, National Bureau of Standards, as well as ANSI B1.1, Unified Screw Threads so as to be familiar with the data presented therein, for future reference.

Our concern here is with the suitable measurement and checking of screw threads to determine their conformance or non-conformance to specifications.

The Mating Part as a Gage

Possibly the most elementary method of measurement lies in the use of a so-called mating part. Years ago, you would see a nut of the proper size wired to the lathe where a man was turning a screw thread. Or hanging on the wall near any tapping operation would be a screw or stud that was judged to be the proper size for checking purposes. If all the screws being cut fitted a certain nut, if a selected stud would readily run into the holes being tapped, that was all that was required. Even

461

today, more often than might be suspected, batches of screws, nuts, or tapped holes in workpieces are checked with some selected matching part. However, as will be seen, a much more exacting inspection is usually called for.

Factors to be Checked

There are several factors that will affect the assembly and proper fit of mating external and internal threads. In addition to the rather obvious visual defects of burrs or slivers and stripped, upset, rough, and malformed threads, measurements and checks are made of pitch diameter, major and minor diameters, lead, angle, and thread form. A less obvious visual defect, but important, is the so-called drunken helix. Then, too, the seemingly ridiculous circumstance may arise when the screw, stud or bolt, for example, has a $\frac{1}{2}$ inch-13 thread and the nut or tapped hole has been tapped with a $\frac{1}{2}$ inch-12 thread. A check of the number of threads per inch is all too often overlooked, ignored, or assumed.

If a complete job of thread measurement is to be done, the following checks should be made in about the order shown.

Threads-per-inch count or pitch measurement.
Visual inspection for burrs, slivers, stripping, upset thread, and
    drunken helix.
Pitch diameter.
Major or outside diameter.
Minor or root diameter.
Lead.
Thread angle and form.

The test for thread count is simple. Lay a steel rule against an outside thread and count the number of peaks in 1 inch. Even better and faster is the use of a screw-pitch gage, a template such as is illustrated in Fig. 1. This type of gage is ordinarily necessary for counting internal threads.

Visual Inspection of Threads

The visual inspection of the workpiece thread, probably with the aid of a glass, should be one of the first steps in screw inspection although most people try to use a gage first and then stop to examine the screw to find out why it doesn't fit the gage. In other words, get rid of the variables of dirt, chips, splinters, malformed threads, or a drunken helix before wearing valuable metal off the ring gage or before

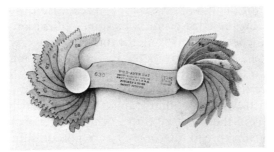

Fig. 1. Commercial type of screw pitch gage.

deciding that the thread is oversize, tapered, or that lead error is present.

Thread Micrometers

For measuring or checking the pitch diameter of an external screw thread, a thread micrometer may be used. The end of the spindle of a thread micrometer is pointed as shown in Fig. 2 to a 60-degree cone for Unified threads and an accurate 60-degree V is ground in the anvil. This anvil is free to rotate so as to adjust itself to the helix angle of the thread being measured. The sharp tip of the spindle point is ground off. Likewise, flats are ground on the peaks of the V and the root of the V is ground out or "cleared." This is done to make sure that only the pitch diameter is measured by the thread micrometer and not the root

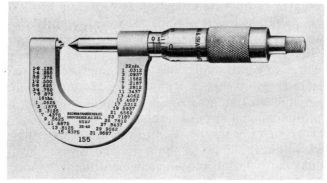

Fig. 2. Screw thread micrometer.

diameter or major diameter of the screw thread. Before using a thread micrometer, screw the cone point of its spindle down into contact with the V-anvil and check the micrometer thimble's zero reading.

A thread measuring micrometer is designed to measure threads within a certain range of pitches. Thus, one thread micrometer may be used to measure the pitch diameters of threads in the range of 48 to 64 threads per inch while another thread micrometer is required to measure threads in the range of 8 to 13 threads per inch.

Since any given thread micrometer is required to measure a range of threads of different pitches, each of which may cause a slight variation of the anvil position on the thread, small errors in measurement are sometimes introduced. For this reason, the best procedure to follow in using a thread micrometer is to first measure the pitch diameter of a standard thread plug gage of the same size as the thread to be measured (if one is available), and to note the possible error. This error is then compensated for when checking the workpiece thread. As an example, suppose a batch of workpieces having a 1″-8 NC-3 thread was to be checked for pitch diameter with a thread micrometer and that the known pitch diameter of the available 1″-8 NC-3 thread plug gage was 0.9188 inch. If the thread micrometer measurement of the plug gage is 0.9183 inch, then the error of 0.0005 inch (0.9188 — 0.9183) must be added to the micrometer reading when a workpiece thread is measured. Thus, if the thread micrometer reading for a workpiece is, say, 0.9180, the actual pitch diameter is 0.9185 (0.9180 + 0.0005).

Classification of Thread Gages

Thread gages may be classified into two broad groups; in one group are the gages used to check the product and in the other are the gages used for reference. In the first group are the *working* gages that are used to check the product as it is being machined, and the *inspection* gages that are used to determine the acceptance or nonacceptance of the product. In the second group are the *setting* or *check* gages that are thread plug gages to which adjustable thread ring gages, thread snap gages, and other thread comparators are checked for size, and *master* or *basic* gages that are thread plug gages representing the physical dimensions of the nominal or basic size of the part.

In the first group of gages, the *working* gages are sometimes set to limits that are within the limits of the *inspection* gages. This practice ensures that any part which is passed as being within tolerance by the working gage will also be passed by the inspection gage thereby reducing the possibility of a disagreement between the machine operator and the inspector in borderline cases, a disagreement that often arises when working and inspection gages are set to identical limits. The principle involved may be summarized by use of an analogy: If a 1″ ball fits into a 1.1″ hole, surely it will fit into a 1.2″ hole.

Working, inspection, setting and *master* gages differ with respect to the accuracy with which they are made. A *working* gage is made to the widest tolerances and is, therefore, the least accurate, while a *master* gage has the narrowest tolerances and is the most accurate. *Inspection* and *setting* gages lie in between. Gage makers manufacture gages to *working, inspection,* or *master* gage tolerances, depending upon the intended application.

Screw thread gages are also classified according to accuracy as W, X, and Y, the W being the most accurate. The dimensions and applications of each of these three classes of gages are covered in *Machinery's Handbook,* the National Bureau of Standards Handbook H28, Screw Thread Standards for Federal Services, and in ANSI B1.2, Screw Thread Gages and Gaging.

Thread Ring Gages

While a thread micrometer might be a natural first choice as an instrument for checking pitch diameter, in most shops thread ring gages are available and commonly used. The ring gage is the modern, wear-resistant, accurate counterpart of the old-fashioned nut hung by a wire on a machine. Illustrations of thread ring gages — also thread plug gages — appear in Fig. 3.

Basically, of course, a thread ring gage is just that — a single threaded ring. It can be a "Go" ring or a "No Go" ring. As Fig. 3 indicates, thread ring gages are usually supplied in pairs. In most cases, a "Go" ring or a "Go" plug alone is insufficient for making a suitable check of the conformance of a screw thread, and the "No Go" member of the team also needs to be employed.

The first step in using a set of ring thread gages is to read the legend stamped on the rings. One ring should read "Go." The legend *should* give the thread size and pitch as, for instance 7/16-20 NF, which, translated, means the workpiece should be 7/16 inch major or outside diameter; there should be 20 threads to the inch (National Fine Thread Series). The pitch diameter, upper limit, should be .4050 inch which should also be stamped on the gage. The "No Go" member should show the same legend except that the pitch diameter (lower limit) would be .4024 inch. Sometimes the class of fit appears, but comparison of the pitch diameter tolerances with handbook thread tolerance tables will classify the fit. For example, the pitch diameter limits just

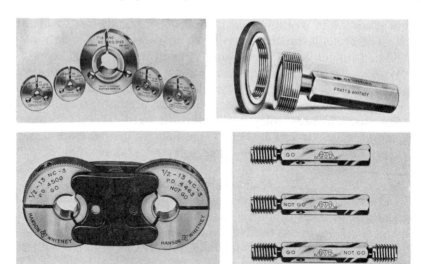

Courtesy of Colt Industries, Pratt & Whitney Machine Tool Div.
Courtesy of Hanson-Whitney Co.
Courtesy of Greenfield Tap & Die Corp.

Fig. 3. Typical thread ring and thread plug gages.

mentioned are for a Class 3 fit. The gage legend should correspond to the work specifications.

There is a strong tendency to assume that so long as a thread ring gage is on the job it is automatically an accurate gaging device. Where the shop systematically checks its thread gages as a routine or where the inspector is sure the gage has recently been tested, he can forego his own check of it, because the gage should withstand a number of gaging operations without appreciable loss of accuracy. Where this is not the case, the pair of rings should be checked on setting plugs if the inspector knows nothing of their history, career, or existing condition. A setting plug is a very accurately made thread plug that is used to facilitate the adjustment of a thread ring gage to its proper pitch diameter. One such setting plug is required for each size of "Go" ring gage, and one for each size of "No Go" ring gage. These setting plugs, therefore, are used to control the accuracy of the ring gages that are used to check external threads.

Because of three radial slots — one of them a through slot — cut into the body of the thread ring gage, see Fig. 4, the effective pitch diameter of the gaging section can be reduced or increased within a restricted range (from .002 inch on small sizes to .010 inch on large rings). The assembled view in Fig. 4 illustrates how this can be done.

Turning screw No. 3, pushes the head of the precision ground sleeve No. 2 against the shoulder in the left-hand segment of the gage and bends to open or spread the gage. Unscrewing No. 3 reverses the action and withdraws the sleeve. When screw No. 1 is tightened, the gage segment shoulder is tightened against the sleeve No. 2 while, simultaneously, the split screw No. 3 expands in its tapped hole and the whole mechanism clamps into position.

Courtesy of Taft-Peirce Mfg. Co.

Fig. 4. (Left) Assembly view of a thread ring gage. (Right) Ring gage with adjusting elements removed.

In most shops, a gage that has been suitably checked and set on a correct setting plug is officially "sealed" by pouring wax in the two adjusting screw holes. Where such a system prevails, an inspector should at least suspect and probably not use a thread ring gage if either or both of these wax seals are missing. Workers and others without authority frequently set thread ring gages to suit some particular circumstance or variation of their own.

Thread ring and plug gage tolerances, and other standard gage-makers' tolerance should be mentioned and understood in this context of using these go/no go devices. Once again, these measurement gages are used based on the need for inspection of a certain type, based on the particular customer requirements that are displayed for a booster rocket or a piece of farm machinery.

Obviously, one will not suffice in all situations. Many categories and grades of standards are available, based on need, based on requirements that drive the inspection. A clearly understood relationship should exist between the standard that is used and the requirement. And the more precise and refined standard will cost more — cost the customer more, as well as the inspection group buying the gages.

Setting an Adjustable Ring Gage

The technique of setting or checking a thread ring gage by means of a setting plug involves one or two peculiarities of its own. In terms of fingertip gaging pressure, the setting plug should screw into the ring all the way so that the end of the plug is flush with the face of the ring gage under certainly not less than $\frac{1}{2}$-pound pressure nor with more forcing than, say, 2 pounds. It is not necessary to "wring" a setting plug in. However, in screwing the plug in, there should be no local binding, catching, hanging up, chatter, nor grating of metal surfaces. A slight surfacing of the ring gage threads with thin oil is permissible. With the setting plug in position, try the gage thoroughly for a snug fit; try to rock it and particularly note if any movement occurs when longitudinal or end thrust is applied. If the ring gage is being adjusted to the correct pitch diameter, test the setting with the master plug after the ring's clamp screw has been finally tightened up. (The master plug is an extremely accurate gage that is used only for checking, never for setting, a ring gage.)

It goes without saying that setting plugs should be used as infrequently as possible and should be checked for size practically every time after use. Where wear up to .0001 inch to .0002 inch is detected, the setting plug should be discarded. (Sections farther on describe the three-wire method of checking setting and master plugs.)

Proper Care and Use of Thread Ring Gages

While the radial slots in a thread ring gage form natural, so-called dirt grooves that tend to clean the threads of a workpiece, it is wise to clean the screw being measured so that it is free from grit, chips, or sludge. The ring itself should be frequently purged in solvent, kept coated with a rust-preventive when in storage, and, in general, treated like a precision gage. If not set down on its face, a thread ring gage will readily roll off a machine or bench. If this happens, check it again on its setting plug; the blow may have distorted it or changed its original setting.

Don't use a ring gage as a paper weight, convenient tack hammer, or as a vise to hold a workpiece. Its radial slots make it, in effect, a threading die and many workers perform the criminal act of using the gage to size their defective work. If more than the customary two pounds gaging pressure is required to screw a thread ring gage on a workpiece, consider the screw thread oversize, defective in some respect, and reject it. Some shops have a rule that a ring gage may bind over the first thread or two on the assumption that the first or lead thread on a screw may be slightly malformed. If the ring goes part way on the screw and then begins to bind, consider the possibility of taper in the screw or a lead error. Incidentally, be sure the legend on a thread gage corresponds with the screw specifications before rejecting a sample workpiece. It is easy to attempt to gage a $\frac{1}{2}''$-13 thread with a $\frac{1}{2}''$-12 gage.

A thread ring gage cannot ordinarily be used to analyze the individual errors present in a screw. If the pitch diameter is oversize, the ring will not engage, of course. Neither will it turn on over an oversize major diameter. If the minor diameter is "off," if the screw teeth roots are filled with dirt or filleted from a worn die or lathe tool, the gage will bind. Excessive screw thread lead error, like taper, will ordinarily be detected after a few turns of the ring gage. Where very fine threads are being gaged, care must be exercised not to "cross thread" the ring gage on the screw.

Some people do not wind a watch or an alarm clock by turning the winding stem. No, they hold the stem and turn the clock. The same thing is unconsciously done with a ring gage. On the whole, it is better to hold the gage firmly and screw the workpiece into it. A more accurate, a more delicate test is secured and much less wear is ordinarily imposed on the gage. If, however, the workpiece is large and heavy, the reverse rule prevails.

How "No Go" Gage Checks the Pitch Diameter

A major or minor diameter may be undersize to a limited extent and not affect the assembly of the screw or possibly not affect its strength in use enough to count, but a pitch diameter must be "on the button," within tolerances, and certainly not undersize. Hence the "No Go" gage. Figure 5 shows the main difference in thread form between the "Go" and the "No Go" thread gages. The "Go" gage will be affected by conditions other than pitch diameter variations. Its roots only are "cleared." But on the "No Go" member not only are the roots cleared, to a greater extent than in the "Go" member, but also the peaks of the gage threads are purposely truncated. The intent is that the "No Go" member shall check pitch diameter only and be unaffected by these other conditions.

If a screw thread will enter a "No Go" ring thread gage by more than one turn and a half, the pitch diameter of the thread being tested is considered undersize.

Ring-gage accuracy can be affected by temperature variations. With many threaded pieces to be checked, the tendency is to hold the ring gage in the hand for a long, continuous period so that its temperature is raised appreciably. This could be avoided by holding the gage in a vise or stand.

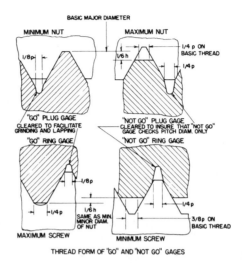

Fig. 5. Thread form of "Go" and "No Go" gages. The "No Go" gage checks only the pitch diameter.

Thread Plug Gages

The subject of the mating-part type of gage is naturally not complete without taking up thread plug gages. This is a single purpose type of gage — one gage to a size — and it is not subject to adjustment at all. More so than a ring gage, a plug gage must be made for a specific thread size, class, and tolerance. A plug gage with "Go" and "No Go" member tolerances for Class 1 threads, for instance, is of no use for Class 3 fits. Wear cannot be compensated for by adjustment, the gage must be replaced after it has worn down several "tenths'" although some plants renew them temporarily by chrome plating them ostensibly back up to size.

In general, the same rules apply to the care, maintenance, and manipulation of plug gages as for ring gages. The "Go" plug ring gage should be screwed in to full engagement with a gaging pressure not exceeding two pounds. The "No Go" end must not enter more than a turn and a half, else the tapped hole is too large. The plug gage may reject pieces for variations in pitch diameter, major and minor diameters, and lead error or for combinations of these, but provides no basis for analyzing the cause.

Although most thread plug gages are provided with dirt grooves, where precision measurement of tapped holes is required, it is preferable that the holes be thoroughly clean and that the gage shall not have dirt grooves. The grooves have a tendency to make the gage act like a hand tap and to size the work. Also, if dirt grooves are absent, the touch or feel of gaging is more sensitive.

One other too common practice with thread plug gages should be avoided. Where the work being tapped is held in a lathe chuck, some workers fail to wait for the motion of the machine to cease before introducing the gage. They want the machine to wind the workpiece upon to the gage. But this is not gaging; all sense of feel is lost.

While it is better to introduce the workpiece to the ring gage, the reverse is true of the plug gage. If you will imagine yourself handling a pencil (except in the case of the big diameter double-handled thread gages), you will probably do a more satisfactory job of measuring.

As will be seen in a section or two farther on, other types of external thread measurement apparatus can well be used in place of the ring gage, but up to this time no better general instrument than the thread plug gage has yet been devised for checking internal threads. One or two types of indicating gages are available, but they must generally be used on hole sizes above $1\frac{1}{4}$ inches.

Roll Thread Snap Gages

One of the objections to the ring thread gage is that the thread size cannot be checked while the workpiece is between centers in the machine. To overcome this difficulty, thread snap gages of the type illustrated in Fig. 6, have been devised; another name for them being *roll* thread snap gages.

The gage "jaws" are really pairs of free-turning rolls that also have lateral freedom or play. There are two sets of rolls: A "Go" pair and a "No Go" pair arranged in the conventional manner of the ordinary "Go/No Go" snap gage. The "Go" roll width — short or long — usually compares with the average length of engagement of the screw threads to be measured. The thread forms on the rolls are annular rings. In other words, the rolls are not cut in a helix like a screw thread.

The ribs of the "Go" rolls have the full thread profile and also some root clearance, whereas on the "No Go" rolls they are truncated and have extra root clearance so that the "No Go" rolls check only the pitch diameter as shown in Fig. 5.

The major attribute of roll snap thread gages is their ability to get at work in the machine. They are also much faster to use for checking a batch of workpieces because the inspector does not have to slow down to screw the workpiece into the gage to full engagement as in the case of the ring gage.

Their major drawback is the ability of the rolls to "roll" literally over a slightly oversize workpiece. For this reason, the recom-

Fig. 6. Roll-thread snap gages shown being checked with combination gage block master, with a setting plug, and being used for gaging workpieces in and out of the machine.

mendation is often made that they be deliberately set from .0002 to .0005 inch undersize. No more than a pound of fingertip pressure should be used in gaging the work, otherwise oversize pieces will surely be passed. Roll thread gages will not readily catch excess ovality in the workpiece unless the inspector takes the trouble to revolve the workpiece in the gage jaws for this purpose.

Other Types of Thread Gages

A variation of the roll snap thread gage appears in Fig. 7, a snap gage whose threaded anvils are rigid but adjustable to a setting plug. While a gage like that shown in Fig. 7 *looks* very satisfactory, it must be used expertly. The anvils are really segments of threads ground with the proper helix angle on cylinders of large diameter. In making the measurement with such a gage, it must be canted or twisted a little — juggled just a trifle on the workpiece threads — to align it with the workpiece thread helix angle.

Courtesy of Taft-Peirce Mfg. Co.

Fig. 7. Thread snap gage in which the threaded anvils are rigid, but adjustable to a setting plug.

The solid-jaw thread snap gage is subject to very ready wear. Its setting should be checked frequently.

The ribbed roll idea has been applied to indicating type gages of the sort shown in Fig. 8. In the type of gage at the left, a single ribbed roll is used on the upper or sensitive (measuring) contact and a double ribbed roll on the lower, solid, reference anvil. The tooth form of the ribs is somewhat truncated and the roots and the bottom of the flanks are "cleared" or ground back so as to make sure that the gage's reading will be affected by nothing more than variations in workpiece pitch diameters. In the type of gage at the right, three rolls are used. The two lower rolls are mounted in a stationary position, while the third roll, mounted on a preloaded armature, is swung into contact with the workpiece. The comparator shown has two-rib rolls for checking pitch diameters only, but the same type of gage is also available with multirib rolls for checking cumulative errors in lead, angle, and pitch diameter.

The indicating thread gage is mastered on a suitable basic or master

Courtesy of Federal Products Corp.
Courtesy of Colt Industries, Pratt & Whitney Machine Tool Div.

Fig. 8. Indicating types of thread comparators used for checking external thread pitch diameters. Initial setting is established with a setting plug.

thread plug gage. The trick in mastering it and in using it subsequently on workpieces is to be sure the back stop is set at the proper height, angle, and depth so that the pair of gaging thread rolls contact on the true diameter of the workpiece and not on a chord. As usual, the back stop can be positioned by watching the gage's indicator as the master is manipulated between the rolls. The indicator's maximum reading denotes the proper position for back stop setting.

Still another fast, accurate, method of checking external threads makes use of the type of gage illustrated in Fig. 9. This upper gage is essentially a functional ring thread gage split in two halves or segments which pivot on studs. The segments literally wrap around the workpiece thread and the indicator shows the comparison between the workpiece and the setting plug with which the gage is mastered. Actually, the indicator reading is the "assembly size" of the screw. The lower gage (comparator) checks the pitch diameter only. The elements are annular single ribbed cone and vee rolls with profile for "No Go," flank contact only. The lower pair of rolls is mounted on a pivoting cradle for maximum ease and speed of operation.

An additional advantage of indicating-type thread gages is their ability to detect out-of-round and taper conditions if the piece is rotated and moved longitudinally. A tapered screw with a maximum pitch diameter that is within tolerance but with a pitch diameter at the other end of the taper that is undersize, may produce an undesirably loose fit in a straight tapped hole especially under strain or vibration.

Fig. 9. A pair of indicating-type gages mounted for "Go" and "No Go" inspection of a screw thread.

Checking Tapered Threads

Screws and nuts with deliberately tapered threaded sections are sometimes encountered. Pipe fittings form the most common example. In considering the measurement of extreme tapered threads, the inspector needs to keep the geometry of tapers in mind. Two of the dimensional variables of a taper must be known by measurement before its conformance can be determined.

What is meant appears, as review, in diagram A of Fig. 10. If diameters *a* and *b* and length *c* are known, then the amount of taper is known. An indicating gage can of course be "mastered" with a correctly tapered piece, its reference anvils *r* and *r*, see Fig. 10-B, and its indicators properly set with the master against a suitable stop *s*. The indicators will then register the deviations of the workpieces from the master.

The same geometry applies to threaded tapers if it is remembered that the pitch lines, *p* and *p* of sketch C, are used for checking the taper. Indicating gages can be made to perform the function hinted at in sketch C. Tapered threads also can be checked by the three wire method — see farther on.

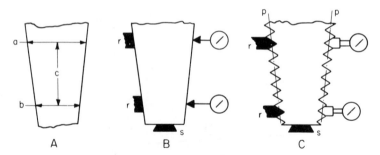

Fig. 10. The same geometry that applies to the measurement of plain tapers is used
in the measurement of tapered threads.

A common and suitable gage for checking tapered threads appears in Fig. 11-A, a gage with a suitable reference stop or base and with triple rolls whose thread ribs are made on a suitable taper. The major diameter or outside diameter is similarly checked by a gage with three tapered rolls that make contact only with the crests of the workpiece thread as shown in Fig. 11-B.

Ring thread gages are also made with suitable internal tapers. The technique is to turn the ring gage onto the tapered thread until the face of the gage is flush with the end of the threaded piece. If the gage will not turn flush, the workpiece is oversize; if the ring gage sinks below flush under normal gaging pressure, the workpiece thread and taper are underisize.

The same technique is used for gaging internal tapered threads with a tapered plug gage. The gage has a flat ground in it and/or scribed lines. The gage should screw in just to the mark in a properly tapered tapped hole.*

Aside from the plug gage, the inspector has no other means of checking tapered threaded holes except by making a plaster of Paris of sulfur cast of the hole (a technique about to be described) and checking this cast with an optical comparator. If the tapered hole is of large enough inside diameter, indicating gage equipment can be designed to solve the problem.

The method of making plaster of Paris casts is fairly simple. Lightly coat the internal thread with oil or grease. Make up the usual thick mixture of plaster of Paris and water and tamp it thoroughly into the tapped hole. If the hole is large enough, insert a pair of wires or a thin

---

*Where the inspector works in a shop making more or less of a specialty of tapered threads as, for instance, on hose couplings, pipe fittings, "dry seal" joints and the like, he should make a special study of the subject in *Machinery's Handbook*, in the Bureau of Standards Screw Thread Handbook (H28), and in the applicable American National Standards Institute publications.

Courtesy of Colt Industries, Pratt & Whitney Machine Tool Div.

Fig. 11. (A, left) Roll-type gage for checking tapered threads. (B, right) Roll-type gage for checking the major diameter of a tapered thread.

steel strip into the soft plaster to reinforce the cast and to provide a means for unscrewing it when hardened. A little practice will indicate how to get casts without cracking or crumbling them plus obtaining sharp, accurate impressions of the internal threads. The cast can then be used like an external thread for various measurements and observations.

Three-Wire Method of Thread Measurement

The fundamental standard method for checking the pitch diameter, taper, and ovality of externally threaded pieces, including setting plugs and plug thread gages and casts of internal threads, is the "three-wire method." Knowledge of the measuring-roll technique for taking measurements on tapered parts facilitates an understanding of the geometry behind the three-wire system.

The pitch diameter may be checked very accurately by this method. It is especially useful in checking very accurate work, such, for example, as thread gages. It usually would not be employed in checking parts in connection with ordinary manufacturing practice because thread gages require much less time and are preferable for shop measurements. The three-wire method, however, is so generally used for precision work that it should be understood. Three wires or pins of the same diameter (within very close limits) are placed in contact with the screw thread, as illustrated by the diagram, Fig. 12. Two wires are

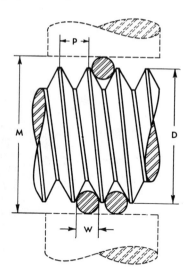

Fig. 12. Method of checking pitch diameter by the three-wire method.

placed in contact with the thread on one side and a third wire on the opposite side. When the micrometer is in contact with all three wires, this ensures measuring perpendicular to the axis of the screw thread. The following simple formula is for determining what the pitch diameter of the American Standard screw is for a given measurement M:

$$\text{Pitch diameter } E = M + (0.86603 \times P) - (3 \times W)$$

where P is the pitch of the thread and W is the pin diameter. As an example, assume that a $2\frac{1}{2}$-inch American Standard screw thread (Coarse-thread Series) has a measurement M over the wires of 2.556 inches, using wires of 0.1443-inch diameter. What pitch diameter does this measurement M represent? Applying the formula, we have

$$E = 2.556 + (0.86603 \times 0.25) - (3 \times 0.1443) = 2.3396$$

Similar three-wire formulas for the measurement of other forms of screw threads such as Acme and British Whitworth are given in *Machinery's Handbook* and other standard handbooks.

If this screw thread is in the Class 3A Series the maximum pitch diameter is 2.3376 and the minimum 2.3298 inches; hence, in this case, the formula shows that when measurement M is 2.556 inches, the pitch diameter is 0.002 inch larger than the maximum allowable pitch diameter of 2.3376 inches.

Measurement M over the wires may be made by using an ordinary micrometer. Special measuring fixtures of the micrometer type have also been developed for use with the three-wire method. These fixtures provide convenient means of holding the wires in position and the micrometer is mounted so that it can move freely either parallel or perpendicular to the axis of the screw thread that is held in a horizontal position between adjustable centers. Some of the measuring devices used are shown in Fig. 13.

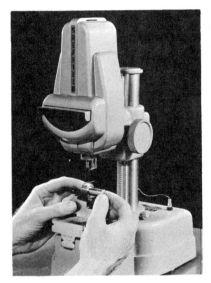 

Courtesy of Sheffield   *Courtesy of Federal Products Corp.*

Fig. 13. Some of the measuring devices used in the three-wire method of measuring pitch diameters.

Determining the Wire Size

In checking screw threads by the three-wire method, any wire diameter W, Fig. 12 may be used provided the wires are small enough to enter the thread and contact with the sloping sides and are large enough to project above the top or crest of the thread, thus permitting proper contact with the micrometer or other measuring instrument. It is preferable, however, to use wires of the size required to make contact at the pitch line or mid-slope of the thread, because then measurement of the pitch diameter is least affected by any error in the thread angle. The term "best size" is commonly applied to wires making pitch-line contact.

To determine the best size wire, divide one-half the pitch by the cosine of one-half the included thread angle in the axial plane or by the cosine of 30 degrees for a Unified thread form. The best size may also be obtained by multiplying one-half of the pitch by the secant of one-half included thread angle. For the Unified or other 60-degree threads, this rule may be simplifed as follows:

Best size wire for Unified Screw Thread = 0.57735 × pitch

Effect of Lead Angle on Three-Wire Measurements

If the lead angle is large as in the case of many multiple screw threads such as are found on worms, quick-traversing lead-screws, etc., the ordinary rule or formula for checking the pitch diameter by the three-wire method is inaccurate and the effect of the lead angle on the position of the wires should be taken into account. This effect depends not only upon the size of the lead angle, but to some extent upon the degree of accuracy required in checking the pitch diameter. The formula given is sufficiently accurate for practically all three-wire measurement of standard 60-degree single-thread screws which have lead angles not greater than about $4\frac{1}{2}$ degrees. (The error in measurement M is about 0.0005 inch when the lead angle is $4\frac{1}{2}$ degrees.) For lead angles, above $4\frac{1}{2}$ degrees, formulas that compensate for the effect of lead angle on the wire measurement should be used. Such formulas may be found in standard handbooks and reference works.

Pitch Diameter Measurement of Drunken Threads

When the thread is correct in form, either the thread micrometer or the three-wire method will give equally good results; however, if the thread is not of the correct shape, but is "drunken," as indicated in the illustrations Figs. 14 and 15, then only the anvil type of thread micrometer shows this variation, while the three-wire method would not indicate any error, provided the thread angle is correct. This is because the three-wire system measures the grooves cut by the thread tool, which is always at the same depth and is unvarying in shape; hence, the error, if any, would not be detected. The same condition is met with in the ball-point micrometer. (See Fig. 16.) The lower anvil point of a regular thread micrometer, however, since it spans the abnormal thread, as shown in Fig. 14, instead of making contact with the sides of the adjacent threads, indicates the irregularity by giving an increasing reading for the pitch diameter.

This does not mean that the three-wire method of measuring pitch diameters is unreliable for ordinary use. With the methods used for

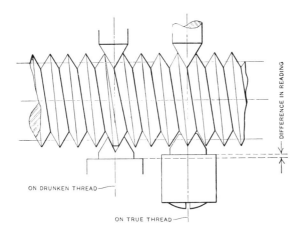

Fig. 14. The anvil type of thread micrometer gives varying measurements on true and drunken threads.

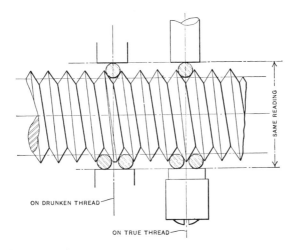

Fig. 15 The three-wire method of thread measurement makes no distinction between true and drunken threads

accurate thread cutting in general, a drunken thread is seldom produced. If the thread is drunken, the thread micrometer will indicate this defect, but the three-wire system nevertheless measures the pitch diameter correctly under all circumstances, since the principle of its use depends on the bearing of the wire on the sides of the thread groove.

The ball-point type of micrometer (Fig. 16) is especially useful for comparing the pitch diameter of a tap or screw thread with that of a standard thread plug gage. Since the purpose is to compare pitch

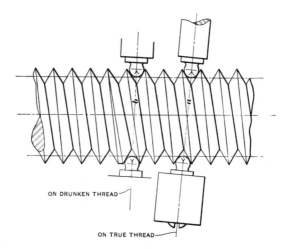

Fig. 16. The only difference in measurement on true and drunken threads made by the ball point micrometer is due to the very slight difference in inclination, too slight to be appreciable.

diameters instead of measuring them, an exact relation between the pitch and the diameter of the ball points is not necessary. An approximate relationship, however, is necessary, since the ball point must be small enough to enter the thread groove and bear on the angular sides. If the thread is a Unified the ball diameter for pitch-line contact is 0.577 × the pitch, but this diameter might vary anywhere from 0.6 to 0.8 × the pitch.

Accuracy of the Three-Wire Method of Measuring Screw Threads

It is possible to check screw thread sizes very accurately by this method; however, the degree of accuracy depends very much upon the accuracy of the wires used as well as the accuracy of the measuring instrument. The measurement may also be affected appreciably by the amount of contact pressure against the wires in measuring. If the accuracy of the pitch diameter of a screw thread gage is to be checked within 0.0001 inch by the wire method, it is necessary to know the wire diameters to within 0.00002 inch. Each wire should be round within 0.00002 inch and should be straight within the same amount over any quarter-inch section. A set of three wires should have the same diameter within 0.0003 inch; moreover, this common diameter should be within 0.0001 inch of the "best size" for any given pitch. As previously explained, the "best size," as it is commonly called, is one which makes contact at the pitch line or at one-half thread depth where any errors in the thread angle will have the least effect upon the measurement over the wires. Tests made to show the effect of contact pressure were made by measuring a 24-pitch thread gage of the plug type. The measurement, with a contact pressure of 5 pounds, was 0.00013 inch less than with a pressure of 2 pounds. If proper precautions are taken regarding wire accuracy and contact pressure, it should be possible to check plug gages within an accuracy of 0.0001 inch. If the wire diameters are accurate to only 0.0001 inch, then the pitch diameter measurement is not likely to be more accurate than 0.0003 inch. This, however, may be accurate enough for many classes of work.

Optical Comparators

Within certain limitations, the optical comparator is readily used to check external threads. The shadow, the sharp silhouette of the thread on the projector screen can be compared with a fine line outline drawn on a supplementary glass placed on the main screen. An advantage of optical projection is the ability to observe outside diameter, pitch diameter, root diameter, form, thread angle, and lead error all at once. The disadvantage is the observer's inability to read and trust measurements where the tolerances are less than .0005 inch.

Indicating Gages

As has been said, the plug thread gage is about the only practical gage for the rapid checking of internal threads, especially on small holes. (At least an hour can be used up preparing a plaster of Paris cast of a tapped hole and checking it either with three wires or by optical projection.) However, when the inside diameter is large enough, indicating gages like those pictured in Fig. 17 are very satisfactory. A master must be used, of course, and great care should be exercised to see that the gage is centralized. The criterion of correct measurement here would be repetition of readings. The contacts or anvils of the indicating type gages, one of which is manually retractable, have thread forms of the proper pitch and type generated on them.

Checking Major and Minor Diameters

The major diameter of screws can be checked with caliper-type gages — vernier, micrometer, or indicating in practically the same

Fig. 17. (A) A portable type of indicating thread gage for checking large internal threads. (B) A bench type of indicating thread gage for internal and external threads.

manner in which the outside diameter of a cylinder is measured. The minor diameter measurement, however, brings up special considerations. For this purpose "chisel" shaped anvils must be used on the gage, anvils whose flanks are so cleared they will not make contact at the pitch diameter or some other diameter along the flanks of the thread tooth.

A diagram of the gage anvil appears in exaggerated form in Fig. 18, which shows at *a* and *b* the cleared flanks of the gage's chisel shaped anvils. The two anvils also should be (theoretically at least) offset by an amount equal to one-half the pitch to accommodate the lead of the

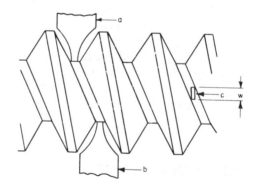

Fig. 18. Schematic diagram showing how the anvils of caliper type gages should be proportioned to prevent contact with the sides of the thread.

thread helix and their width *w* in the other direction, as diagrammed at *c*, should not be so great as to touch the sides of the thread.

Checking Lead Error

The lead error in a screw thread also should be checked. For this purpose the optical comparator or the toolmaker's microscope is used. If the measurement must be made to very close tolerances (as in the case of checking the lead of a plug gage), a measuring machine is necessary. At the inspection bench, gage blocks and special points can be combined as indicated in Fig. 19 or the type of gage appearing in Fig. 20 can be secured. Care must be used in the latter two methods illustrated, not to tip or twist either the gage or the workpiece or in

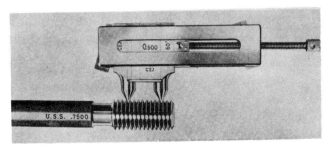

Fig. 19. Checking the lead of a thread gage by using a combination of gage blocks and special points.

Courtesy of Federal Products Corp.

Fig. 20. Checking the lead of a thread gage with an indicating type of gage.

other words make sure that the axis of each measuring point is on a true diameter of the screw.

There is no ready or satisfactory instrumentation available for checking the lead error for internal threads unless the hole size is large enough so that some method similar to that used for external threads can be devised. Where it is necessary to analyze internal thread conditions, the plaster of Paris cast method is satisfactory.

A common manufacturing-plant mixup occurs when screw threads or nuts are measured for conformance and accepted, then plated or galvanized and finally used for assembly. The unplated screws are approved as being to size, but they will not assemble after plating. The answer is obvious, of course; the plating has built upon the screw flanks and changed the pitch diameter (also the major and minor diameters, though the increase in pitch diameter is the more troublesome ordinarily). The question arises, what allowance should be made, in threading screws or tapping holes on parts that are to be subsequently plated? The answer, of course, must come from the shop itself and from some knowledge of how thick a plate or galvanize coating is put on the threaded parts on the average. A series of checks can be made (perhaps by the three-wire method or with an indicating gage like that shown in Fig. 8 on screws and on casts of tapped holes) on unplated components and on the same components after they have gone through the normal plating process until a good average answer is reached. One bothersome feature, however, is the fact that plating (especially cadmium) does not necessarily go on evenly or it may flake off and clog the threads of conventional ring and plug gages. Hence, in many plants, an empirical and arbitrary allowance of 0.0005 inch for plating is established and all dies, taps, thread rings, and plugs to be used on unplated pieces are bought .0005 inch under the standard screw specification and .0005 inch over the standard nut specification.*

Practically everything written in this section applies also to the inspection of 29 degree Acme threads, British Whitworth 55 degree threads, metric threads, buttress and square threads except, of course, that certain mathematical constants are different. Conventional plug, ring, setting, basic, and master gages can be secured for these types of threads as well as indicating and special design gages.

---

*The Unified and American Screw Threads Standard Class 1A and 2A External Thread Limits provide allowances in the maximum dimensions which accommodate plated finishes or coatings.

Checking Visual Defects

Last, but far from least, in connection with screw thread inspection, is the matter of visual defects. Unless the screw is large and the pitch coarse, the visual examination should be made under a four- to ten-power magnifying glass. Such an examination may show metal slivers, thin helical strips of metal partially removed by the die and left attached in the roots, along the flanks, and on the peaks of the teeth. Rough, flaky surface finish may be seen, where the metal has knurled and chattered into tiny humps along the flanks of the teeth. These conditions, especially in the fine thread series and Class 3A and 3B Limits, can well prevent otherwise dimensionally correct screw threads from assembling.

If the die or tool is dull, the roots will be filled; or if the die itself has been improperly made, the peaks of the screw thread will come out too sharp as shown in solid outline, Fig. 21-A.

Where the machine has "crowded" the die — forced it too rapidly over the rod from which the screw is being cut — all sorts of weird shapes may result as shown by the solid outline in Fig. 21-B. The thread angle, lead error, sharp peaks, and filled roots may not be apparent enough under the ordinary glass and usually the optical projector is relied on for such analyses.

While the shadowgraph will disclose deformed threads, it cannot bring out undue surface roughness and seldom uncovers that other culprit commonly known as a drunken helix. The machine and die used

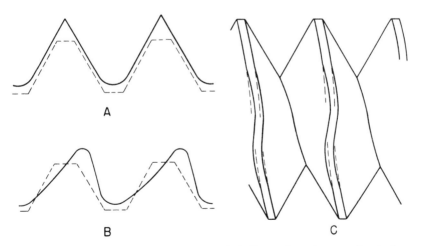

Fig. 21. Visual defects in a screw thread: (A) Sharp thread crests due to improperly made threading die; (B) the result of forcing the threading die too rapidly over the rod from which the screw is cut; (C) drunken helix produced because of intermittent end thrust in the threading machine.

to cut a thread may produce dimensionally correct thread form, surface, lead, pitch diameter, outside diameter, and minor diameter, but because of a short cycle of end thrust in the machine, a wavy, jittery effect known as a drunken helix is produced on a small portion of the thread much as illustrated in Fig. 21-C, the dotted lines indicating the true helix.

If a tapped hole is deep or under 1 inch in diameter, it is impossible to detect directly the sort of visual defects mentioned above. The usual procedure is to reject a tapped hole or nut simply because a plug gage will not enter it. If it is necessary to find out why certain tapped holes are continuously rejected by a plug gage, especially when the taps themselves have been carefully checked, it is necessary to take plaster casts of the hole for more complete observation and analysis. Many times a tapped hole can be readily sectioned or a nut cut in half, thus exposing the threads for visual observation.

CHAPTER 14

Special Measuring and Inspection Problems

In addition to measuring thickness, depth, length, height, outside diameter, inside diameter, radii, and such conditions as taper, ovality, squareness, or eccentricity, the inspector is called on to check angular relationships. Many times the angle between one surface and another is not too important from the point of view of precision; on other occasions it must conform to specifications with an accuracy of minutes or fractions of a degree, a relationship nearly comparable to tenths of thousandths in linear measurements.

Angles and arcs are measurements based on a circular scale or plane of reference. The increments of this circular plane of reference are degrees, of which there are 360 in a circle. These degree increments can be further broken down to minutes (1 degree equals 60 minutes), and seconds (1 minute equals 60 seconds). Some angle measuring tools are capable of measuring to greater than 1 degree precision, while some are not.

Angles and arcs are critical to physical descriptions and blue print reading. Without them, we cannot specify certain measurable relationships. Like other measurements, the need to measure precise levels of angles and arcs varies according to the applications; the pitch of a roof as well as a precisely milled component of a close-tolerance machine are described by angles. In all cases accuracy is important, but the degree of accuracy attained varies with the need and the cost.

Several general methods are used to measure an angle. One is the equivalent of laying a template against the work, much after the fashion of using a radius gage or a curvature plate. The try square is a device of this type, a template literally, for checking a 90-degree angle. Corresponding to the try square, but for measuring angles other than 90 degrees, there is the adjustable instrument called a bevel protractor.

Less precise angular measurements can be obtained through the use of the bevel protractor without a vernier scale, or a simple steel protractor with an angular scale attached to a straight steel rule (Fig. 1). Angle gages are available in various increments up to 45 degrees for rough assessments of angles on a workpiece. Vernier scale protractors and sine bars, as well as some modern electronic angle-measuring

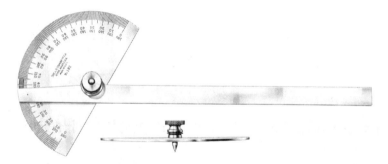

Courtesy of Starrett Co.

Fig. 1. Steel protractor.

devices, can be used for measurements where more precision is demanded.

Angle, slope, and taper measurements are also secured optically, with rather exacting precision, through the cross hairs and transparent protractor of a toolmaker's microscope or on the shadow screen of an optical comparator, which have been previously described.

For superior accuracy, angles are determined by means of precision gage blocks, cylinders, and sine bars as will be seen in sections farther on. Special mechanical indicating or air gages or electric comparators can be designed, constructed or contrived for precision checking of angular relationships.

Angles

Electronics has brought an amazing tool of angle measurement to the world of inspection in the inclinometer/level/protractor that is available in several sizes and applications on the market (Fig. 2).

Sometimes angles cannot be measured or assessed directly. When this is the case, angle translation devices must be employed. A common device intended for this purpose is the bevel, which consists of a sliding and pivoting edge that references from another straightedge, providing the reproduction of an infinite number of angular spacings.

These bevels assess an angle and are then referenced against a measuring device for an angular dimension, and are available in three styles: the plain bevel, the combination bevel, and the universal bevel.

Fig. 2. Inclinometer.

Bevel Protractors

Figure 3 shows a commercial bevel vernier protractor and several examples of its use.

In using a vernier protractor make sure it is bearing correctly against the workpiece as illustrated in Fig. 4-A. It is easy to clamp a protractor against a workpiece and have the blade tip away as Fig. 4-B indicates or for the blade to ride up on the slope as sketch C in Fig. 4 shows. Similarly, the base of the protractor can be tipped or canted as shown in Figs. 4-D and -E. The protractor must feel solid against the work; at the time the angle measurement is made the protractor must not be capable of rocking. Many protractors are equipped with fine adjustment knobs that enable the inspector to revolve the blade readily to a "homing" position through the last fraction of a degree.

If, with the bevel protractor, the principle of reference and measuring surfaces is recalled, there is unlikely to be an error. First, be sure the gage's base, as illustrated at A in Fig. 4, is solidly against the

Courtesy of Brown & Sharpe Mfg. Co.

Fig. 3. Commercial bevel vernier protractor and several examples of its use.

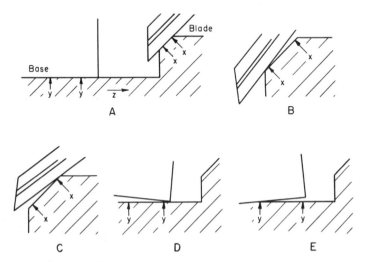

Fig. 4. In using a vernier protractor, contact between the blade and the workpiece should be as shown at (A) not as shown at (B, C, D, and E).

workpiece reference surface at two points as at y and y. Hold the base firmly in this position and turn the blade until it contacts evenly, as points x and x of Fig. 4-A suggest, but be sure the pressure used on the blade does not exceed the pressure used on the base. Guard against inadvertently sliding the whole instrument in the direction of, say, arrow z in Fig. 4-A.

Having set the blade in proper position in relation to the base — having correctly used the protractor as a template — tighten the clamping knob, check the feel of the protractor once more against the work, and then read its scale and vernier.

Reading the Bevel Protractor

Reference to Figs. 3 and 5 will show that the circular clamp of the vernier protractor is divided into 90-degree quadrants and that it is also equipped with a vernier. The 0 of the main scale divides it in either direction; likewise the 0 of the vernier scale allows readings up to 60 minutes (1/60 degree is 1 minute) in either direction.

In reading the protractor scale and vernier it is necessary to note the direction of revolution the *vernier* 0 is taking in relation to the scale 0. If it is traveling to the right of the scale 0, as in Fig. 5, read the vernier scale, 0–60, on *that* side of the vernier 0 for its coincident line. If the vernier 0 moves away to the left of the scale 0, read the vernier scale to the left of its 0. In Fig. 5, for instance, the vernier 0 checks 17 degrees on the main scale in the upper illustration, while in the lower picture the vernier 0-line lies beyond 12 degrees on the main scale. Reading along the vernier in the same direction, to the right, the 50-minute vernier graduation is coincident with a main scale division. Hence, the reading in the lower illustration is 12 degrees, 50 minutes.

Care must be exercised in using the vernier protractor not to register the complementary angle, 90 degrees *minus* the angle you really mean to read. Such a mistake is especially easy where the workpiece angle is close to 45 degrees. If the actual angle were 43 degrees, for example, it is easy to get reversed and register 47 degrees (90 degrees minus 43 degrees = 47 degrees).

Care of the Bevel Protractor

As for maintenance, the bevel protractor should be kept rust-free and clean. Reasonable care should be exerted to prevent its being nicked. More than reasonable care should be observed to see that the blade does not become bent. When the tongue and clamp screw arrangement wears so that the blade cannot be adequately tightened, the instrument should be returned to the manufacturer for repair or

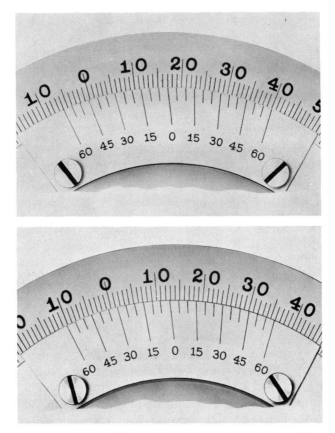

Fig. 5. (Upper) Close-up view of bevel protractor vernier scale. Protractor is set at 17 degrees. (Lower) Protractor set to 12 degrees 50 minutes.

renewal. It is wise to check the vernier reading on a known angle (on a sine bar, for instance) and, by means of the vernier plate screws, adjust its accuracy.

The bevel protractor has an ordinary accuracy, discrimination and precision to 5 minutes or 1/12th of a degree. This discrimination and the ability of the inspector to manipulate the gage correctly compare with the precision of the steel rule, which measures to .015 inch. When greater accuracy in measuring angles or tapers is required, the inspector must resort to more precise equipment such as the sine bar.

Sine Bars

A sine bar is a carefully machined tool steel bar that is used with two properly spaced cylinders that may or may not be fastened to it. In its simplest form, see Fig. 6, the oblong steel bar is 4-square, the cylinders are round and free from taper to close tolerances. In particular, the axes of the cylinders are parallel to the adjacent sides of the sine bar and are located at a definite distance apart, usually 10 inches. The bar comes with holes and slots through it so that work-pieces can be more readily clamped or bolted to it. As Fig. 6 shows, sine bars come in other and special shapes, widths, and thicknesses, also with clamping, revolving, and base attachments.

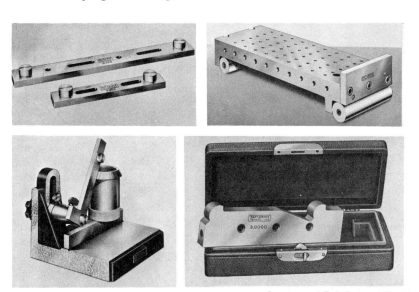

Courtesy of Taft-Peirce Mfg. Co.

Fig. 6. Various types and sizes of sine bars.

Principle of the Sine Bar

The principle of the sine bar can be understood from an examination of Fig. 7. In this illustration, the sine bar is tilted and rested on the slopes of the workpiece whose angular position, the angle a, is required. If measurements h_1 and h_2 are determined, the value of leg b of the right triangle (Fig. 7-B) can be calculated since it is equal to $h_2 - h_1$. The hypotenuse of the triangle is 10 inches because the sine bar is made that way. And by trigonometry, the sine of the angle a equals $(h_2 - h_1)/10$. In other words, after h_2 and h_1 are measured, the sine of the angle becomes 1/10th of the difference between them. (Some sine bars are made with the cylinders 5 inches apart; a few are made with c ..nders spaced 20 inches. For these, the sine of the angle becomes 1/5th of $h_2 - h_1$ or 1/20th of $h_2 - h_1$, respectively.) To

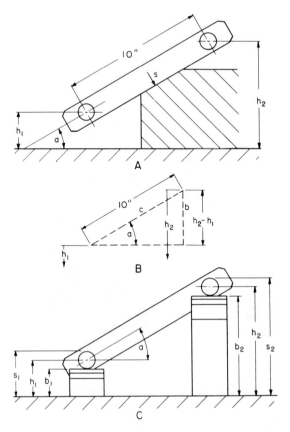

Fig. 7. Use of the sine bar is based upon simple trigonometric relationships.

determine what the angle a is in degrees, minutes, and seconds, the value of $(h_2 - h_1)/10$ is found in a table of sines opposite the corresponding angle or by using a calculator.

Dimensions h_1 and h_2 are ordinarily secured by a vernier height gage or, where extra precision is required, by stacks of gage blocks. Figure 7-C illustrates this diagrammatically. The centers of the cylinders or sine bar plugs do not have to be located, since the diameters of the two plugs are carefully made to be alike. Hence, $h_2 - h_1$ or $b_2 - b_1$ or $s_2 - s_1$ are equal.

Suppose the gage block stack b_2, Fig. 7-C is 7.2657 inches and b_1 is 2.625 inches, the difference between the two ($b_2 - b_1$) is 4.6407 inches. One-tenth of 4.6407 or ($b_2 - b_1)/10$ is .46407. The sine tables show .46407 as 27 degrees, 39 minutes, which is angle a.

A more expert way of handling a sine bar, perhaps, is illustrated in Fig. 8. By one contrivance or another the tapered piece is clamped or

Fig. 8. One method of using a sine bar to determine the included angle of a workpiece.

held firmly to the length of a sine bar (in the case shown in Fig. 8 this is accomplished with the aid of a standard surface plate magnetic block) and the sine bar is tilted at an angle by means of unequal gage block stacks until the upper surface of the tapered piece seems level. A height gage indicator is run along this upper surface and the height of one gage block stack is altered until there is no change of reading on the traversing indicator dial. Then the angle of taper of the workpiece is equal to the angle between the sine bar and the surface plate and can be calculated from the h_1, h_2 relationship of the gage block stacks supporting the sine bar.

A sine bar, it goes without saying, should receive the care and maintenance properly accorded gage blocks. Dirt, grease, grit, sweat, scratches, nicks, and corrosion are enemies of sine bar accuracy, and a sine bar should be inspected regularly for such defects.

Measuring Angle of an External Taper

The angle of taper of a tapered piece can also be measured and calculated as indicated in Fig. 9-A. The tapered piece is set upon a surface plate or machinist's flat. Two measuring rolls or cylinders, c_1 and c_2, are secured. (For this purpose the standard cylinders are usually

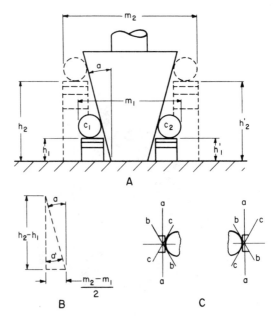

Fig. 9. Schematic diagram of the setup and the dimensions required to determine the angle of taper of a workpiece by the measuring roll method.

about 2 inches long and about $\frac{1}{2}$ inch in diameter. Within gagemakers' tolerances they are free from taper and out-of-round and they are usually equal in size, with surface roughness generally not worse than 5 microinches.)

Equal-height gage block stacks are assembled for h_1 and h_1'. These heights can be almost any arbitrary and convenient choice, although they are usually selected as being about one-quarter the total length of the workpiece. The rolls rest on top of the stacks, as Fig. 9-A indicates, tangent to the sides or walls of the tapered workpiece. Measurement m_1 is secured with micrometers, vernier calipers, or an indicating gage, depending on the discrimination and precision desired.

The gage block stacks are then extended to height h_2 and h_2' which, if possible, should be in the order of two or three times h_1. The rolls are again used and measurement m_2 is made.

Subtracting the h_1 and m_1 measurements from the h_2 and m_2 readings, respectively, gives in effect a triangle — see B in Fig. 9. In that triangle, the tangent of the angle a' is $(m_2 - m_1)/2(h_2 - h_1)$ and angle a', as sketch B indicates, is equal to angle a, which is the angle of slope of the tapered workpiece.

In the measuring technique connected with the above, the jaws of the measuring instrument must be centered through the measuring rolls on the maximum, the true, diameter of the workpiece if it has a circular cross section. The measuring rolls must not swing under measuring pressure to the sort of position lines b-b in sketch C of Fig. 9 or lines c-c suggest in contrast to the true-diameter position indicated by lines a-a. Likewise, the measuring instrument surfaces must bear on the true horizontal center lines of the measuring rolls themselves. If contact is accidentally made below the centerlines, the rolls may lift up off the gage block stacks and give a false instrument reading.

Other Measurements on External-tapered Part

A tapered workpiece like that shown in Fig. 9 must be measured in other respects if the conformance of the piece to specifications is to be adequately checked. Determining the taper or the angle a is only one part of the inspection, since there must be complete correlation between angle a, the length h, and the diameters d and e shown in Fig. 10-A.

Technically, it is impossible to measure accurately diameters like e and d directly as indicated in sketch B of Fig. 10. (Workers often accept such direct measurements as these, but machined intersections — "feather edges" — like x and y are too unreliable to allow accurate direct measurement.) Therefore, the measuring rolls (c_1 and c_2 of Fig. 9) are used again as follows.

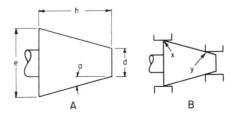

Fig. 10. A satisfactory check of a tapered workpiece should show close correlation between the angle a, the length h, and the diameters d and e shown in (A). It is impossible to measure, accurately, diameters like d and e directly as indicated in (B).

The tapered workpiece is mounted truly vertical on a surface plate as shown in Fig. 11. The rolls are placed against the workpiece as shown and measurement m is taken. Before finding diameters d and e, other dimensions must be known. Some of these can be measured directly, others must be calculated. The dimensions obtained by direct measurement are m, h, and r, r being one-half of the diameter of the measuring roll. Angle A can be determined in the same manner as was

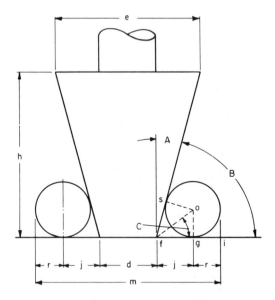

Fig. 11. Once angle *A* has been determined, diameters *d* and *e* can be calculated using dimensions *m*, *h*, and *r*.

angle *a* in Fig. 9; angle *B* is equal to $90° - A$ and angle *C* from the geometry of the figure is *B*/2.

Dimension *j* is therefore equal to $r \times \cot C$

Diameter *d* can now be found:

$$d = m - 2r - 2j$$

Diameter *e* can also be found:

$$e = d + (2h \times \tan A)$$

Checking Dovetails

Another problem in inspection involving angles and tapers is the checking of dovetails. The cross section of a female dovetail is shown in Fig. 12. If, in the length of a dovetail track, the angle *a*, height *h*, width *b*, or width *c* vary too widely, the corresponding male or sliding member will bind or wedge in it. So, again, the familiar measuring rolls or cylinders are used. In this case, the measurement *m* is an inside or width measurement. Ordinarily, in the case of dovetail checking, the angle *a* is determined with a bevel vernier protractor and not with the gage block, cylinder technique suggested in Fig. 9.

$$e = d + (2h \times \tan A)$$

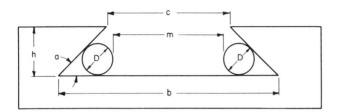

Fig. 12. Measuring-roll setup used to check an internal dovetail.

Knowing angle a, Fig. 12, dimension b can be determined by using the following formula:

$$b = m + D \left(1 + \cot\frac{a}{2} \right)$$

where D is the known diameter of either of the equal-size measuring cylinders. Dimension c is then equal to

$$b - (2 \times h \times \cot a)$$

Similarly, the dimensions for a male dovetail, Fig. 13, are obtained by measurement m, the diameters of the cylinders, D, angle a, and the following formula:

$$b = m - D \left(1 + \cot\frac{a}{2} \right) + (2h \cot a)$$

While the use of standard measuring rolls or cylinders has been described previously, occasionally an experienced inspector prefers to use precision balls. Geometrically, the application is the same, and there are measuring situations where the use of balls is less awkward or more accurate than the use of rolls.

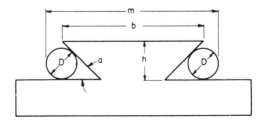

Fig. 13. Measuring-roll setup used to check an external dovetail.

In a shop manufacturing daily many pieces of a dovetail shape, specially designed indicating gages may be employed. The retractable anvils of such gages have measuring rolls permanently fastened to them. The gage is mastered to a correct size dovetail (a workpiece perhaps carefully checked by the slower, detailed method described above) and from then on conformance of the regular production dovetails can be quickly checked directly by the gage.

Measuring Internal Tapers

A practical method of determining internal tapers also involves the use of balls.

Figure 14-A shows a tapered hole of known or measurable dimension h. It is required to measure angle a and diameters s and t. Two balls are selected; a small ball of diameter d such that it falls nearly to the bottom of the tapered hole as shown in Fig. 14-B, and a larger ball of diameter D, which nearly comes up to the top of the hole as shown.

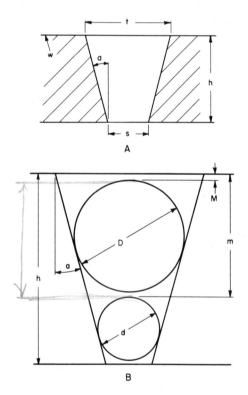

Fig. 14. A setup involving the use of balls to determine internal tapers.

Measurements M and m are secured by setting up a depth gage arrangement with surface w as the reference surface or, if w for some reason is not a suitable surface for reference (it may be too small to support a depth gage), the workpiece may be set on a surface plate, the surface of which then becomes the reference surface.

Angle a can now be calculated using the formula:

$$\csc a = \frac{2(m - M)}{D - d} - 1$$

Figure 15 shows the set-up used to determine diameters s and t. The small ball which appears in Fig. 14-B is not shown in Fig. 15 since it is not used. Dimensions M, D, h, and angle a are known; therefore, we proceed as follows to find t.

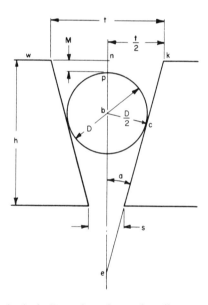

Fig. 15. Setup with single ball used to determine diameters s and t of the taper shown in Fig. 14-A.

In triangle enk, $t/2 = ne \tan a$. Length ne is made up of three other lengths, $M + D/2 + be$; M has been found by measurement, $D/2$ is known, and length be, which is the hypotenuse of right-angled triangle ebc, can be calculated:

$$be = \frac{D}{2} \csc a$$

then $t = \left(M + \dfrac{D}{2} + \dfrac{D}{2} \csc a\right) \times 2 \tan a$

Dimension s can then be calculated:

$s = t - 2h \tan a$

Special Gage Setups for Taper Checking

The techniques just described for measuring tapered plugs and holes not only require surface plate setups but they also take a lot of time. If there are many pieces to be checked — if, as is so often the case, the pieces are from daily production runs perhaps in the thousands — much faster means for inspecting and measuring them must be devised.

For pieces with outside diameter taper a special gage may be built which will work on the principle shown in Fig. 16-A. The indicators are mastered on a piece having the correct taper, probably a hardened taper plug measured and checked with gage blocks and rolls as described in preceding sections. Thereafter, workpieces can be run into the indicating gage about as fast as the inspector can handle them. If either indicator, or both, varies from the "master" readings beyond tolerances, the piece is rejected.

Somewhat similarly, inside diameter tapers are checked in quantity by having an indicating gage made with a mandrel or matrix that is the duplicate of the prescribed internal taper as sketched in Fig. 16-B. The two sensitive contacts recede into sockets in the tapered gage mandrel and in so doing operate two indicators through special internal gage mechanisms. This gage is also mastered first, of course. Rather than a mechanical indicating gage and mandrel, the tapered gage plug may be

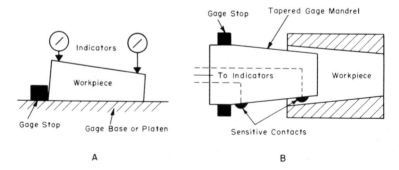

A   B

Fig. 16. (A) External taper measuring setup in which two dial indicators are used. (B) Internal taper measuring setup that utilizes two dial indicators.

an air gage plug with air jets substituted for sensitive contacts which in turn are connected to an air gage column or indicator.

Inspection of Gears

To more or less complete the discussion of the field of measurement, instrumentation, and gaging, at least brief mention should be made here of the special sort of techniques required for ensuring conformance of gear teeth and gears. On the whole, gear and gear-tooth checking are in a highly specialized field. The inspector who works in a shop manufacturing, purchasing, assembling, or using gears in its products should make a special study of the subject, for it is much too complicated for this text to handle in its entirety. Theoretical and practical information can be secured from library technical sections in the form of books, and the manufacturers of gear cutting machinery and equipment can supply inspection, gaging, and checking information. In a few paragraphs here, an attempt will be made only to furnish a guidepost to gear inspection with the description of a few of the simpler techniques.

As on the screw thread, most measurement relationships on a gear hinge about the pitch diameter. (For the sake of brevity here it is assumed the inspector has studied on his own the elements and theory of gears and gear-tooth forms.) The outside diameter and the root diameter are important as well as tooth form, and tooth thickness at the pitch line. Likewise the number of teeth and their even spacing around the gear periphery must be taken account of.

Checking Conformance of Gear Hub or Hole

Consistent with other inspection routines, it is a good thing to establish an order of events in connection with the examination of gears. Perhaps the first inspection step, one commonly overlooked, would be to check the conformance of the gear hub or the hole in the gear. Many times this inspection is made at the place of manufacture of the gear blanks before the operation of generating the gear teeth, but it should not be overlooked in considering the finished product. Is the hub or hole of the gear oval, tapered, or barrel or hour-glass shaped? Is the axis of the hub or hole parallel to the gear axis or, more specifically, perpendicular to the face or plane of the gear disc? Is the diameter of the hole or hub within tolerances? Finally, if the gear's hub is to run in bearings or if the gear is to turn on a shaft, is the surface finish satisfactory?

The effect of some of the difficulties described above on meshing gears is to produce what is sometimes called wobble. In theory, the line of contact across the teeth of mating gears is supposed to be parallel with the axis as illustrated at A in Fig. 17, the side view of the contact line being illustrated at B. If either or both of the gears wobble or weave as they revolve, if the hole or hub of one of the gears is not perpendicular to the disc plane, for instance, then the line of contact will be tilted as exaggerated in sketch C of Fig. 17. Just a little of this effect will produce noisy gear trains and cause unnecessary tooth wear. It is possible that the wobble could be so bad that the interlocking gear teeth could bind or lock. Always remember the purpose of toothed gearing is to transmit motion and power smoothly at highest efficiency and, ordinarily, with as little noise as possible.

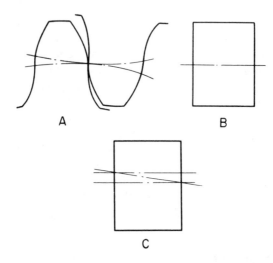

Fig. 17. (A) A pair of mating gears for which the line of contact across the teeth of the gears is properly parallel with the axis. (B) Side view of line of contact described in (A). (C) Side view showing improper line of contact resulting if either or both of the gears wobble or weave as they revolve.

The gear hub or hole must be concentric with the pitch circle. When it fails to be, the condition commonly known as run-out exists. Then gears clack, rattle, run noisily and wear out sooner. The outside diameter of the gear should also be concentric with the center axis. If not, noise and wear result because of imbalance and unequal centrifugal forces tugging at it, conditions that become increasingly important at high speed.

Checking the Pitch Diameter or Tooth Thickness

Having examined the gear hub or hole and having located its deficiencies, the next step should be to check the pitch diameter. One way of doing this is similar to the three-wire method of screw thread measurement except that two measuring rolls or wires are used after the fashion shown in Fig. 18. The wires are located in diametrically opposite tooth spaces, if the gear has an *even* number of teeth, and as nearly opposite tooth spaces as possible, if the gear has an *odd* number of teeth.

For this purpose, a preferred wire or pin diameter for an external spur gear is equal to $1.728 \div P$, and for an internal spur gear to $1.44 \div P$, where P equals the diametral pitch of the gear to be measured. The correct measurement of M over wires for an external gear or between wires for an internal gear of a given number of teeth and pressure angle can be found in convenient handbook tables. Figure 18 shows measurement M, as taken over two wires located in the tooth spaces of an external gear.

In Fig. 19 the wires are held in place in the internal gear tooth spaces by a metal support strip. They extend beyond the ends of the tooth spaces so that instead of having to take the measurement between wires, a more convenient measurement over wires can be made. Before comparing the measured value taken in this way with that given in a table for measurements *between* wires, however, twice the diameter of the measuring wire must be subtracted from the observed measurement.

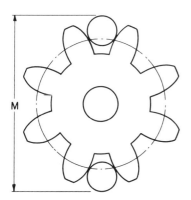
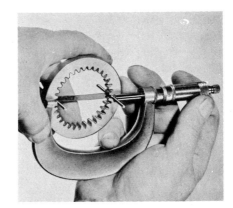

Fig. 18. (Left) Setup used when the pitch diameter or tooth thickness of a gear is to be checked with measuring rolls or wires. Fig. 19. (Right) When wires are used in measuring an internal gear, they may be held in place by a metal support strip as shown.

Figure 20 shows one wire in place for a check on a helical gear. The other wire is placed in the opposite tooth space and the measurement *M* is taken over both wires as in the case of a spur gear.

Although measurement over wires is commonly referred to as a check on gear pitch diameter, it might be more properly designated as a check on tooth thickness at the pitch diameter. If the teeth of an external gear are too thick, the wires or pins will ride higher in the tooth spaces and measurement *M* will be too large. If the teeth of an external gear are too thin, measurement *M* will be too small. Tables showing the difference in tooth thickness for a given difference in measurement *M* are also given in some handbooks.

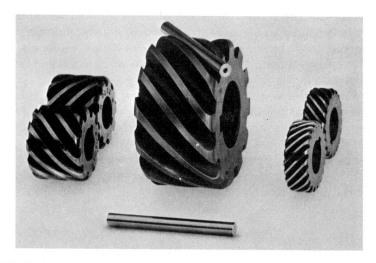

Fig. 20. Position assumed by a measuring roll when it is used to check a helical gear.

Figure 21 shows a vernier caliper designed to measure the thickness of a gear tooth at the pitch circle. The vertical scale is set so that when the caliper is supported by the top of the tooth as shown, the jaws of the caliper will contact the sides of the tooth on the pitch circle. Such a caliper does not measure the thickness along the pitch circle, but the *chordal* thickness *T*.

Measurement of tooth thickness also can be made on an optical comparator where the shadow of the gear, if the gear is small enough, can be projected on a suitably accurate outline of the theoretically correct tooth form and pitch diameter. Such a comparator is also an excellent means for checking on tooth spacing and tooth profile.

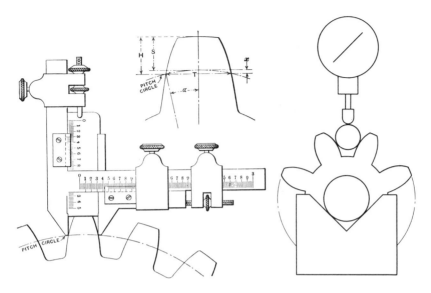

Fig. 21. (Left) Vernier caliper designed to measure the chordal thickness of a a gear tooth at the pitch circle. Fig. 22 (Right) Setup used in checking the concentricity of the pitch circle and the axis of the hub.

Checking for Ovality and Eccentricity

Where a gear has a suitable hub, it can be mounted in a V-block and a check of the concentricity of the pitch circle with the axis of the hub can be secured, using one wire and an indicator as shown in Fig. 22. If the gear has a hole through its hub, the hole can be "plugged" and the plug used as a hub in the V block. This latter method may be used when the hub is unsuitable for mounting in a V block.

The outside diameter should be checked for ovality or eccentricity, especially where the gear is to enter high-speed service, because either of these errors produce centrifugal and inertia effects that result in noisy and worn gears. The outside diameter should be within tolerance or else there will be too little clearance, resulting in interference at the roots of the mating gear teeth.

The eccentricity and/or ovality of the pitch circle is frequently checked by operating the gear against a master gear in the type of equipment shown in Figs. 23 and 24. Such a test will also indicate wobble or side play.

In using such apparatus, be sure to place the two gears at the theoretically correct center distance. An inspection of this sort will

Fig. 23. Gears are frequently checked by operating the gear against a master gear in the type of equipment shown.

determine composite errors but does not differentiate between eccentricity, wobble, ovality, indexing error, hub ovality, poor tooth form, or excessive surface roughness.

Checking Backlash Allowance

When a gear is mounted so that it can run in natural fashion in proper mesh with a master gear, as in Figs. 23 and 24, the "backlash" allowance for the gear can be checked. This is done usually by locking the master gear and by setting an indicator point against one flank of a tooth of the gear being checked. Figure 25 shows the setup with the amount of backlash appearing at *b* in exaggerated form. If the gear is rotated back and forth as arrow *a* indicates, the indicator will register the amount of backlash.

While the short discussion of gear inspection in the paragraphs above has been confined to spur gears, the same general geometrical principles apply, of course, to checking the elements of bevel, helical, herringbone, spiral, and worm gears. But here, the physical shapes of the teeth and gears naturally present special problems of measurement as compared to ordinary, simpler spur gears. It is necessary to return to the suggestion made earlier in this section where the inspector is referred for further study, instruction, and coaching to special treatises on gearing and to the manufacturers of gear cutting equipment or to other gear specialists.

Courtesy of Fellows Gear Shaper Co.

Fig. 24. Gear checking equipment with provision for recording the composite errors.

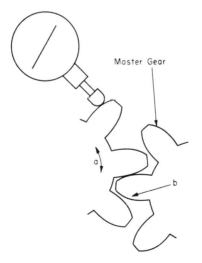

Fig. 25. Checking the amount of blacklash allowance at a given center distance.

Visual inspection of gear components should not be neglected of course. A gear is no different than any other machine-cut product so far as the effect of surface roughness, burrs, slivers, nicks, dents, blow holes, and corrosion are concerned.

Surface Finish Measurement

A little attention has already been given to measuring or comparing surface roughness by means of sample workpieces selected as standards or by means of replica blocks. There are, however, many circumstances where the eye or the fingernail is nowhere near accurate enough to establish the conformance of surface finish to specifications. Since the analysis of surface roughness has been reduced to terms of microinches, surface measuring apparatus of the general type and form of that illustrated in Fig. 26 is coming into more common use. In principle, there is a fine pointed, diamond tipped stylus for tracing the surface irregularities. The stylus, or an extension arm, is moved back and forth in uniform strokes across the workpiece surface (a limited area or path on it) by means of a motor-driven mechanism. Usually the measuring instrument and the work are supported on a solid, flat smooth surface like a surface plate.

The relatively minute vertical motion of the stylus point, as it moves over hill and dale, is transmitted by means of an electric crystal and electronic amplifier to a recorder pen and chart.

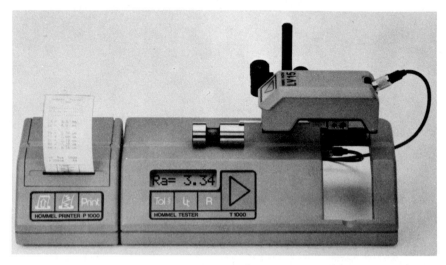

Courtesy of Valmet Inc.

Fig. 26. Surface finish measuring unit.

Figure 27 shows a chart obtained from checking the surface conditions of worm gear flanks. Each heavy longitudinal line on the chart represents 10 microinches, each fine line 2 microinches, and the curved lines cover increments of time or chart speed rate.

While the proper technique for operating a surface recorder is better obtained from manufacturers' catalogues and instruction books, a couple of pointers can be offered here. Remember you are trying to take measurements in millionths of an inch. You should use the same care you would naturally exercise if you were trying, for instance, to measure a diameter to millionths of an inch with a comparator. The workpiece surface should be truly free from dirt, lint, grit, chips or grease. The workpiece should be clamped or held securely enough so that the pressure and movement of the stylus will not deflect or move it. The instrument's measuring arm, motor unit, V-block, and surface plate must be clean, smooth and flat, for the workpiece must not tip, rock, or vibrate under the delicate impact of the stylus.

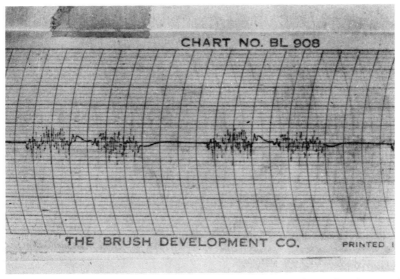

Courtesy of The Brush Development Co.

Fig. 27. Record of the surface condition of a worn gear obtained with apparatus similar to that shown in Fig. 26.

CHAPTER 15

Gage Checking and Calibration

Some inspectors run into the relatively special problem of checking gages and measuring instruments, which calls for observing several basic precautions no matter what type or form of gage or apparatus is being checked.

In the first place, to do a real job, the operation should be unhurried. Someone once remarked that the atmosphere in a gage lab should correspond to that in a library. An attempt should be made to do the work itself in a location — on a bench — where there is as nearly as possible complete freedom from vibration, noise, loud talk, drafts, and glaring light. Try to do a job of gage checking at a time when you are unlikely to be interrupted. A gage checker should lead a cloistered life when actually at his work.

The calibration and checking of measurement tools and instruments are critically important because of the nature of most types of inspection and gaging. Obviously, the reliability, accuracy, and precision of go/no go type inspection measurement relies on an on-going correlation between a gaging device and the master or standard. Especially in a manufacturing world that relies heavily on in-process measurement and control by operators, the calibration of their measurement tools is critical. Many times modern quality control methodology places the burden of measurement and control on the on-line operator. In any case, the go/no go and the indirect or transfer-and-compare methods of inspection will provide worthless or misleading information if a rigid routine gaging calibration and control is not in place. And the very principles of direct-measurement devices and machines are undermined if they are not continuously calibrated to the (ultimate) standard.

Indirect measurements are used in every aspect of inspection. The comparison of indirect measurements to a standard requires calibrated standards *and* calibrated comparison devices. As measurement devices turn to measurement machines, and these machines become more technologically sophisticated, the notion of calibration also becomes more critical. With the increasing complexity of measuring machines comes a built-in distance between the machine and the inspector and the operator. Most operators and inspectors do not understand all the measurement and discriminatory variables of their machines. A confidence level must be maintained between the measurement machine and their operators that can only come with a reliable calibration program.

So calibration and standards-control programs have come to mean many things in industry and inspection today. The surveillance and maintenance of all sorts of measurement and control devices are included in this category. It may be that airborne contamination measurement and control is important to a factory. In such a case, so is the calibration of that system. Software control is also primarily essential in measurement and inspection routines: its regular assessment exists as a form of calibration, since it may be directly and indirectly involved with the measurement and data recording process.

Highly Accurate Equipment is Needed

The next general observation is that the equipment and apparatus used to check shop gages should have accuracies and discriminations greater than the gages to be tested. A vernier depth gage, for example, should not be checked with an indicating depth gage whose dial divisions read no closer than .001 inch. No, check the vernier with a dial indicator whose dial reads to .0005 inch, .00025 inch, or .0001 inch.

Closely allied to the above is the suggestion that the master equipment used for gage checking should not be used for any other purpose. A set of gage blocks used continuously out in the shop, for instance, should be avoided for gage checking in favor of a conscientiously maintained, seldom but carefully used "inspection set" assigned to that purpose only. Take time to keep them clean, shining, rust free, unscratched, and free from nicks and dents.

If you use gage blocks, anvils, platens, V-blocks, flats, knees and similar apparatus for gage checking be sure the relevant surfaces are truly flat. If you use cylinders, rolls, or wires, be sure they are truly round and not oval, that they have no taper. In general, surface roughness on checking equipment probably should never exceed 5 microinches, and if you can get the apparatus lapped and polished to fractions of microinches, so much the better. Be sure such blocks, cylinders, rings, and the like are suitably hardened.

There should be a regular schedule for sending basic equipment such as gage blocks, optical flats, and master rings to the manufacturers for recalibration, refinishing, or replacing. Depending on the frequency of equipment use, this should be done perhaps quarterly, perhaps twice a year, certainly once a year.

Gage checking should be done in an area where the temperature can stay in equilibrium for long periods of time — away from doors or windows that are frequently opened and closed. Don't do gage checking in direct sunlight through windows, or near radiators, registers, steam pipes, or heating apparatus. (A recording thermometer is handy apparatus in a gage checking area.) It is a good thing to get into the habit of using gloves, wooden or plastic tongs, or felt pads when handling gages to be checked, or the checking equipment, so as to minimize the effect of heat transfer from the hands. It takes time frequently to set up the gages to be checked along with the rolls, rings, and gage blocks to be used. When the setup is complete, rehearse the gage checking operations once. Then let the gage and checking apparatus lay idle for, say, 20 minutes to equalize all temperatures. Finally make the check or calibration as rapidly as possible with a minimum of handling.

If electric or electronic comparators and measuring apparatus are used, try to have some means of regularly checking the shop line voltage; in the same manner have a pressure gage on the air line if air gages are used as checking apparatus.

Establish a Definite Checking Procedure

Once more, a definite order of events should be adopted. The first step to be taken with the gage to be checked is to clean it thoroughly. Eliminate the variable of dirt. Second, try the gage, its anvils, its accessories, its clamping devices, and all such for tightness. Test for end shake in a micrometer for example. Anything loose or wobbly, any part that will move or deflect when it shouldn't is taboo on gages. Third, perform a visual inspection. Clean off any rust or corrosion. Examine carefully for nicks, pits, scratches, burrs, and the like. Consider, for instance, loose carbide or diamond surfaces on indicator contact and anvil surfaces. Hunt for cracks and weak spots in gage frames and bases. (Don't forget how measuring apparatus gets dropped onto concrete floors out in the shop, an event that seldom receives publicity.)

Where the type of equipment or gage demands, test next for flatness, with an optical flat, or for parallelism, taper, or ovality.

If a contact point is supposed to be spherical, project its enlarged shadow onto an optical projector screen and look for worn or flat areas. Check the surface finish, perhaps. In other words, narrow down one-by-one all the kinds of variables and errors that cause measuring apparatus to go wrong until you get to the actual setting or to the actual calibration of indexing, graduations, vernier adjustment, and the like. Before you reach the final step, there may be an amount of repairing, refinishing, replating, relapping, or replacing. You may decide to ship the equipment or tool back to the original manufacturer.

The gage checker uses his geometry instinctively. He locates on the true diameter of a roll orring and not on the chord; he is careful not to tip, cramp, or cant a gage, the checking blocks, or cylinder; the conceptions of parallel, perpendicular, or flat are clear to him. He makes certain of never springing or deflecting the equipment he is testing by the use of excess gaging pressure, yet he realizes that a wringing contact is necessary in particular instances.

Of course, for extra finesse in gage checking many shops have built dust- and vibration-free, temperature-controlled rooms (constant temperature rooms held within one degree of 68° F) and require that the gages to be checked shall have lodged in the 68° F temperature room at least 27 hours before they are checked.

Effects of Temperature

Last but certainly not least in this brief treatment of measurement and gaging some consideration should be given to the effect of temperature changes both in the workpieces being gaged and in the metal makeup of the gaging apparatus. For the most part the gaging of a workpiece is a matter of linear measurement and the length or linear dimension of any metal piece changes with the temperature. An inch length of the average ferrous material will alter by about .000006 inch (6 millionths) with each degree change of its temperature. A similar figure for aluminum is .000013 inch (13 millionths) and for copper and copper alloys about .000009inch (9 millionths).

To get a practical picture of how a steel piece, for instance, can expand without your realizing it under average temperature increase conditions, suppose you pick up a bar 10 inches long which has been at the ordinary room temperature of, say, 70° F and hold it in your hand for 10 minutes. Generally speaking, by that time the temperature of the bar will have gone up some 10° and it will be .0006 inch longer:

$$10 \times 10 \times .000006 = .0006$$

In short, the normal handling of a 10-inch steel bar under average factory conditions can increase its length a half-thousandth inch without any one realizing it.

Furthermore, while a piece of stock or a measuring instrument will heat up rapidly, it takes it a long time to cool down to its previous (room) temperature. The chart in Fig. 1 shows what happened to a 24-inch caliper held in two hands for 5 minutes and then set down to cool off. (The dotted line, c, shows the temperature rise where the same caliper was handled with $\frac{1}{4}$ inch thick felt pads insulating the hands.)

Where modern assemblies are requiring tolerances and fits to .0001 inch (all of which makes for a smoother, quieter running and much longer

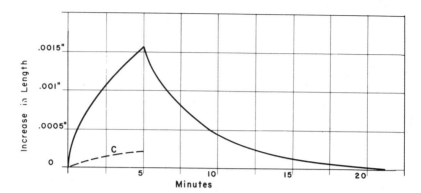

Fig. 1. Chart showing gage expansion due to handling. Solid curve: 24 inch caliper held in hands for 5 minutes and then set down to cool. Broken line: temperature rise for same caliper when handled with $\frac{1}{4}$-inch thick felt pads.

lived automobile power plant or refrigerator unit, for example), not only must surface roughness under 10 microinches be achieved but also attention must be paid to the effects of temperature on sizing and on measuring the conformance of metal pieces to the size required. Hence, the modern technology of turning, boring, milling, grinding, lapping, and honing with refrigerated and temperature-controlled coolants. The inspector must enter the game by realizing what errors his hands can produce, not only by incorrect manipulation of his gage but also by changing the accuracy of his gages through the effect of handling temperatures.

To be practical about it, it is unusual or unlikely for a sufficiently noticeable error to creep into the average measurement if the instruments and workpieces are handled as much as possible with the fingertips and if the measurement is completed within a minute or so. But to stand enveloping the frame of a portable gage in your fist for ten minutes and then turn to make a precision measurement is another story.

As a standard temperature for measurement, 68° F has been selected in the United States as the temperature at which any linear measurement will be considered correct. If the work is measured at a lower temperature, it will be, of course, smaller and larger at a higher temperature. Again to be practical, the actual room temperature seldom has too much effect (even though it might be 90° F) because it is possible for workpiece, gage, and master all to be at the same temperature.

The Need for Gage Control

It's a pretty safe gamble that eight out of ten micrometers owned and used by machine shop operators are "off" by some .0005 inch either in zero setting, end play, calibration, or anvils worn out of parallel. Disc and ring masters, which are used to zero comparators and dial bore gages, become forgotten, neglected, and seriously worn. Maybe the psychologists can explain why a worker will set his dial bore gage to a rusted ring master he unearthed almost anywhere and proceed fatuously and complacently confident of his ability to ream exact 1-inch holes because the legend stamped on the ring says 1.000-inch. The finer the tolerances the more serious such a complex becomes.

Lately also, the discovery is being made in many factories that gage elements wear away much faster than has been the common belief. Traditionally, a gage wear allowance up to 10% of the tolerance has been permitted — which was all right where the tolerances were plus or minus .010-inch or even plus or minus .005-inch. But when manufacturing tolerances narrow to tenths or less, a broad wear allowance goes out the window.

Under some conditions it does take a gage a long while to show error-producing wear. In others, the gage surfaces might wear back several thousandths in a single day. That's where the rub comes in; too often nobody knows how fast the gages do wear. Hence, the need for gage control.

There is no set formula for the rate of gage wear. A man may check brass pieces steadily for a week and have his micrometer show no appreciable wear. The next day he may be measuring sand castings and have a tenth or so chewed off his mike anvils in a matter of hours. Heavy-handing a thread ring gage onto just a few oversize threads may spoil its accurate setting. The best formula is to know the time elapsed and the number of gagings on what type of material and parts during the intervals between which the gage is checked for wear.

No type of gage seems to be immune to wear, whether it has flat or spherical contacts, whether it is an indicating comparator platen, a ring or tapered plug, or whatever. Gages with roller anvils usually require less maintenance. Heavy versus light gaging pressure has less effect than might be suspected. Other parts of gages should not be forgotten; check the skirts of air plugs, and calibrate indicators and gage meters because internal gear teeth, pivots, and jewels also wear.

The material a gage contact is made of is important. Anything softer than tool steel hardened to 50–60 Rockwell C seems useless. Chrome plate on hardened surfaces has withstood from 10 to 100 times as many gagings, without appreciable wear, as plain hardened steel, while tungsten carbide anvils have been known to stay in shape through 100,000 gagings. Despite such performance records, however, never quite trust a gage unless it has been checked or unless a valid wear record is available. But inspectors and workers can't always be stopping to check a measuring instrument; hence again, the desirability of gage control.

It used to be that no factory bought any measuring instruments. As part of his trade, the mechanic, machinist, and inspector furnished his own micrometers, verniers and test indicators. He made whatever plug and flush pin gages were needed. Whether or when they were ever checked for accuracy depended much on his pride of craftsmanship. A trace of this old system still clings even in many modern plants where the individual often buys and uses his own, rather than company-supplied, 1-inch micrometers.

An in-between system still flourishes in a great many factories where the worker owns some of the instruments he uses but the plant supplies the remainder, which make up the majority. Often, too, the factory-bought instruments — indicators, height gages, fixed gages, gage blocks, etc. — are put right out on the floor with no more record of them from then on perhaps than the annual inventory. How or when they were checked was secondary. In many plants gage control still embraces some of these traditional "hang-overs."

However, during the last decade definite company policies have been developed concerning the type, design, application, selection, procurement, maintenance, and complete control of any and all gages, measuring instruments, size control equipment, and testing apparatus. The ideal setup is perhaps yet to be reached, but progressive organizations are working toward it. The new approach means systems for issuing, maintaining, and returning gages plus up-to-date records of every move.

There are about as many kinds of gage control systems and records as there are establishments or individuals using them, all much alike and differing mostly in details. The situation resembles buying men's suits — outwardly each garment seems to be the duplicate of the other but any suit selected from the rack must be tailored, or altered at least a little, to properly fit the buyer. It would be impossible to outline here a universal gage control plan that would precisely meet the requirements of all manufacturing areas; the system for any one plant or department requires study and trial. However, there are certain fundamentals which can be touched upon.

Planning for Gage Control Raises Questions

An adequate gage control system should be able to answer two fundamental questions correctly:

1. Where is the gage right now?
2. How accurate is it?

There are other basic questions, for instance, plant management wants to know how much money is tied up in measuring equipment. But just getting the answers to 1 and 2 raises further detailed questions. To whom, where, and when was the gage issued? What was its condition then? When was it last checked? How long has it been on the job? On what product, part, or material is it being used? How often per hour or per day? Is it due to be returned? Is it missing?

As questions like these are being answered, the type, scope, and extent of a record system is being patterned. And then other decisions have to be made. Who will make and keep the records? Where? Will gages be kept in some sort of central storage and issued as required or will "gage control" simply maintain some patrol system of constant surveillance out on the factory floor? Or a mixture of both? Who will check the gages and who will be responsible for gage control? Some answers come from examining existing systems.

Gage Control Programs in Use

In certain large manufacturing companies Gage Control is a responsibility of the Inspection and Quality function. It is a separate department under that heading with its own supervisor. Physically it operates the gage lab, a records office, a central gage crib, and the disbursement of gages through tool cribs or similar facilities strategically located in manufacturing areas. In addition to gage maintenance, Gage Control is charged with the analyzing of size control and the measuring requirements of Production and with the designing, selection, and procurement of the measuring equipment that is needed. Gage Control specifies the gages to be used for each operation and job. These duties are carried out with help from Production, Methods and Tool Engineering. Management holds Gage Control responsible for the inventory of measuring and testing equipment and for meeting company accounting requirements of budgeting, depreciation, and obsolescence.

In a number of smaller plants, also, the custom has grown of placing the gage selection and maintenance responsibility with the chief inspector. Such a setup usually functions under much more restricted and improvised conditions from the standpoint of gage control facilities than does the large plant organization.

Traditionally gages and measuring instruments have been classed to a considerable extent as small tools along with jigs and fixtures, reamers, milling cutters, and the like. Hence gage control has been a routine of the tool crib in many establishments. Usually under such circumstances Manufacturing and Tool Engineering specified the type and extent of gaging and its equipment for production with Inspection selecting many of its own requirements.

One occasionally found variation requires all gages to be requisitioned from a tool and gage crib and returned there either after immediate use, at shift's end or surely by week's end. The crib does not toss the gages back on the shelves but sends them to Inspection for checking, correction, and repair, whereupon they are returned to the crib. In other words, the crib's source of supply for gages that it is allowed to issue is the inspection department.

At the far end of the spectrum is the factory where Manufacturing, Engineering, and Inspection each procure and provide whatever gaging and testing equipment each thinks it needs, an arrangement usually accompanied, as already intimated, by little or no organized attempt to check, maintain, or replace measuring devices.

Every so often, especially in the smaller establishment, some one inspector may take on to himself an unofficial, individual responsibility for gage checking in the area he serves. One way or another he will secure or gain access to equipment — a set of gage blocks, a comparator, a sensitive dial bore gage perhaps — and find time to examine not only his own instruments at regular intervals but also the micrometers, gages, and equipment used by adjacent machinists.

In between the extremes of practically complete gage control and none at all a large variety of patterns can be found in industry. The trend lately has been to pay more attention to the subject and to invest more money in gages, instruments, and size control equipment and alsoin means for maintaining and controlling them. This occurs as a plant grows in size; again where its product machining becomes more diverse and complex; and finally where competition or product requirements (missile parts are one example) demand closer tolerances.

Gage Control Records

Records are probably the essence of a gage control system. In industry they may vary from simple card files to complex ledgers and punch card systems to computer databases. Most of them display common factors an enumeration of which might help the inspector in contriving or revising a plan if he were faced with the problem.

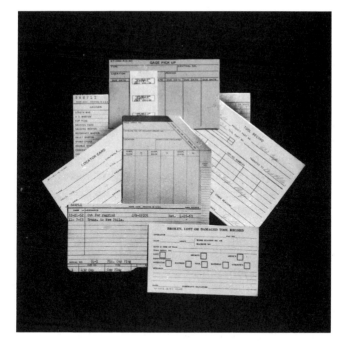

Fig. 2. Samples of the wide variety of cards, tally sheets, ledger forms, and other records used in various plants as part of their gage control record systems.

Every gage, instrument, or size control unit should have its own serial number. Such a series of numbers may be completely independent or it may be related to an existing plant register system. One rule in connection with gage serial numbers seems to establish a necessary safeguard. Once a serial number has been issued for a gage, that number is never used again. Even when a gage is replaced by a facsimile, give the latter a new number. Old gages that have been ordered out of commission and scrapped have a knack for reappearing on the job. Who fishes them out of junk piles and sneaks them back into use is one of the unsolved mysteries of gage control.

Usually a basic record includes the gage's serial number, date of purchase and from whom (or date of manufacture if made in the company tool room), its general description including make and model number, and some word concerning its purpose or use. If it is a special design, some reference is made to an engineering file, correspondence, or purchase order. For inventory purposes the record usually contains the original price or cost and, often, the up-to-date depreciated value.

For each gage, a continuous record is kept of its location. A column for dates is provided paralleled by space in which to note whether it is in stock — unissued — or to whom and where it has been given out on that date. ("To whom and where" includes sending it out or away for repair.) Another column provides space for noting the date of return. The same record form often provides space for notes about the product, part, or material the gage is being used on.

Another must is some form of receipt covering the issue and return of the gage. These vary from a simple brass tool check to a carbon copy slip containing sufficient legal phraseology to ensure a successful suit against the gage borrower in the event of failure to return the gage. In the better programs the gage receipt contains provision for marking in most of the data required to recapitulate the sort of information described in the foregoing on the main record forms.

Maintaining a rather careful, detailed record of gage condition is customary. For example, a micrometer might be issued although it is known to have a zero setting error of minus .0002-inch and anvils out-of-parallel by two light bands (23 millionths). Upon its return another record is made of its condition to see if it has worn more. Such a record also registers who made the gage checks and when the gage was finally repaired or replaced.

It will be found that the successful gage control setup depends more on the care and conscientiousness — the dedication — of some one person in gage control who manages the records than it does on the physical makeup of forms or routines established by executive order. Good gage control means constant attention and rather rigid discipline. Whoever keeps the records is best placed to pull the right levers on time to assure that discipline.

Gage Surveillance

To maintain accurate, up-to-the minute records is seldom enough, however. Gage Control people find it necessary to adopt supplementary methods for keeping track of gages because gages are apt to disappear, not show up on time to be checked, or be subject to other dilatory tactics of the users.

It pays to mark the serial number on gages, instruments, and size control equipment rather prominently, the psychology being to deliberately call attention to them, to attach importance to them, and thus perhaps remind the user that they need to be checked frequently. Some plants use brass or plastic discs for this purpose, others paint the numbers on.

The biggest trouble is to get gages back in on schedule to be checked. This means having some idea, in the first place, of about how long a gage can safely be left on the job. In some situations gages wear fast enough to warrant calling them in at the end of every shift. Others need policing once a week, some once a month. The correct scheduling of gage examination comes from experience, and especially from closely following and analyzing the gage condition records mentioned in a foregoing paragraph.

Also to successfully recall gages for maintenance, it is usually necessary to carry a stock of duplicates. to deprive a worker of a frequently used gage for several hours, while it is being tested and repaired, meets with resistance and also involves too much costly, emergency repairs. Exchanging the worker's gage with an accurate alternative solves the problem.

Most gage control systems provide for flagging the gage record when a gage is issued, the flag serving to signal the date it should be returned. The records are combed each day with orders sent out for the return or exchange of the gages involved.

In a few factories, where gage use is regular, uniform, and not too severe, a system of "color coding" has been found effective. A number of the gages are painted or striped red for instance; their duplicates are colored blue. Red gages are for use, say, during the first and third weeks of the month and the blue ones during the alternate weeks. Each week, Gage Control brings one group up-to-date. If a red gage is found in use during a "blue week," it is suspect and automatically recalled. Color coding also includes monthly control intervals for some gages — green, say, for January, March, May, etc., and yellow for February, April, and June. Usually color code cards, posters, or calendars are displayed in manufacturing areas advertising which gage color is legitimate or official for the particular period.

A gage control staff usually needs to do a certain amount of detective work. A percentage of workers for some reason resort to various artful dodges to avoid the gage control, recall, and replacement routines. Then, too, human frailty enters — some lapse in completing gage records, some neglect to return a gage that is due in. It pays, especially in larger operations, to have someone from Gage Control more or less comb the manufacturing areas from time to time, making a sort of inventory and hunting especially for culprit gages that belong back in custody for reexamination and rehabilitation.

The effort and cost of framing, installing, and maintaining a practical gage control program almost inevitably pays off in an upsurge of product quality, a reduction in scrap and rework, and lowered assembly expense, because required dimensions are more regularly met. A measurement is never much better than the reliability of the instrument making it.

CHAPTER 16

Measuring in Millionths

Many inspectors today are faced with the problem of measuring size differences of 50, 20, or 10 millionths of an inch or even less. Blue prints now appear with tolerances in the millionths. Methods for machining and measuring in terms of 10 millionths of an inch are not uncommon. This chapter will focus on the relatively new extra-precise technology of millionth measurement and provide a few tips on ways and means of improving measurement techniques to help develop a respect for that elusive millionth of an inch.

The Person Who Measures Millionths

The man (or the woman) who can measure to a millionth of an inch should have an "even temperament" and must have a high degree of emotional stability. He should take pride in his work and be always striving to improve his skills and his job knowledge. There are many training courses available. These are offered by professional societies and by the manufacturers of the measuring instruments, which the inspector must use. They will help him to solve many of his measurement problems and improve his ability to use his gages to the best advantage. He should have a good background in basic mathematics, which he can augment with technical courses in such subjects as physics, metallurgy, statistics, and mechanical engineering. There are many publications and reference books available for the millionth inspector, and he should try to read as many of them as he can so as to keep himself up to date with the continuous progress being made in the millionths measurement field.

The millionths measurement inspector is considered to be a perfectionist, because he is working in an area that many people regard as perfection in the art of measurement. However, if being a perfectionist is believing that one can always find exact answers to all measurement problems, we would suggest that an inspector with that philosophy will be doomed to disappointment. Millionths measurement has many limitations and unknowns — which we will explore in this chapter.

526

Good judgment is a quality that one expects to find in any inspector, but the millionths measurement inspector should have this attribute highly developed. His judgment will be strengthened by his training and experience and by his deliberate and careful attention to his craft. He must be neat, clean, and painstaking, but should restrain himself from chasing the will-o'-the-wisp of an impossible perfection. The millionth measurement inspector should be a very special person but he must, at all costs, keep a sense of humor — about himself and the nearly impossible situations he is sure to find himself involved in as he chases the elusive millionth of an inch.

Measuring in millionths can mean one of two basic approaches to the inspector. Either the inspector is pushing himself, his own skills, and his equipment to the limit of their abilities (as on many common surface plate techniques), or he is relying on an incredibly sophisticated machine to do most of the work. In the first case, a basic understanding of measurement techniques and variables is needed, and accuracy, reliability, and precision really do depend mainly on the inspector. When new machines are used to perform the millionth, or micron, or even angstrom-level measurement, the inspection routine can be fairly fast, straightforward, and even simple. In this case the machine — the electronics, the laser, whatever — does the work. The role of the operator is to understand and support the proper functioning of the machine.

Gaging Equipment

Before attempting millionths measurement, the inspector is urged to become fully acquainted with the gages and equipment available. He should evaluate the manufacturer's claims, follow instructions meticulously, and, above all, check the accuracy and ability of the apparatus he plans to use.

Modern measuring machines, of the contact or noncontact variety, electronic, optical (laser-equipped), computerized, air-cushioned, or whatever are constantly changing and getting better — more accurate, precise, easier to use, faster. Support and calibration devices, such as surface plate flatness measuring devices and other surface plate equipment such as electronic height gages are also evolving and rapidly improving. Whether the operator is "standing back," in support of department's new multimillion dollar machine, or is intimately engaged in the setup and execution of the surface plate apparatus that is being pushed to its limit, many principles of precision measurement remain the same.

Mechanical indicators with dial readings in "half-tenths" (0.000050 inch) have been on the market for many years. Most of them magnify spindle movement through a gear train, but some make use of a reed mechanism or a twisted wire principle.

To gain exacting results with a mechanical dial indicating gage, it should be used on a sturdy comparator over as short a range as possible. The size difference between the master standard and the workpiece probably should not exceed 0.000050 in. In addition, the accuracy and repeatability of the indicator should have been carefully tested and the amount and location of any instrument error known. Other precautions applying to millionths measurement, discussed in detail later, also need to be observed. If the job requires the reading of size differences in increments of less than 0.000050 inch, air, special optical or electronic equipment should be used.

Developments in air gages during the past decade not only have made them highly suitable for hole measurements, but have also led to their frequent selection when size differences between 5 and 50 millionths of an inch were to be measured. Those with dial or column graduations in increments of 5 and 50 millionths are favored. Air calipers, air cartridges, test set heads, and air rings provide many shops with the solution of gaging problems in the 5- to 50-millionth tolerance range where thickness and outside diameter measurement were involved.

Magnification in air gages is attained by employing certain combinations of internal master jet diameters and plug jet diameters, and reducing the clearances between the plugs and the workpiece. Air gages have been marketed that register — theoretically, at least — size differences of 5 millionths of an inch or less. Although the inspector may apparently see size differences of a millionth or two on the scale graduations of an air gage, he should probably accept such readings with reservation. Barometric pressure, temperature, and humidity may have to be considered.

An air gage is strictly a comparator that is "set" to a master or some size standard. The validity of its measurement depends very much on the known, exact size of the master. Hence, it would not be justified to assert an internal diameter in terms of a particular and exact millionth-inch size with an air gage "set to" a commercial master ring gage that has its dimension certified only to within 10 millionths of an inch.

The exact level of the meniscus of a mercury column is difficult to determine where millionth-inch observations are required. At high magnifications the pointer of a dial-type air gage, or the float in a column type, is liable to flutter enough to affect the guess about that last millionth. In the present state of the art it may be better to confine the use of air gaging to measurement within 10 to 50 millionths of an inch and turn to electronic apparatus or interferometry for detecting size differences between 1 and 10 millionths.

In the case of electric and electronic gages, the position and motion of calipering metallic contacts, spindles, jaws, or anvils are translated through variations in voltage, current, resistance, reactance, frequency, magnetism, or capacitance to meters. Magnification is accomplished through what are essentially transformers so that a millionth or so will "look" as big as, say, 1/16 inch on a meter scale.

Many commercial models of this type gage provide switching mechanisms so that different magnifications and ranges can be selected. A gage might, for example, provide a plus or minus 0.001-inch scale for "rough" work with a value of 0.0001 inch for each of its 20 dial divisions. It could then be switched to a 10 times greater magnification, with each dial division registering a 0.000010-inch size change over a total range of plus or minus 0.0001 inch. A third magnification could allow a total scale range of plus or minus 0.000010 inch, with each dial division marking 1 millionth of an inch.

Before measuring with any make of "millionth comparator" that uses internal electronic circuitry, the inspector should fully understand the manufacturer's instructions. He should have a calibration chart of its accuracy from one end of the dial to the other. Since, to some extent, these gages are sensitive to supply voltage fluctuations, a knowledge of these variations at the outlet is valuable. Too much voltage deviation may affect gage calibration. The inspector should find out from the gage supplier how and where to master the gage — whether (1) at each end of the scale with minimum and maximum masters or (2) at the zero center of the scale with one master. Both methods of setting may be required.

All types of ultra precise gages can fail the user for some or all of the reasons previously listed. The professional standing of the inspector as well as the reliability of his measurements is in proportion to his conscientious observance of precautions against uncertain measurement.

The Effects of Environment — Temperature

The expansion and contraction of metal from temperature changes is figured in millionths. An inch of steel will lengthen about 0.000006 inch with a 1° F rise. A general, similar figure for copper and brass is 0.000009 inch, and for aluminium, 0.000013 inch. The international standard inch is exactly 2.54 centimeters at exactly 68° F. If an inch of steel is measured at 69° F, it would be 1.000006 inch long. Obviously, in millionths measurement an error of 0.000006 inch is large.

A piece of steel held in the hand absorbs heat faster than we think. A 10° temperature rise in a matter of 5 minutes or so is not impossible. Then it may take hours, literally, for the metal to cool off and contract to the original length. The professional technique in millionths measurement is to handle gage blocks, masters, and workpieces by remote control with insulated forceps or tweezers (see Fig. 1) with plastic pads or gloves.

Courtesy of Federal Products Corp.

Fig. 1. While peering through a plastic shield, this inspector uses tweezers with insulated tips to place parts under the gage. Previous to measurement, parts are stored on the heat sink (lower right) until they are stabilized at the controlled room temperature.

In industrial situations the usual first thought is an air-conditioned, temperature-controlled room (see Fig. 2) for millionths measurement. Such a room is valuable and handy — occasionally absolutely necessary — but there are circumstances where it may fail and others where some different and far less costly arrangement will do as well.

An air-conditioned gaging laboratory must be free from drafts. Maintaining a general 68° F temperature is not enough. There have been installations where a stream of 60° F air from an inlet blew directly on a sensitive gage while a dead spot over in the corner of the room stayed at 72° F. Ideally, the cool air circulation should be so deflected and gently diffused that a match flame would not flicker and yet a thermometer would read 68° F anywhere in the room.

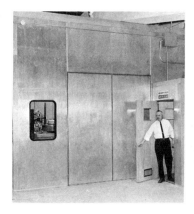
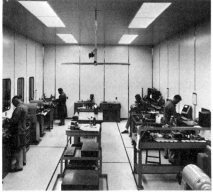

Courtesy of Sheffield

Fig. 2. (Left) Exterior view of a typical controlled environment room. (Right) Controlled environment room being used for working and measuring to millionths of an inch.

Avoid the natural tendency to place a gage near a window. Infrared rays of direct sunlight strike the gage and tend to heat it and the work, although the surrounding temperature is kept assiduously at 68° F. Similarly, radiant body heat affects the work. Plexiglas barriers serve quite satisfactorily in shielding the instrument and the work from convected body heat and the inspector's "hot breath." A practical and inexpensive commercial setup for eliminating many of the effects of convected body heat is illustrated in Fig. 1. In certain situations the gage or interferometer is entirely boxed in with transparent plastic, and the operator manipulates the parts with long-handled, insulated forceps introduced through self-sealing rubber portholes.

In addition to providing a stable atmospheric environment and guarding millionths measurement from stray and erratic temperature variations, probably the most thorough method of securing temperature stabilization is "togetherness." This technique is, in fact, the essence of temperature control. In simple words, place the masters and workpieces on the gage anvil in metallic contact with each other and leave them thus for 20 minutes at the very least, or much better, for one, two, three hours or more. Such a "soaking" procedure is especially essential where gage blocks have been wrung together.

Often it is handy to use a "heat sink" that is usually a slab of steel or aluminium of considerable mass with a clean smooth surface placed beside the gage as in Fig. 1. Gage blocks, masters, and workpieces are placed on the heat sink (after they are cleaned) as a convenient means of obtaining common metallic contact for equalizing their temperatures.

Can millionths measurements be reliably made in the ordinary heat of manufacturing areas where there is neither air conditioning nor temperature control? The answer for size differences from 0 to 5 millionths is no; and from 5 to 10 millionths, probably not safely. But differences between 10 and 50 millionths can be measured with considerable assurance if a few precautions are observed.

Measurements do not necessarily have to be made at 68° F. Accurate results can be secured at 76° F, or any other reasonable temperature. The trick is to have masters, workpieces, and gages all at the same temperature (and, of course, be made of the same materials or those whose expansion coefficients are alike). First, free the gage from convected air currents and sudden drafts. The best way probably is to almost box the gage in. Secondly, stabilize the job temperature by nestling masters and workpieces together on the gage anvil, leaving them to interchange heat for an hour.

Before leaving the gaging setup for a period of temperature stabilization, be sure that the masters and parts and gage anvils are clean, that gage blocks are wrung, and that all other preparations for measuring are completed. This should include setting and mastering the gage and even going through the motions of a tentative measurement or two. By so doing, the final measurement can be taken adroitly and quickly.

Having a reasonably accurate thermometer nearby is always reassuring in precision measurement. A scratch-pad record of hourly temperature variations in the vicinity of the gage gives the inspector an idea of the uniformity, or otherwise, of the ambient temperature at the gage. An accurate thermometer is a must where size differences of a millionth or to are to be measured in a temperature-controlled environment.

Errors Due to Vibration

To register consistent, repetitive readings, a gage should be subject to as little vibration as possible — ideally, none. In most manufacturing and urban areas, a constant tremor exists to which everyone is so acclimated that no notice is taken of the frequency at which the surroundings pulse and vibrate. When trying to read a precision instrument, the value of stillness is realized. Under the majority of circumstances, vibration, thankfully, may not bother measurement much. However, when tremor is sufficiently intense to make a meter hand flutter, it is time to take action. Because instrument vibration is largely indiscernible, one of the better clues to its presence is lack of repetition. When an inspector fails mysteriously to get consistent readings, particularly repeat readings, vibration can be one of the causes.

Usually the first attempt to eliminate vibration is to slip cork, felt, or rubber pads under the gage, an expedient that is effective probably more times than it is ineffective. More elaborate measures have included mounting the gage pedestal or even floor sections on tar mastic.

The solution may be as simple as moving the gage a few feet to another location. (Plant layouts often locate delicate instruments close to aisles where fork trucks travel.) Again, a rigid bracket or shelf lagged to a brick wall can provide a steady support. One plant solved the problem by placing its gages over a heavy floor beam until the maintenance foreman, also looking for solidity, installed an air compressor on the same beam line, three bays away. Distant punch presses also have a faculty for making meter hands move to their rhythm. Concrete piers brought up from the ground through the floor are good unless there is heavy truck traffic nearby.

If there is anywhere near a single solution to vibration proofing, it is gained by weight and massiveness, by something "solid" whose natural vibration frequency differs substantially from the immediate surroundings. Following this reasoning, one should try putting a gage on a surface plate resting in turn on a heavy table.

Odd as it may sound, one group succeeded in stilling the transmission of vibration to a gage by setting it on a pedestal of 16- by 20-inch chimney cinder blocks. The blocks were carefully stacked one on the other without masonry (which is part of the trick), and they were not fastened to the floor. As a safety measure the uncemented cinder-block pier was encased in a solid wood box cleated firmly to the floor. The sides of the box came as close as $\frac{1}{4}$ inch to the cinder blocks but did not touch them.

A Constant War on Contamination

Contamination might be labeled enemy No. 1 as far as gaging is concerned. It is ever present and so much a part of our accustomed environment that it is unnoticed, forgotten, or ignored. But when the accuracy of measurements in millionths is at stake, the inspector becomes as acutely conscious of it as a surgeon.

Those skeptical of the effects of resident dirt could perform the following experiment. Leave a cleaned workpiece and millionth-inch comparator untouched for a day or overnight. Then measure the piece, taking pains not to clean it or the gage in any fashion. Next, carefully clean the piece and the gage, and take a reading. The size difference between dirty and clean will be a couple of millionths and probably more. If the inspector then gives the clean gage anvil and the workpiece the traditional final wipe-off with the palm of the hand, he will probably find the workpiece seemingly up to 10 millionths thicker.

Grubby hands have provoked many a precision measurement error. Add as other causes of error lint from clothing, wiping cloths, and bench and instrument covers. The latter should be of plastic and frequently cleaned. Paper towels and tissues are lint breeders, though not as prolific as cloths. Some inspectors use clean chamois; others prefer a soft artist's brush for last-minute cleaning. Special wiping tissues, almost completely lint-free, can be purchased commercially. With oil and moisture as vehicles, resident dirt has an insidious faculty for creeping into an ultrasensitive gage to foul spindles, plungers, and pivots, and make it sluggish.

Climatic conditions require the use of rust inhibitors in most of the country. These add to the cleanliness problem because they attract dirt. The usual routine is to spray or bathe the coated surfaces with a suitable solvent — but observe one precaution: Be sure the solvent is filtered clean.

Some gage laboratories adopt the expedient of maintaining low-humidity environment through the use of electric dehumidifiers or silica and calcium-chloride dehydrators. They also observe the rule of never touching clean workpiece surfaces or gage anvils with moist fingers. Thus rusting and the need for rust inhibitors are eliminated. Incidentally, an electrostatic dust precipitator in the lab or in the air ducts assists the air filters.

The point to remember is that the controlled environment is needed not only for proper functioning of the measurement machines, but also for protection to whatever is being measured, and the effects of environmental variation on the dimensions and even functionality of those parts and assemblies. Temperature, humidity, electrostatic discharge, as well as ambient atmospheric contamination levels, are all controlled in a modern "white" room or clean room atmosphere.

Precision measurement requires this control to extend to the air; to the surfaces of all parts, tools, and gage equipment that are brought into the room; and, not the least, to the people who come into the room and work. The room control ensures a clean ambient air flow — forced air and HEPA filters properly balanced with the room size and return air system ensure air cleanliness. Many parts that have to be measured and/or used in modern production require extreme levels of cleanliness themselves, and are cleaned in special and ever-changing systems. This surface cleanliness level is an entire inspection field in itself — the quantification of contaminants on a given component surface, often to the microinch level. And people and other guests to this controlled environment must also be shielded from exuding contaminants.

Millionths of an inch inspections are not your run of the mill process assessments, but rather special measurements, at the first article or incoming level, and often in a functional mode, to describe the dynamic interface between two critical parts, for instance. The setups for these operations are costly and time-consuming, and the results have to be accurate; usually, important decisions ride on these assessments, and the ability to maintain and control variation at such a level of discrimination often dictates what the inspection consists of.

Metallurgical Effects

One effect of metallurgy has been considered in the discussion of temperature coefficients of expansion and other effects will be dealt with in subsequent sections on penetration, deflection, and wear. But there are still additional reasons why an error of several millionths may occur because the inspector failed to realize the weak points of different materials.

Workpiece surface conditions vary according to their composition. Hardened tool steels can be given finishes so smooth that surface roughness is not obvious. On the other hand, it is safer not to trust a smooth-appearing surface. For millionths measurement at least, try to determine the amount of surface roughness. The gage almost invariably measures the peaks. If the workpiece is to be subjected to any metallic contact, rubbing, friction, or wear, the peaks will soon disappear.

Stainless steels (and occasionally chrome plate) may look very smooth, but sometimes tiny flecks of chrome expand bubble like or loosen up at the surface to disturb millionths measurement. Tungsten carbides present brittle-hard, smooth surfaces, but beware of nearly invisible pits. Tiny as they are, they may accept a spherical gage contact and "throw off" a millionths measurement. Aluminum surfaces oxidize, leaving a nearly impalpable, white, powdery coating of aluminum oxide with which to contend. Aluminum oxide is very hard and extremely abrasive. Although it may not directly affect accurate measurement, it has been known to swiftly lap a millionth or so off the surface of a gage anvil or contact.

In trying to distinguish size differences in terms of 1 or 2 millionths of an inch, it will sometimes be observed that the metal "grows." This is frequently true of work that has just been hardened or tempered — in other words, where it has been subjected to considerable internal stress. A piece will measure a certain number of millionths. An hour, a day, a month later, the same dimension may be several millionths greater. If the few millionths inch of size increase is important, as where tolerances are in millionths, some means must be adopted to stabilize the metal before final sizing. Sometimes it is difficult to convince the manufacturing division that a measurement discrepancy is due to an unstable, internal metallic structure.

There is a possibility in millionths measurement that the size, area, and weight of a workpiece will create an error in obtaining a true dimension. For example, in a test made by the National Bureau of Standards with a 4-inch gage block, the length was measured (a) in a horizontal position, and then (b) while standing upright, and finally (c) while vertically suspended. The results are shown in Fig. 3. When upright, the block shrank or compressed to become 0.00000008 inch shorter, but it stretched 0.00000008 inch when suspended, as at (c). Elongation and shortening varies as the square of the length. The 0.08-millionth-inch error in measurement of a 4-inch block, or a smaller error obtained in gaging a shorter one, is negligible for all practical purposes. However, if millionth-inch tolerances and measurement are required on a part whose basic dimension is 18 inches in length, thickness, or diameter, the compression or elongation of the metal due to gravity may be pertinent.

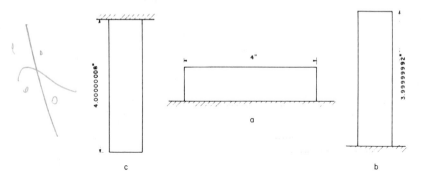

Fig. 3. Stretching and compressing of a 4-inch gage block due to its own weight. The changes in size, shown at (b) and (c), will vary with the square of the length.

Contact Point Penetration

At first thought, the phenomenon of penetration seems nothing more than an illusion, which the inspector has never observed or considered as a source of error until he attempts measuring those last few millionths. Penetration is the bending, depressing, deforming, and yielding of the surface of the workpiece under the pressure of the gaging contact. Its effect is illustrated in greatly exaggerated fashion in Fig. 4.

A succession of tests have shown that with a diamond-tipped, spherical contact point of 0.125-inch radius pushing down on a standard, hardened steel gage block under only a 6.4-ounce pressure, the penetration amounts to 10 millionths of an inch. Even with a contact pressure of as little as 1 ounce (27 grams), the penetration amounts to about 3 millionths of an inch.

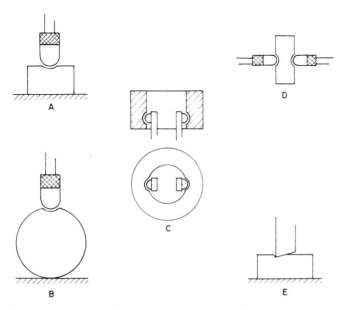

Fig. 4. Exaggerated examples of penetration, the deforming of the workpiece surface under the contacting pressure of a gaging point. It is an important factor in millionth-inch measurement.

The penetration effect varies with the metals being gaged, depending mostly on Young's modulus for the material. Penetration into tunsten carbide is often negligible because of its greater rigidity. In comparison, brass yields readily. Hence the inspector needs to be alert to penetration when he uses a carbide block as a master and then measures a brass cylinder. Probably, penetration in the surface of a cylinder is a little greater than that in a flat surface, which in turn is greater than in the bore of a ring. The penetration error doubles when the piece part is being measured with outside or inside calipering jaws, as seen at C and D in Fig. 4.

Some might say immediately that if the spherical contact were not so hard, or diamond-tipped, the penetration would be less. Perhaps this is true, but if the spherical contact point is relatively soft, might not it flatten a trifle? Again, others might recommend a flat gage contact, since the penetration should be less (depending on the contact area). It is extremely difficult, however, to obtain perfect parallelism between the plane of a flat contact and the gage anvil. Consequently, penetration like that illustrated at E in Fig. 4 would most probably occur.

Another thought might be that an error from penetration could be nullified by the use of extremely light gaging pressure. The problem, however, is not solved quite so simply, as will be seen from discussions of deflection and of gaging pressure versus internal friction.

To escape penetration error from metallic gage contact pressure, some argue that an interferometer or an air gage might be used for millionths measurement. In neither case is a mechanical contact pressed against the workpiece.

But the light waves of an interferometer do "penetrate" in their fashion, or act as if they do. The phenomenon, known optically as change of phase, accompanies reflection at the surface of a light-absorbing material. The change of phase for fused quartz is negligible; for a properly polished steel surface the reading error would not ordinarily exceed 0.7 millionth of an inch. The used gage block or workpiece surface, however, can have enough lapping marks or other surface defects so that the change-of-phase error can mount up to some 3 millionths of an inch. But new machines, and proper setups of laser and other light-derived measurement systems, do hold a great deal of encouragement in terms of noncontact measurement and all of its advantages. Some of the most simple and precise of all measurement systems in use today employ the relatively simple principles of measurement through noncontact light beam assessment or interferometry.

Air-gage proponents contend that neither optical nor mechanical penetration occurs when air is used. But for one thing, they forget or ignore the equivalent effects of the surface roughness of the workpiece or master. (A discussion of surface-roughness error, explaining how air tends to measure the "pitch diameter" of the surface ridges, is included in Chapter 11.) If air gages are to be used for the sort of extremely close work implied here, the masters and the work should have no discernible roughness. Where 1 microinch of surface roughness can be measured, expect an air gage to err by an amount close to 1 microinch from this cause.

Air-gaging systems display other inherent, error-producing characteristics that are deceptive and difficult to detect and correct. As yet, some professional metrologists are unwilling to rely on air gaging for detecting size differences less than 5 millionths.

Contact Pressure Steals Millionths

Since contact pressure interferes with those last few millionths of measurement, extra sensitive gages exert pressures varying from a few grams to several ounces, depending on the design. Some makes of electronic gages permit the operator to regulate gaging pressure between an ounce and 4 ounces. Another manufacturer advertises contact pressure of 1/10 ounce ($2\frac{1}{2}$ grams).

For detecting size differences of less than 100 millionths (0.0001 inch), gaging pressure is seldom trusted to human touch. Contact pressure in most precision gages is controlled by spring action, as in the case of dial indicators. In a few designs, gaging pressure is exerted by weights and counterweights.

No doubt the ideal condition is to use as little pressure as possible in order to neutralize penetration deformation, and deflection; but at the same time, there cannot be any wavering of contact against the workpiece. Contact must be certain. Invariably there is some internal force opposing gaging pressure — dirt, friction in pivots and bearings, "hysteresis" in springs, or the pull of a magnetic field. Uncertainty in measurement comes when the gaging pressure is reduced to a value just about equal to any internal reaction force, and a sort of null point is reached. Vibration is also a major cause of uncertain readings with light gaging pressure. The usual symptoms for the trouble are fluttering, unstable meter readings or lack of repetition.

Deflection "Sneaks" into Measurements

The simple experiment illustrated in Fig. 5 could be enlightening to anyone incredulous about deflection. A micrometer is mounted as a snap gage and set to the workpiece diameter with the customary finger pressure on the thimble. The spindle is then locked and the workpiece is withdrawn. A 10-thousandth-inch indicator is also set up as shown in the illustration to register the vertical motion of the micrometer frame and barrel. When the workpiece is moved in and out between the locked jaws, the indicator shows the micrometer frame deflection. (This deflection and the indicator reading can be increased by screwing the micrometer spindle a little more tightly down on the workpiece.) If the workpiece is rigid enough so that its deflection can be neglected, its diameter can be obtained by adding the indicator reading to that of the micrometer.

Deflection is about the stealthiest of all the error-producing conditions. It "sneaks" into measurements, yet is hard to believe present because the unaided vision cannot discern it. Deflection errors

Fig. 5. A setup which demonstrates the effect of gaging pressure on the deflection of the gage frame. The indicator shows gage deflection when the workpiece is set in place.

often are mistakenly included with those of temperature, penetration, and looseness. It is ever present if there is contact gaging pressure of any amount. In some setups, merely an ounce or two of pressure can produce deflections measurable to several millionths of an inch. The basic reason why gage frame deflection interferes with correct measurement is illustrated in Fig. 6, which shows a spring-loaded contact exerting downward pressure on a workpiece with a force F.

Simultaneously, an equal and opposite reaction force R becomes set up exactly equal to F. This opposing force tends to make the gage bracket bend upward, as at D. But force R operates with a leverage or moment arm L. Such mechanical advantage not only increases the immediate deflection D but adds to the deflection of the gage upright D'. Without going into Newton's law of equal and opposite reaction, or the mathematics of moment arms, or Young's modulus, Fig. 6 shows

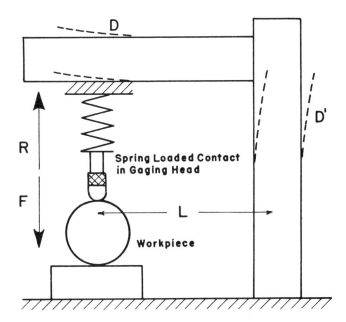

Fig. 6. Diagram showing how the mechanical advantage gained by the gage frame
moment arm amplifies the deflection due to contact pressure.

that reaction R, though equal to force F, will more readily "spring" the gage frame because of leverage L.

The cure for this type of deflection error is to (1) have F as small as possible; (2) make the throat depth L as short as possible; (3) make the gage frame rigid and of adequate cross section; and (4) securely tighten any gage clamps or adjusting devices. The sturdy, virtually massive gage construction seen in Fig. 7 helps reduce deflection errors to negligible quantities.

The obvious protest might be offered that if the gage is used as a comparator, its deflection error (as well as penetration) is in effect nullified. The deflection of a gage set to a master should remain the same when the latter is replaced by the workpiece, a contention that proves true in many circumstances. On the other hand, deflection, when friction is present, is an unpredictable variable. If a comparator is inherently prone to this combination, inexplicable differences in readings may appear, especially in millionth-inch measurement.

Lack of repetition in results can thus point to the presence of deflection. Often deflection errors are uncovered when a piece happens to be measured in two different gages. The gage that fails to hold a zero setting for a period of time might be offering a clue to excessive deflection.

Deflection errors creep into inside-diameter calipering measurements and are especially troublesome where master rings are to be calibrated to within a few millionths of an inch. In the present state of the art, most master-ring checking is done on "measuring machines" more or less typical of the one seen in Fig. 8. The "machine" jaws are mastered or zero-set with gage block stack calipers as shown. Then the ring is centered over the jaws and the difference between the readings, in millionths, is read on the meter. Probably both the stationary and the sensitive jaws bend and penetrate a trifle.

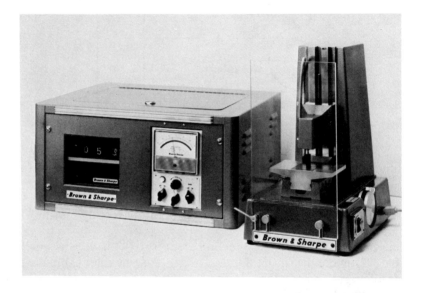

Courtesy of Brown & Sharpe Mfg. Co.

Fig. 7. The main frame of this millionth-inch comparator is designed with a "built-in" web to resist deflection and prevent errors from this source.

In this type of measurement, the belief that the same amount of jaw deflection would be transferred without error during the comparison of the workpiece and the master could turn out to be erroneous. The diagrams in Fig. 9 show why. While zero-setting a gage block caliper, the solid reference jaw R and the spring-actuated sensitive jaw S might deflect as illustrated at A. The direction of possible deflection of each jaw is indicated, although the amount is greatly exaggerated. Another deflection possibility is seen at B, where the workpiece has replaced the master. There are many such jaw-deflection combinations.

Some of the difficulty comes from the friction (F in View B, Fig. 9) between the workpiece and the gage platen. Being spring-loaded, the

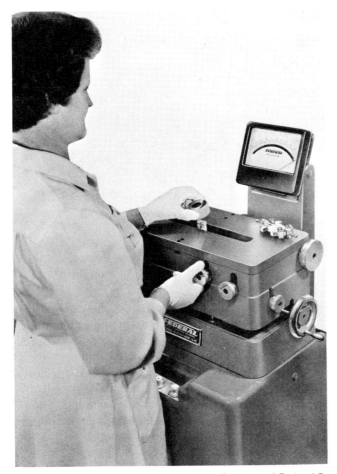

Fig. 8. An electronic internal measuring machine has been set to caliper jaws spaced with gage blocks prior to the checking of master ring gages.

sensitive contact S tries to drag the work into final gaging position. To reduce deflection produced by this effort, free-rolling cylinders or balls are sometimes used under the work, as suggested in View C. One make of measuring machine has a cam device, that, in effect, locks the calipering jaws in gaging position while the master is replaced by the workpiece. Jaw deflection is not eliminated by this scheme (nor penetration), but the error is virtually nullified because the arrangement keeps the jaws positioned precisely alike for each measurement.

Jaws of precision calipers for internal diameter gaging are made of rigid material with cross-section areas large enough to reduce bending

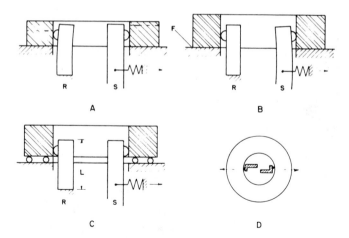

Fig. 9. Views (A) and (B) show two of the many possible ways internal gage jaws can deflect. Cylinders or balls reduce friction drag of a workpiece, View (C). View (D) shows how jaws are made to overlap for measuring small holes.

errors to negligible amounts. Gage designers shun any jaw length (L, View C) greater than $1\frac{5}{8}$ inches in order to minimize bending leverage. When the hole size is less than 0.750 inch, however, jaw cross section must be reduced because of the lack of space in the hole. (View D, Fig. 8, illustrates how jaws overlap to fit into a small hole.) Jaw deflection becomes a fairly important problem in holes smaller than 0.250 inch.

The demonstration in Fig. 10 verifies the deflection potential in jaws of small cross section. Such a jaw is shown solidly clamped on a jig, and an electronic transducer gaging point is making contact with it. The connected meter was set to read zero when no weight was hanging on the end of the jaw. But when, as in the illustration, a 20-gram weight was hung on the jaw, the electronic indicator registered a deflection of 5 millionths of an inch.

Some Sources of Looseness

A first cousin to penetration and deflection as an error breeder in millionths measurement is looseness. It displays the family trait of being hard to detect and is often unobserved. Its symptoms — meter flutter, lack of repeat readings, and inaccurate calibration — are also characteristic of other gage troubles and thus contribute to making looseness difficult to diagnose or isolate.

Parts that make up a gaging assembly often mysteriously work loose at inopportune times. Probably a systematic check for looseness

Fig. 10. This arrangement demonstrates gage jaw deflection. The electronic gage head revealed a 5-millionth-inch movement when a 20-gram weight was hung on the securely mounted jaw.

should be made periodically. One way would be to set the gage contact on the gage anvil and zero the meter. Finger pressure or a light tap can then be applied to each location where looseness might appear, and the meter will reveal its presence. There are a number of typical places where looseness can be expected.

Be sure the gage anvil is clamped down securely to the gage frame. Seemingly rigid, it can float on a film of air or it can teeter a few millionths of an inch in its holder. If the gage post is round, remember that most cylinders are turned or ground a trifle oval and any hole or socket will be bored out of round by a similar few millionths. All too often the major axes of ovality are 90 degrees to each other and the post then readily pivots in the hole. Metal chips or plain dirt between post and clamp can produce the same pivoting effect. Two flat, joining bracket surfaces tend to yield a trifle (no matter how hard they are tightened up) if either surface is not completely flat or a ridge of dirt has crept between them. If the gage has spring reed mechanisms, check and tighten the assembly of the springs to the gage frame and parts.

A fairly complete knowledge of the gage meter may help in a search for loose pivots, worn bushings, and plunger side play. The gage contact point should be tight in the spindle. A diamond or sapphire insert can work loose in its setting. Having checked and corrected as far as possible for sources of looseness, the inspector is wise to recalibrate the gage throughout its range, also testing for repetition at each calibration station.

Don't Ignore Wear in Gages

Perhaps because wear is the most common cause of gaging error, it is probably the most ignored; peculiarly too, because it is about the most obvious and easily checked of discrepancies. Although wear crops up in instrument pivots, bushings, and plungers, it is usually more troublesome on anvils and contacts. What occurs under typical conditions is illustrated in View A, Fig. 11. A comparator is mastered with either a flat gage block b or a cylindrical master m and then an attempt is made to check the diameter of a cylindrical workpiece w. When the flat surfaces of the contact and anvil are not parallel or hollows are worn in them, an accurate comparison cannot be obtained.

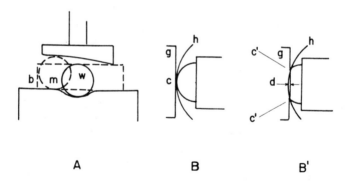

A   B   B'

Fig. 11. Typical errors produced by anvil and contact wear. View A shows effects of wear when comparing work (w) with masters (b) and (m). A difference (d) in measurement, as in Views B and B' may be due to worn contacts.

The error produced by a spherical contact which has been allowed to wear flat is shown in Views B and B'. When an unworn contact is used to caliper an internal diameter h and then is used to compare it to the flat, plane surface of a gage block caliper g, there is no loss of dimension at the point of contact c (View B). But when the spherical gage contact is worn (View B'), it bears against the gage block as shown and against the hole at points c and c'. The dimensional loss due to wear — the error in obtaining the true diameter of the hole — is equal to the distance d.

The extent of wear hollows in anvils and lack of parallelism can be readily measured with optical flats. The spherical contact can be checked by examination with a microscope. The contribution to measurement error from internal instrument wear is shown if the instrument is calibrated and lack of repetition is observed.

Wear and dirt inevitably appear together. One way to control wear error is to keep gages, masters, and workpieces clean. However, even with perfect cleanliness, some infinitesmal amount of wear takes place with each gaging. Some materials increase wear. Carbide masters are wear resistant themselves, but can raise havoc with hardened steel anvils. Cast-iron workpieces present the problem of sand particles and glass-hard cooling checks. All surfaces should be completely clean and free from lapping and buffing compounds.

The question is frequently asked: How often should a gage be checked for wear — how many gagings can be made before the error from wear becomes appreciable? Making steady checks for wear, keeping a record of them plus a count of the number and type of gagings completed between checks — recorded experience, in other words, and not the personal opinion of an individual — becomes the answer.

Where measurement is a matter of detecting a size difference of a millionth or two, contacts, anvils, gage blocks, and masters have to be replaced or renewed the instant any wear at all is discernible. However, for most commercial precision measurements in the neighborhood of 5-, 10-, 20-, or 50-millionth increments, a "wear allowance" of up to 5% of the tolerance is often permitted. Where the amount of wear is constantly known, measurements may be corrected for it arithmetically to arrive at the final correct figure. Corrections for instrument or meter wear are usually made in this fashion, especially in cases where careful calibrations have, in particular, shown the extent, location, and direction of such errors.

Gage anvil and contact material is also important. Unhardened surfaces are almost completely unreliable. Chrome-plated parts seem to withstand 10 times more wear than those of unplated, hardened tool steel. (When chrome is used against chrome, however, there is a possibility of galling.) Carbide blocks, masters, anvils, and contacts usually repay their initial higher cost by providing from 100 to 1000 more wear-free gagings than those of hardened and tempered steel. Sapphire-tipped contacts, although subject to breakdown, usually wear about as well as those of tungsten carbide, but diamond contacts, of course, set the endurance records. It must be remembered, however, that even diamonds do wear down, and the inspector cannot be complacent about checking the condition of such contacts periodically.

The Importance of Correct Manipulation

The effect of manipulation in breeding errors, or in correcting them, has been implied in several of the preceding cases. For one example, forceps or tweezers, and gloves or pads, are used to keep the heat of hands away from the work. Intelligent checking for looseness and constant, thorough cleaning are other important phases of manipulation to be carried out in millionth-inch measurement.

No doubt, the first rule of correct manipulation would be to use the hands as little as possible or, to put it more positively, take the hand off the work as soon as possible. Let the gage do its own work! For some reason, inspectors tend habitually to use one hand, or both of them, as a vise and never let go while sizing a part. Most gages are designed to hold the work properly, and once they are placed in the prescribed position, they will be more likely to give the correct measurement.

Cramping is the term often used to describe this tendency to hold the work rigid. The mechanic who uses an indicating micrometer for the first time quickly realizes how often he may have forcibly, though unintentionally, held the work and the micrometer at a slight angle to each other, enough sometimes to produce a half-thousandth-inch measurement error. Another example of potential cramping is seen in Fig. 12. As long as the inspector's fingers cling to the ends of the cylindrical master, he will be inclined to tilt it, twist it, or hold it out of true geometrical position and thus introduce several millionths of an inch error in his reading.

True, the cylinder shown in Fig. 12 must be positioned so that its full diameter and not a chord is fully under the gage contact, but when it seems to be properly centralized, both hands should be removed while the indicator or meter is read. The operation should be repeated several times for confirmation of true centralization and the reading. Some inspectors have formed the worthy habit of lightly tapping the cylinder back and forth under the gage contact with a pencil, until repeat readings indicate it is exactly centered.

Keeping a grip on the workpiece is sometimes excused because the workpiece overhangs the anvil — the unbalanced weight might cause it to tilt indiscernibly against the gage contact pressure and result in measurement error. This is good thinking except that it is probably better to secure a wider anvil. Hands invariably tremble, but the workpiece must not flutter. Often it pays to design and have available some sort of auxiliary holder for the workpiece. Clamps and magnets are handy devices.

Since gage blocks are widely used as masters, their continually correct manipulation is important. Experience has shown that carelessness about wringing gage blocks together and to anvils readily produces

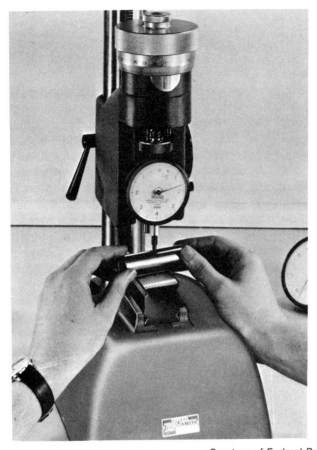

Courtesy of Federal Products Corp.

Fig. 12. Hands often unconsciously tend to "cramp" work out of correct gaging position. Both hands should be removed from the workpiece while the indicator or gage is read.

errors as great as 100 millionths (0.0001 inch). On the other hand, experiment indicates that proper wringing may possibly create an error no greater than an immeasurable fraction of 1/10 millionth inch.

Perfectly flat, perfectly clean, dry gage blocks are sometimes difficult to wring. Many inspectors apply a little "wrist oil," but this expedient should be positively taboo where less than 5 millionths of an inch are involved. An extra light, filtered clean, kerosene like oil, coated on the block or anvil so thinly that it is iridescent, produces no discernible error. A 1-inch, smooth cube of basswood kept soaking in a covered container of clean kerosene can serve as an applicator.

Prior to use, it should be lifted out with tweezers, grain end up, onto a piece of clean tissue or cloth to let the excess oil drain and evaporate. A gage block rubbed once across with the end grain of the oiled basswood cube will have almost exactly the right film of lubricant on its surface for error-free wringing.

A man could not be expected to work at precision measurement day in and day out without occasionally dropping a gage block, master, or workpiece. Such an accident is bad enough, but humans seem compelled to conceal personal clumsiness. Consequently, the fallen object is retrieved as rapidly as possible and put to work with an innocent air of its never having been dropped. After such an accident, the fallen part should be carefully examined for dirt, nicks, scratches, dents, or distortion. Any injury should be repaired or the piece replaced before making further millionth-inch measurements. Gage extensions, accessories, and anvils roll or slide readily off benches. If an attachment contains a diamond or sapphire insert, it should be checked with a microscope for breakage.

Clumsiness and awkwardness, although unintentional, are the opposites of a certain adroitness and dexterity required in millionths measurement. It is good practice to complete a setup and take one or two readings simply as a sort of rehearsal. Such a custom not only permits a critical analysis of the measuring method, but also gives the inspector extra practice toward the manual deftness desired.

The word *flinching* is frequently heard in inspection circles. While one dictionary definition describes it as a loss of nerve in the face of an unpleasant duty, the "trade" use of the word signifies more a subconscious tendency to make a gage read what is desired rather than the true dimension, especially in marginal cases. The finger pressure exerted on a micrometer spindle readily varies to accommodate a more desirable reading. The added leverage of tweezers makes it easier to cramp a piece into a more favorable position under the comparator contact. Inspectors readily miscount a dial division or two to get a subconsciously desired reading. Flinching is an unintended human observation in measurement that lies in the borderland between manipulation and approximation.

The trends in today's modern inspection equipment design is as much to increase the repeatability and reliability of the machines, as it is to bring on the discriminatory capabilities. The discrimination is here — there is nothing that cannot be measured physically, that exists in the physical plane. The problem now is more one of ease of measurement, and of course those measurements' reliability. Most of today's sophisticated measurement machines "hold" the part themselves, and position and compensate for "flinching" and other inspector-related variables through automatic and powered sensing and adjustment systems. If the inspector has to tap or compensate physically for positioning and movement through the inspection, the effectiveness and efficiency is going to be severely impacted.

Approximation Should Be Avoided

Many like to guess, estimate, and round off numbers, and these traits tend to subvert the exactness required in millionth measurement. In Chapter 5, the estimating of an exact size is discussed. The text suggests the use of a scale of finer discrimination or, to use more modern terminology, of greater resolution. If a micrometer reads 0.494 inch plus something, use the vernier scale to read that exact 0.0004 inch. It is better to give the complete, correct reading of the dimension as 0.4944 inch, rather than guess that the final figure is, say, 3 "tenths," or something else.

The minute a marker reads between graduations on any instrument, the desire to estimate its position becomes practically irresistible. In millionths measurement, it is safer to name the graduation next higher or lower, the one closer to the meter hand, rather than to try to estimate half a scale division. In Fig. 13, which would be the correct decision — to call the market halfway between *a* and *b*, or to say it is more on the side of *a*? Probably few would call the measurement as size *b*; the majority would vote for *a* plus a half; the others, battle-scarred from many measurement errors, would refuse to estimate but would arbitrarily settle for graduation *a* as the size, especially when each graduation represented a difference of only 1 millionth of an inch.

At the present state of the art, no one (including the National Bureau of Standards) is yet 100% positive of accurately measuring a

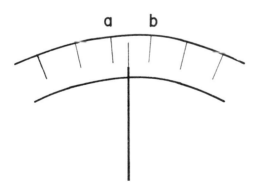

Fig. 13. An opportunity for approximation. When each graduation represents a millionth of an inch, estimating to a fraction of a division is unwarranted.

size difference of a fraction of a millionth. An instrument might seem to show such a variance, but instrument error can be greater than the actual size difference. A millionth can be "blown up" to look like an inch, but all the inherent errors are seen in proportion.

Another misleading habit is to average a group of readings and call the result the true workpiece size. If there were, for instance, two readings of a gage block thickness, one of 0.120001 inch and the other of 0.120002 inch, the tendency would be to call the block size exactly 0.1200015 inch. Almost without exception, measurements reported to the seventh decimal place represent an arithmetical average and not an absolute measurement.

Currently, the recommendation is to take several successive readings which are then averaged and their standard deviation* calculated. The measurement is reported to be possibly as high as the average plus three standard deviations, and very probably no higher, or as low as the average minus three deviations, and very probably no lower. Where the exact size might lie between these plus or minus limits is purely a guess. A further analysis of a group of figures, however, sometimes influences a statistician to make an "educated" estimate. Repeated similar readings in a series are about the best guide to the exact mesurement, although, to be doggedly pessimistic, a recurring (systematic) error either in the apparatus or in manipulation can invalidate the result.

Anyone regularly reading instruments must understand parallax and appreciate the extent of error its neglect can produce. To avoid parallax, view the meter hand normal to the graduations. Never trust an oblique, careless glance. Nullifying parallax is actually very much a part of correct gaging manipulation.

"Round" Holes Are Seldom Round

In millionths measurement especially, the inspector might as well resign himself to inevitably finding some degree of "relationship" trouble, usually involving a combination of misshapes grouped together on one piece. Luckily any lack of squareness, parallelism, roundness, or other geometric discrepancy can be measured and the direction and amount present counteracted by correcting the final size readings.

Gage blocks whose end surfaces are consistently smooth, flat, and parallel are available. In the case of cylinders and holes, the situation is

---

*The statistical technique of procuring and using the standard deviation is described in *Quality Control, Reliability, and Process Improvement*, N. L. Enrick, Industrial Press Inc. and in other standard statistical reference books.

different. Probably no pieces having these geometric forms are ever-entirely free from some combination of measuring ovality, taper, waviness, or barrel or hourglass shape.

Until tolerances are down in the 10-millionth-inch range, a hole with a condition like that sketched in Fig. 14 does not cause assembly difficulties too often. How would an inspector go about checking for a leaning hole whose diameter was just about perfect, as well as being free from other geometrical errors? How would he be sure of finding the condition illustrated when the error in inches is only a matter of 3 or 4 millionths of an inch or involves only a few seconds in angularity?

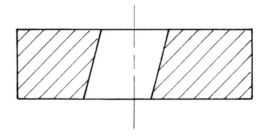

Fig. 14. "Lean" in an otherwise perfect hole is difficult to detect when the error is only a few millionths of an inch.

Millionths measurement reveals another hole condition which may never have been detected in holes made and measured to more ordinary tolerances. The geometric irregularity is triangular out of round, or "cloverleaf." Most inspectors have encountered these peripheral characteristics which are peculiar to centerless-ground cylinders. But how, they ask, do such effects occur in a hole? Nevertheless, triangular out-of-roundness does show up in holes to prevent close-tolerance assembly fits and confuse results in two-point, millionth-inch calipering.

If the quill that holds a grinding wheel, or any related revolving shaft on an internal grinder, happens to have been centerless ground, and if its cloverleaf condition is ignored, the shaft's triangular out-of-round pattern will somehow be implanted to some degree on that periphery of the hole being ground. The condition is seldom detected until control of hole geometry in millionths of an inch is needed. Even then the effect will never reveal itself under two-point, inside-diameter calipering. It can be detected and measured to some degree if the hole is checked over a three-jet air plug or on a three-point internal comparator of a type similar to that shown in Fig. 15. The latter must be equipped with an indicator, transducer, or an air cartridge device having a resoluton of 50 millionths (0.00050 inch) or even finer.

Fig. 15. This gage with three sensing jaws is suitable for detecting three-point out-of-roundness in a hole.

A Discussion of Lobing

The foregoing is merely an introduction to a whole field of measurement geometry that has been given little attention where parts tolerances are never smaller than 0.0005 inch. Out-of-roundness and cloverleaf, along with taper, barrel shape, and hourglass profile on cylinders and in holes, became of greater concern as tolerances narrowed to 0.0001 inch. When size control to within a few millionths inch loomed as a requirement, out-of-roundness error became very important.

Until lately, the term *out of round* in a machine shop has meant almost any condition that might show up on a test indicator — eccentricity, bend, warp — and, in particular, plain ovality. Similarly, an out-of-round hole was considered to be only egg-shaped with respect to the major and minor axes. Nowadays we know that there is more to it.

Lately, also, a sort of double ovality has been recognized where the workpiece seems to have two major and two minor axes. In Fig. 16, simple ovality is seen at A and ovality occurring in two directions is illustrated at B.

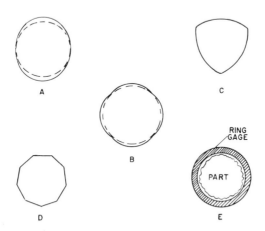

Fig. 16. Diagrams illustrating a few of the actual profiles of machined cylinders or holes often supposed to be perfectly round. Lobing and waviness are greatly exaggerated in size.

From the observation of such conditions has grown the terms lobe or lobing. In fact, industry is recognizing that no hole, cylinder, or sphere probably ever possesses a geometrically perfect, symmetrical round shape. Nor does it but seldom assume a perfect oval (two-lobe) shape. The profile of the average machined cylinder or hole (and sphere) varies, in terms of millionths, from something completely distorted and irregular to, usually at best, a multiple-lobed, fairly symmetrical outline. Even then it often exhibits much waviness, chatter, and surface roughness.

The familiar three-lobed shape common to centerless grinding is seen at C, while diagram D pictures the sort of multiple lobe shapes that are far from uncommon. Finally, sketch E illustrates cylindrical surface waviness and surface roughness of a workpiece, as well as some out-of-round distortion and with possibly a trace of lobing also being present.

The diagrams in Fig. 16 have one necessary exaggeration. The actual dimension of the amount of lobing seen at A, B, C, and D is usually mostly a matter of millionths. Occasionally, it can equal 0.0001inch, 0.0002 inch, or more. (Sometimes the distortion, ovality, or lobing is visible to the naked eye.) The cylinder or hole diameter is shown nearly full size, but the lobes are magnified some 10,000 times to make them visible. In other words, the diameter might be, say, 3 inches but the lobing might be only 30 millionths inch though it looks to be sizable in proportion to the diameter.

Although lobing (ovality), distortion, waviness, and roughness may change the periphery of the shaft or hole from the desired perfect circle by only a few millionths, it is still these very conditions that cause hum, vibration, heating up, and wear. These effects, in turn, increase to cause noise, rattle, and finally the failures much too common in industry.

Some of the Causes of Lobing

Probably the centerless grinder did more to focus attention on lobing than any other single factor. But out-of-roundness can be produced in many ways. Drive shafts and bearings in lathes and grinders, if they are not truly round, to some degree impress their patterns of peripheral irregularity on the work, whether a hole or a cylinder is being machined. Centers that are out of line may produce similar effects. Three-jaw and multiple-jaw chucks no doubt cause many workpieces to become out of round. In recent years, magnetic chucks are coming more into use, since they are often able to grip workpieces without distorting them. The springing of metal under cutting pressure also probably contributes to lobing. All causes of lobing are not known at present. The important matter, however, is to detect, analyze, and measure lobing as a first step toward eliminating or minimizing it.

One more point is important. If distortion, out-of-roundness, or ovality, or lobing (even waviness sometimes) has been established on a cylinder or in a hole or ring, subsequent operations cannot be depended on to eliminate any of these conditions completely. The distortion pattern set up by the roughing lathe will usually show up after grinding, although the degree of error may be reduced to millionths of an inch. Lobing is, perhaps, the most persistent in this respect.

Measuring Lobing

The plain elliptical form of out-of-roundness (what might be called two-point lobing) is readily detected by turning the piece on the platen of a comparator under some form of indicator with adequate resolution. For that matter, any type of indicating caliper or snap gage may be used. The oval hole responds to the dial bore gage or internal measuring machine as seen in Fig. 8. The rarer case of four lobes (or any even number of lobes) can be checked by the same sort of two-point measurement.

Many inspectors, however, have experienced the mystery of precisely "miking" a shaft only to discover it would not then assemble to a hole or ring gage that is very close to the same diameter. The reason is that measuring devices having diametrically opposed contacts do not measure the "envelope" or complete circumference and cannot, therefore, detect odd-numbered lobes, as well as some forms of distortion and waviness. A ring gage does "envelope" the periphery of a cylinder (View E, Fig. 16). A plug gage acts similarly in a hole. Plug and ring gages will tell whether a shaft and bearing will assemble without interference, although there may not be a very good fit from the point of view of wear and noise as View E suggests.

Why a shaft with lobes will not enter a truly circular hole of similar diameter is illustrated in Fig. 16. Also — and this is often hard to believe — a shaft with the shape shown at C will roll just as smoothly on a flat surface as a perfect cylinder. Multiple-lobe pieces show the same general characteristics and their lobing is equally unmeasurable by two-point methods if there is an odd number of lobes and if the lobing has a symmetrical geometrical arrangement.

To detect lobing and count the number of lobes, use of a V-block is recommended. Ordinarily, the standard, 90-degree included-angle V-block will do. But to measure the exact amount of the lobing protruding, $R - r$ (Fig. 17) requires a V-block with a special angle. This correct angle A can be calculated from:

$$2A = 180 - \frac{360}{n},$$

where n equals the number of lobes. For a three-lobe configuration, A becomes 30 degrees and the V-block used should have a 60-degree included angle. The measurement M, which is obtained by revolving the "round" piece in the V-block under the comparator contact, can be converted to a measure of the radial variation in cylinder contour by means of the formula: $M = (R - r)(1 + \operatorname{cosec} A)$.

If the discrepancy from true form is to be analyzed in terms of millionths, attention must be paid to the V-block. The two-plane sides should be checked for flatness and wear from time to time and surface roughness of the two areas should be at a minimum. The vee angle must be known to be unchanging throughout the length of the V-block, and the center line of the vee must be perpendicular to any gage platen on which it rests — truly 90 degrees, as indicated in Fig. 17.

If a cylinder were mounted in a V-block under the contact of a comparator, the latter set to zero on it, and the cylinder revolved, say,

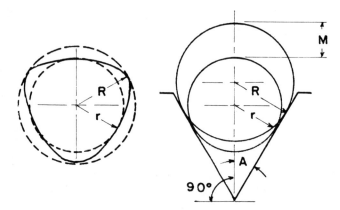

Fig. 17. Measurement of the exact amount of lobing requires a V-block with a special angle (A). Here illustrated is the geometry involved.

5 degrees at a time, a series of plus and minus readings would be observed. If these readings were plotted around a sort of circle chart, a wavy outline would be completed that would be similar to that shown at E Fig. 16 or those in the charts illustrated in Fig. 18. To perform the same measurement on the surface of a hole is another story. For this purpose (and also to get an external profile rapidly) special instrumentation is helpful.

Several types of roundness measurement apparatus are available, and the inspector should be acquainted with the techniques and details involved. Their basic principle is not unlike the mechanism of a jig borer which permits a circular sweep. One type, illustrated in Fig. 19, has a precision spindle, which runs true within 0.000002 inch, as its principal feature. An attached stylus pickup contacts the specimen as the spindle rotates. What the stylus "sees" as it sweeps the exterior of a cylinder or the inside of a hole is transmitted electronically, amplified, and recorded as a polar diagram on a chart. (Sample charts are also shown in Fig. 18.)

This kind of instrument gives no indication of the workpiece diameter; it simply measures and charts the plus and minus deviations of the workpiece periphery from a true circle. Ingenious electronic filters permit separation of roughness and waviness values from those of lobing and general contour, if desired. A caliper stylus and worktable arrangements permit studies of such features as concentricity, parallelism, and squareness. (The instrument shown will also detect and measure the "leaning" hole illustrated in Fig. 14.) In another type of instrument, the "sweep" is gained by revolving the workpiece against the stylus, which is held rigidly by the gage.

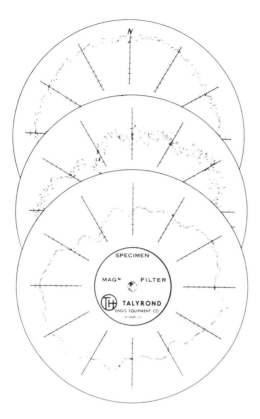

Fig. 18. Specimen charts which show, in millionths of an inch, actual roundness, surface waviness, and roughness. The chart of a perfect piece (which does not exist) would be a perfect, smooth circle. The charts do not record diameter.

The Gaging Setup Should Be Square

An important part of the geometry of measurement is the four-square construction required of gages and gaging setups. A try square is an important tool for the metrologist.

It has been suggested previously that flat anvils must be checked regularly for the exacting flatness required of them. Similarly, if flat anvils are supposed to be parallel to each other, that necessary condition should be appraised by rolling a small precision cylinder or ball between them or, much better, by slipping an optical flat between them. The idea of examining the profiles of spherical contacts has also been suggested.

There is a tendency to overlook the matter of perpendicularity, or the lack of it, and to assume that the gage will surely be at right angles

Courtesy of Mitutoyo/MTI Corp.

Fig. 19. An electronic instrument that detects and records deviations from perfect
roundness.

when placed on its platen. The desired condition is shown at A in Fig.
20. A comparator or gage should be carefully checked for squareness
from several different angles and at either end of its travel to be sure
that the ideal condition prevails — with the gage clamped tight!

In direct measurement, the condition seen at B in the same
illustration is obviously unwanted because measurement h is greater
than dimension u (View A), and the difference between h and u
increases as angle a increases. The relationship between h and u is
shown in View C.

It should be realized that all errors of perpendicularity are not due
to gages. A sizable number of them are the result of crooked setups.
Furthermore, even though the gage head and contact could be perfectly
perpendicular, the reference anvil might be tilted, as indicated in
View D.

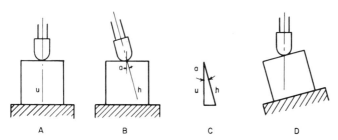

Fig. 20. Perpendicularity in gaging setups should not be assumed in millionths measurement. An angular error of 1 minute can cause a 10-millionth-inch error in 1 inch.

The Surface Plate in Millionth-Inch Measurement

Traditionally, the surface plate has been the basic reference surface, so much so that many inspectors are reluctant to adopt other techniques. Setups of the sort illustrated in Fig. 21 seem natural to them, but to make a similar length measurement with a modern precision calipering gage or even with a comparator is considered not quite legitimate. Although instruments like the one seen in Fig. 21 are available commercially having resolutions of at least 5 millionths inch, there is considerable doubt whether a height gage on a surface plate can take advantage of such fine readings.

An old surface plate could more resemble a strip of rolling prairie when its surface flatness is considered in terms of millionth-inch measurement, and the underside of a height gage shoe is also likely to be no paragon for flatness. Although either condition could be considerably corrupted, too often they are overlooked because surface-plate setups unfortunately give a false sense of security. Even at best, probably only very few, small areas of a surface plate would show as little as 10 millionths inch of unevenness. As a result, workpieces, height gages, and other equipment tend to tip or teeter at least minute amounts.

For ultraprecise measurement, everything should rest stably on the surface plate. The iron surface plate is useful because equipment with a magnetic base will hold tight to it. The traditional custom of sliding the gage back and forth should be avoided in millionths measurement; it is better to move the workpiece under the gage contact, rewringing it where possible.

As a working area, the level, solid surface plate does fill a valuable place in millionths measurement. It can form a dependable foundation on which to fasten precision instrumentation. It is useful as a heat sink; it reduces and often nullifies vibration, it seems to urge the inspector into a more orderly, cleaner, and professional arrangement of his work.

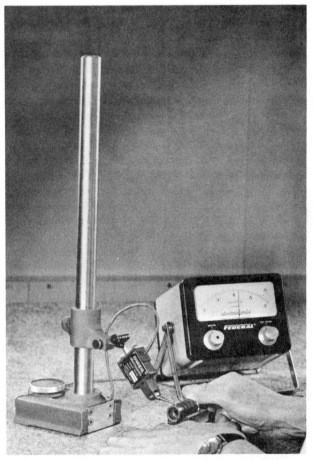

Courtesy of Federal Products Corp.

Fig. 21. A surface plate setup in which an electronic instrument is being used to make measurements in the order of 10-millionths of an inch.

Procedures for Checking Gage Blocks

Gage block checking is almost the most important phase of millionths measurement. The average plant may not yet have too many occasions to inspect parts with tolerances closer than about 20 millionths inch. But gage blocks must be used to master, set, and check the production gages — to check master ring gages, for instance. Gage block checking, thus, becomes a necessary routine. If gage block inaccuracies are unknown, all other measurements stemming from them will be undependable.

Many plants are equipped with electronic comparators capable of measuring to a millionth inch and are especially suitable for gage block calibration. The comparator shown in Fig. 22 has opposed contacts that eliminate the need to wring the blocks to an anvil and are also excellent for checking thin blocks, as can be seen in Fig. 23. A few companies have and use interferometers, the basic equipment for gage block calibration. These instruments are briefly described later.

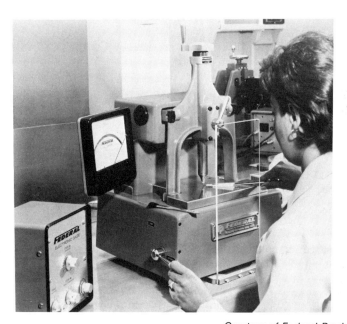

Courtesy of Federal Products Corp.

Fig. 22. This millionth-inch comparator, especially designed for gage-block check-ing, has opposed contacts.

Comparators require less capital investment than an interferometer and are in more general use for checking gage blocks. But there must be available at least one complete set of "master" or "laboratory" gage blocks, with an accompanying and periodically revised calibration chart. This set is used only for recalibrating the blocks in one or more "inspection" grade sets. These latter blocks, in turn, are used for workaday checking of other blocks, of cylindrical and ring masters, and occasionally to calibrate instruments and gages.

Master sets should be sent to the manufacturer (or perhaps to the National Bureau of Standards) periodically, but not less often than once a year, for recalibration. New, calibrated blocks should be purchased to replace any that show much more than plus or minus a

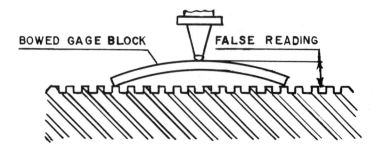

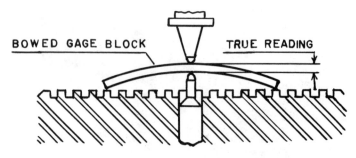

Fig. 23. Dual gaging contacts are advantageous, since thin gage blocks tend to warp when not wrung together.

few-millionths-inch error in length; or an equal or less amount of error in flatness or parallelism. Although a plant may be equipped with an interferometer to do basic calibrations, it is better occasionally to obtain conformation from another source.

Several preparatory steps are needed before actual calibration of any block. First, the block should be clean and grease-free, and its surfaces and edges checked for scratches, pits, and nicks — damage that often means raised metal. An optical flat should be used to determine if any such nonconformities are present.

Some sets of gage blocks come equipped with small, flat deburring stones for smoothing surfaces and edges. A block that needs stoning should not be used. Adequate coaching on how to do stoning without wearing or injuring the gage block is a prerequisite for performing the deburring operation.

The surface roughness of blocks should also be measured and known. Ordinarily roughness is of such slight magnitude that use of a stylus-type surface recorder with a high-resolution probe or an interference microscope is necessary. The purpose is to be sure that peaks of surface roughness are not present to cause damage during wringing and poor wringing. Narrow peaks due to excessive roughness may be flattened down after the blocks have been wrung together once or twice, causing a measurement error. For the latter reason, the surface condition at the tip of the spherical contact of the comparator should be examined with a microscope.

Next, the flatness of both end surfaces of the gage block should be checked with an optical flat. (This is discussed in Chapter 12.) If surface irregularity is apparent from the contours of the fringes, its magnitude should be estimated.

The distance between any two fringes appearing on an optical flat represents close to 11.6 millionths vertically. This is true whether there are many lines close together or a very few wide apart. Characteristic edge wear is illustrated at A in Fig. 24. A straight line can be imagined drawn across the dips in a fringe as at e-e'. (This operation can be simulated physically by laying the straight edge of a transparent plastic rule across the optical flat from e to e'.) The question then is: What proportion of distance f (which represents 0.0000116 inch) is distance d? The latter might be estimated as $\frac{1}{8}$ of f or as a rounding downward at the block edge of approximately 1.5 millionths inch (arithmetically, $\frac{1}{8}$ of 11.6 is 1.45).

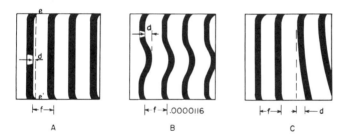

Fig. 24. Estimating the extent of various faults encountered in checking gage blocks by means of optical flat fringe contours. The conditions shown are: rounded edges (A), a depression (B), and a sloping area (C).

A little edge wear on blocks can be tolerated. However, a characteristic wear condition that cannot be overlooked is shown at B in Fig. 24. The depth of the depression d in the sketch at B must be estimated. As drawn, it would seem to be close to half of 11.6 millionths, or about 5 millionths inch. Illustrated at C in Fig. 24 is still another condition. There, the fringes show that the block has a flat area that recedes by an amount d that might be estimated as around 7 millionths inch.

If a contour pattern indicates either hollows or humps on gaging surfaces, it is good practice to sketch their contours on paper for reference.

Where difficulty like that described at B, Fig. 24, appears on both gage surfaces, an enlarged cross section of that block might look somewhat like the condition shown exaggerated in Fig. 25. If such a block were wrung on a comparator anvil, the comparator, depending on where the spherical contact touched the block, would register; (a) at the nominal thickness of the block (*a-a'* in Fig. 25) or (b) at the incorrect thickness *b-b'*. A gage with calipering jaws could register *c-c'* rather than *a-a'*. Hence, the desirability of a contour sketch. It is also convenient to use the manufacturer's legend stamped on the blocks as a means of always obtaining the same orientation and the same end up, when taking a series of measurements of a block.

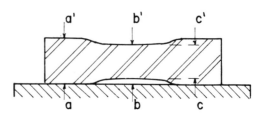

Fig. 25. A possible cross section of a worn block shown greatly exaggerated. Measurement of this block would result in one of three readings representing distances (*a-a'*), (*b-b'*), or (*c-c'*), depending on the gaging method.

Taper or lack of parallelism between the gage block measuring ends is also a condition ordinarily checked as a preliminary to millionths measurement. The usual practice is to move the block in one direction under the millionth comparator contact and then explore through 90 degrees the other way. Any relevant change from a zero meter reading is observed and the direction and the extent of any taper present is marked on the contour sketch.

Since the gage block under question will be compared to a corresponding master block, the latter might be rechecked for flatness and taper. At this time also, all preparations in connection with obtaining proper environment and for avoiding the many types of measurement error previously described should be completed.

The gage block comparator is set to zero by means of the master. This operation is repeated and the comparator reset until the inspector feels confident that it will hold its zero setting while the master is removed and the block being measured is slipped into place.

Another rule for checking blocks is that measurements always should be made at the same point on the block. The dots on the blocks seen in Fig. 26 are located where it is common practice to have the comparator spherical point make contact. The position of thick and thin spots and

Fig. 26. Locations on gage blocks where, in common practice, contact with the spherical gaging point of a comparator is frequently made. All block measurements should be taken at the same point.

taper could make some other points the choice. The choice of gaging location is also influenced by the type of reference anvil used.

Generally speaking, three types of comparator reference anvils are employed for gage block checking. A smooth, perfectly flat anvil of the sort seen in Fig. 1 has many proponents. Some inspectors believe that greatest consistency and reliability in millionth-inch measurement are gained when a gage block is correctly wrung to such an anvil. Their claim is hard to disprove. On the other hand, either there is a possible error from heat transfer (though the block is wrung with gloved hands) or there is a time loss in taking each measurement, while the inspector waits for the block and anvil temperature to equalize.

Others seem to have confidence in a three-ball arrangement like that suggested at A in Fig. 27. Precision bearing balls are carefully inspected for flat spots. Since they are seldom completely spherical, often exhibiting several millionths out-of-roundness, each ball should be explored dimensionally. The three balls are then placed on the anvil, with selected, equal diameters, positioned vertically. The custom is to harness them with a thin fiber plate (View A, Fig. 27) kept in position with a spot or two of sealing wax. Usually the ball group is arranged on the anvil so that one ball (at the

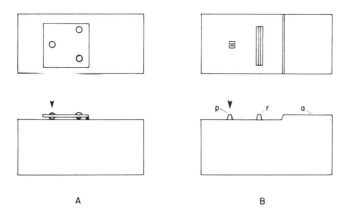

A   B

Fig. 27. A three-ball anvil (A) arranged for checking gage blocks should have one ball directly under the comparator contact as shown by the arrow. Alternatively, pin-and-rail anvils (B) are used.

apex of the triangle) is directly under the comparator contact. One of these anvil setups is shown in use in Fig. 28.

There are some who favor the "pin-and-rail" arrangement such as that seen at B in Fig. 27. The pin does not present a spherical surface for contact but instead has a flat land of about 1/64 inch in width. The rail is also about 1/64 inch wide. Both lands must be straight, flat, and on the same level plane. One practice is to machine away portions of an anvil to provide a profile like that illustrated at B. Initially, and from time to time, surfaces *p, r,* and *a* are lapped flat and level, with surface *a* used as the reference surface for this operation. The spherical gaging point should approach the block surface in a truly perpendicular manner, contacting it immediately above the center of the pin, as seen at B.

Courtesy of Federal Products Corp.

Fig. 28. Three-ball anvils are shown set up in a typical arrangement for measuring gage blocks on a millionth-inch comparator.

Measuring Length with Interferometry

The basic way of determining any gage block size is by a count of a number of light waves by means of an interferometer. The new international standard of length, the meter, is defined as the distance traveled by light in vacuo during 1/299,792,458 of a second.

Interferometers as instruments for measuring length have been developed to a sufficiently practical stage to permit their daily use as an inspection method in the hands of competent people (Fig. 29). The inspector who may be called upon to use interferometric methods is urged to make as extensive a study of the subject as he can. For a start, he can acquire considerable education in the theory, principles, and use of interferometers from manufacturers and suppliers of such equipment, as well as from gage block manufacturers.

Fig. 29. A fringe-count micrometer and electronic fringe counter. The specimen is introduced through a self-sealing rubber porthole in the Plexiglas cover by a remote control device.

One such type of apparatus is generally known as a fringe-count micrometer; and another type (laboratory models of which have been in use for a number of years) is generally known as a gage block interferometer. The fringe-count micrometer makes a calipering con tact with the gage block and includes an automatic electronic fringe counter. No metallic contact with the specimen is made in the more traditional gage block interferometer; measurement depends on the direct impingement of light rays, their displacement, and a slide-rule computation.

A contact type interferometer appears in Fig. 29. Such an instrument is, of course, mounted in a vibration-free manner. Its spherical contact, which has a large radius to reduce penetration error, is mounted in a gaging head that is raised and lowered by a slip-clutch motor drive. The operator brings the contact down to the anvil, and the digital counter is set to zero. He then raises the counterbalanced gage head and moves the specimen block into position. When the contact is again lowered, the gage block size is read in interference bands or fringes and converted to millionths. The customary gaging pressure is about 1 ounce.

It is assumed here that the block has previously rested on the soak plate to stabilize its temperature and that everything has, of course, been thoroughly cleaned. At the moment of measurement the thermometer inside the gage is also read, as well as a barometer hanging near the gage. In addition, psychrometer readings for relative humidity are taken.

Generally, several measurements of the specimen block are taken either to try for repetition or to get an average. The final reading is adjusted — both plus and minus — by several correction factors, including those for temperature, humidity, and barometric pressure. Standard measuring conditions are: water vapor pressure, 7 mm Hg (or approximately 40% relative humidity); atmospheric pressure, 760 mm or 29.92 inch Hg; temperature, 20° C or 68° F.* Correction factors, methods of calculation, operating instructions, and other necessary data are, of course, supplied by the gage manufacturer.

A simple diagram of interferometer optics is seen in Fig. 30. Very briefly the system may be described as follows: A ray of light from the krypton tube is passed through a collimating lens to a beam splitter which breaks it into two components. One beam is directed toward a fixed prism reflector and the other to a prism connected to the measuring tip. The two beams then meet at a combining plate where an interference (dark band) is produced. This, in turn, is directed up to a photomultiplier tube. Impulses from the latter, together with the action of the cathode followers and the digital counter, effect the fringe counting.

---

*In rare cases, the carbon dioxide content of the air is analyzed and a correction is made for the standard condition of 0.03% CO_2 by volume.

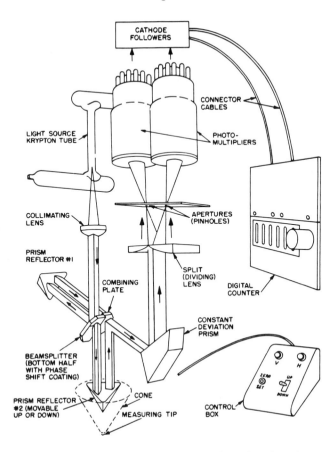

Fig. 30. A diagram which shows the operation of an interferometer.

Arrangements can also be made to add to or alter accessory parts of the fringe-count micrometer so that direct interferometric measurement can be made of strip stock, cylinders, and bearing balls, as well as gage blocks.

The traditional type of interferometer requires no metallic contact with the workpiece, but depends on light beams impinging on the gage block surface. Hence, that surface must be smooth enough to reflect light without introducing a "change of phase" error, the interferometer's equivalent of penetration. Also, all the rules previously given regarding the elimination of such error producers as dirt, vibration, and temperature apply to the nth degree if an interferometer is to give accurate results.

Until lately, interferometers have been strictly laboratory instruments used only by physicists and specialists. Now available are compact models, Fig. 31, equipped with ingenious manipulating

Fig. 31. Bench-type interferometer.

devices with which a competent inspector can secure quite accurate results. However, these still must be used in a temperature- and environment-controlled room.

While information on the construction and application of any such apparatus, as well as the theory behind it, should be obtained in detail from the manufacturer (including, at least, a capsule education in the principles of interferometric measurement), a rough description of the technique is appropriate here.

Internally, most interferometers contain a series of prisms, mirrors, a beam splitter, and other characteristic elements of an interferometer's optical system. These are all mounted on a solid metal base, which, in a fashion, also acts as a soak plate or heat sink. At about the middle of the beam path (under the glass window in the cover) is a card-size steel platform which can be made to move back and forth by means of exterior knobs.

A rectangular optical flat is placed endwise on this platform, at a scale designated location, and the specimen gage block is wrung to it. This setup is shown schematically in Fig. 32. In-phase and out-of-phase light beams from a cadmium source follow their standard interferometric paths. They impinge on the plano optical flat surface P, Fig. 32, and also on surface B of the gage block. The platform — optical flat — gage block assembly is moved laterally until the inspector,

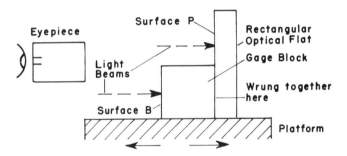

Fig. 32. Diagram of the setup employed in some interferometers. The gage block is wrung to an optical flat placed vertically on a laterally movable platform. Light rays impinged on both the gage block and the optical flat produce the necessary fringe patterns.

looking through the eyepiece, sees sharp fringe patterns simultaneously at *B* and *P*.

What he sees may look like the pattern illustrated at A in Fig. 33, except that the fringes are colored. The instrument has controls so that the red cadmium light ray can be used, then the green, and finally the blue. The pattern seen at A, Fig. 33, would be, say, from cadmium red; at B, from cadmium green; and at C from the blue rays. If for some reason the face of the gage block were not parallel to the optical flat (due to wear or lack of parallelism), the inspector might see a pattern more like that shown at D.

If, for example, the gage block where 1 inch in size and some arrangement were provided whereby the front surface could be moved gradually forward, the observer would see no fringes — just plain light — between each 0.00001267 inch of such movement, when using cadmium red light. After he had counted 78,900.4799 alternate fringes and light flashes he would know that the block surface had traversed a full inch.

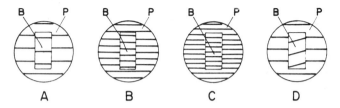

Fig. 33. Fringe patterns as seen through the interferometer eyepiece. Views (A), (B), and (C) illustrate three patterns that might be obtained in measuring a gage block with three wave lengths of cadmium light. The pattern seen at (D) is one which indicates that the gage block face is not parallel to that of the optical flat.

Some such task would be arduous and time-consuming, and probably a counting error would creep in before all 79,000 odd fringes had been tallied, to say nothing of estimating a 0.4799 part of a fringe! Hence the use of the three colors of light and an ingenious scale device.

The fringe spacing for cadmium red is 12.67 millionths inch; for green, 10.02 millionths inch; and for blue, 9.45 millionths inch. This accounts for the differences in fringe spacing suggested by the sketches in Fig. 33. These differences plus an estimate of fringe fractions are employed in determining the length of the block. An exactly 1.000000-inch block would produce only one particular set of patterns and fringe fractions, like those seen in Views A, B, and C of Fig. 33. If the block were 1.000001 inch long, there would be a different set of patterns.

In addition to viewing the fringe patterns, the operator also sees in the eyepiece a ruler-like scale and a cross of hairlines superimposed on the fringes. In making a measurement, he makes a record of the scale readings, for all three colors of light. This information is then transferred to a slide-rule-like device from which the size of the block is calculated. After making the usual corrections for temperature, humidity, and barometer, the accurate size of the block under standard conditions is obtained.

Checking the absolute length of a gage block in an interferometer is somewhat time-consuming, not only in taking readings, but because a waiting period varying from twenty minutes to an hour or so is needed for temperature stabilization after the setup is ready for observation. Some interferometers have two internal thermometers. One registers the temperature of the internal metal (soak plate) and the other, the internal air temperature. When the two thermometers read alike, an observation can be safely made.

CHAPTER 17

Coordinate Measuring Machines

With the advent of numerically controlled machine tools, especially tape controlled milling and drilling machines, demand grew for a means to support this equipment with faster first-piece inspection and, in many cases, 100% inspection. To fill this need, coordinate measuring machines were developed by modifying precision layout machines. Indeed, most coordinate measurement machines "double in brass" because they are used as layout machines before machining and for the checking of hole locations after machining. It has been said that the coordinate measuring machine is not just a refinement in gaging equipment but a major breakthrough in mechanizing the inspection process and in lowering its cost.

Breaking a Bottleneck with a Coordinate Measuring Machine

Let us take a specific example to see why an inspection department is prompted to install a coordinate measurement machine rather than rely on conventional surface-plate inspection methods. In this typical case a bottleneck develops when inspection is confronted with the necessity for measuring 12 hole locations in 50 machined castings where 100% inspection is required. The tolerances are plus or minus .001 inch for location of the holes relative to their reference points on X and Y coordinates. The holes are also located in relation to an edge to very close tolerances. The inspector quite naturally turns to his surface plate, a precision knee, a stack of gage blocks, and a height gage. He mounts the casting on the knee on the surface plate and proceeds very much in the manner described in the chapter on Surface Plate Methods. Of course, he establishes a reference or starting point and, with the height gage and with various plug gages or wires in the holes, he proceeds to measure the hole center coordinates. Now because the basic surface plate technique is always perpendicular to or upward from the horizontal plane of the surface plate, there must be one setup of the part to measure the locations of the holes in one direction (the X direction), and another setup to measure the locations of the holes in

575

the Y direction (90 degrees from the X direction). The surface-plate method is time-consuming and tedious. Also to be taken into account is the possibility of errors in the gaging setup. The inaccuracies of the gage blocks, the height gage and indicator, the plug gages, and the precision knee must all be taken into account. The inspector, too, has to be constantly on the alert in order to maintain accurate readings during 100 different setups on 50 pieces with 600 holes to be located and at least 1200 dimensions to check. The amount of time spent on the measurement of the hole locations in each of these castings might very well average 2 hours and the time spent measuring the batch of 50 parts would then be 100 hours. No wonder the chief inspector looks for another faster solution to this problem!

Now let's take this same measurement job of 100% inspection and instead of using conventional surface-plate technique apply a coordinate measuring machine to the problem. To start with we'll use a mechanical coordinate measuring machine, that is, one that employs dial indicator readout rather than electronic meters. The casting is clamped to the measuring machine table in the same manner that it was held on the table of the drill. The coordinate measuring machine table is usually provided with T-slots or tapped holes or both for hold-down purposes. After the part is staged on the table, the tapered probe of the machine is inserted into each hole in turn, with the X and Y coordinates read directly from the continuous travel dial indicators. Both readings are obtained at each setting. The time to measure the locations of the 12 holes in both directions should not take more than 15 or 20 minutes including the time necessary to set the piece up on the table and establish the datum reference. In this case the probable time saved using the coordinate measuring machine as opposed to conventional inspection methods amounts to more than one and one half hours per part. What is demonstrated by this example is the prime advantage of the coordinate measuring machine which is quicker inspection coupled with accurate measurements. The coordinate measurement machine just described is a mechanical gage without electronics, Fig. 1. This machine measures in the X, Y, and Z directions with a measuring capacity of 18 inches on the X coordinate, 24 inches on the Y coordinate, and 8 inches on the Z axis. The positioning accuracy of the probe is claimed to be within plus or minus .0005 inch. The dials are direct reading and the machine can be bench mounted or fixed on a mobile bench as shown in the illustration.

Coordinate measurement machines (CMMs) Fig. 2 are the most truly modern method of dimensional inspection — the pride of every inspection laboratory, and indeed the way of a rapidly expanding and changing future. Like the multiaxis measuring machines discussed previously, these units have changed manufacturing and inspection or

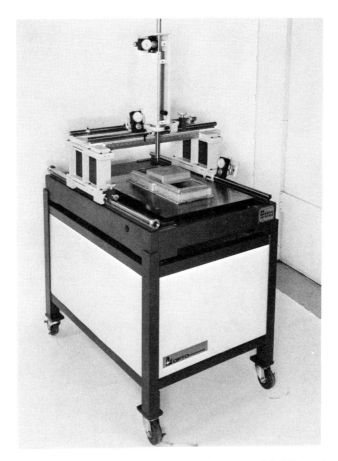

Courtesy of OPTO mechanisms Inc.

Fig. 1. A coordinate measuring machine with travel dial indicators, suitable for small castings or machinings.

measurement parameters. These units also speed up the inspection work, which is very important to first-piece and production runs and workpiece variation assessment. The faster the inspection (and the more reliable), the better the feedback can be to those responsible for understanding dimensional variation and controlling it.

These units are like multiaxis measurement machines in many ways, and they are different in others. First of all, they *do* provide a multiaxis assessment of measurement point of view — a three-dimensional assessment of the workpiece. The machine's probe or measuring device travels along two perpendicular paths that fall in the same plane — these paths represent the X and the Y axis, and the

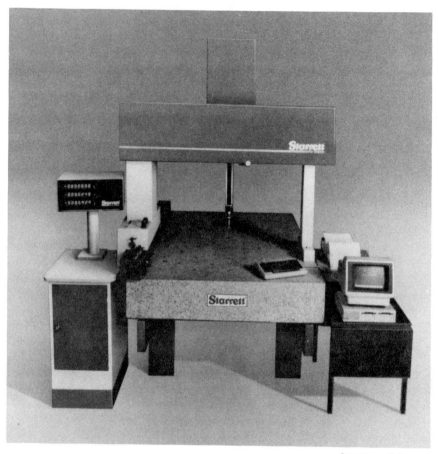

Courtesy of Starrett

Fig. 2. Coordinate measuring machine.

travel along these paths is the basis for the two-dimensional measurements and comparisons that are made. And the Z axis is also defined in terms of the travel of the measurement probe in relation to the other two planes.

The guideways of travel paths that direct the X and Y measurement movements are usually one of two basic types: vertical or horizontal, or bridge-type or cantilever, respectively. These guideways direct a nearly friction-free travel path for the gaging head, which of course performs the actual measurement, referenced against the fine discriminatory system that identifies position and dimension on the guideways.

These CMMs are different from traditional multiaxis measuring machines because they are displacement measuring devices that use a permanently positioned staging table, and a traveling gage head or probe. The probe detects the measurement or displacement between points of reference on the workpiece, positioned on the machine's table. This displacement is then usually displayed in a digital readout in increments of inches or meters.

The performance characteristics and the changing features of these machines is overwhelming. New ideas and applications are being introduced daily that make them more accurate and efficient, as well as versatile. For instance, some units are equipped with a 360 degree rotary table, effectively producing a forth axis of reference with respect to the measuring stylus. Options are endless — video systems, microscopes, other optical adaptations, and a host of accessories make the world of CMMs a field of study unto itself.

A Coordinate Machine with Optical Comparator

Another mechanical coordinate measuring machine, Fig. 3, is equipped with an optical comparator as well as travel dial indicators. A

Courtesy of OPTO mechanisms Inc.

Fig. 3. A mechanical coordinate measuring machine equipped with an optical comparator and travel dial indicators.

machine such as this one can be used to check a wide variety of components and is especially adapted to the checking of hole locations in flat plates, and printed circuit boards. The machine is accurate to ± .001 inch yet capable of the rapid inspection so necessary when over 500 holes must be verified in one piece.

Electronic Measurement and Digital Readout

Just as we have seen continual improvement in other areas of dimensional measurement, there has been constant development of new and better coordinate measuring machines to keep pace with the space age technology of close tolerances, zero defects, and numerically controlled machine tools. Measuring machines with greater accuracy and precision combined with larger capacity and high speed operation are being designed and built. Figure 4 shows a typical three-axis machine, *X, Y,* and *Z* coordinates with accuracy said to be ± .0005

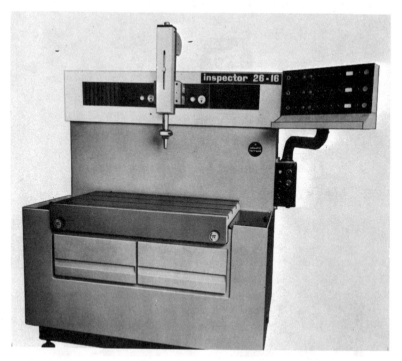

Courtesy of Farrand Controls, Inc.

Fig. 4. A typical three-axis, digital readout coordinate measuring machine. This one uses a measuring element called Inductosyn.

inch and resolution of .0002 inch over all three axes. Measurement is accomplished by electronic means without lead screws and with digital readout. The measuring element is called the Inductosyn data element and uses inductive coupling between conductors separated by a small air gap. Linear accuracy in the order of 50 microinches is claimed by the manufacturer. The element is not subject to wear and so will not become inaccurate, and is impervious to oil, dust and other contaminates. The digital readout has automatic plus and minus indication from a zero reference. The readout is in 5 or 6 digits plus a meter indication.

The workpiece can be aligned quickly with the probe by the means of a swiveling adjustment on the worktable. The operation of this machine is fast and easy for the inspector who has fingertip control of the measuring probe for movement over all three axes X, Y, and Z, of the part being checked. Furthermore, there is no need for reference standards such as gage blocks, templates, or other external devices.

The Moiré Fringe Concept

The digital measuring system used on the coordinate measuring machine pictured in Fig. 5 is unique because it is based on the Moiré fringe concept of measurement. The main element of the system is an accurately ruled grating of the necessary length. An index grating with

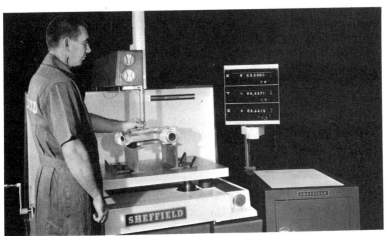

Courtesy of Sheffield

Fig 5, The measuring probe of an XYZ coordinate measuring machine is inserted in a hole in a casting which is staged on precision parallels clamped to the machine table.

the same line structure is superimposed to produce a pattern of dark and light bands. This integrated interference pattern is a Moiré fringe — a greatly magnified replica of the line structure. The light intensity is converted into electrical signals by four photo cells and the outputs from the photo cells are used to produce a digital display equal to the motion. Signals from the photo cells will also give an indication of direction and the digital display instantly follows any change in the slide movement. Although the two gratings are superimposed, they do not touch and continual use will not cause wear. In operation, the workpiece is mounted securely to the table, just as it was, in the earlier examples, using appropriate bolts, clamps, and blocks. The part is aligned with the probe travel by adjusting the table aligning device. With the proper tapered or flat tip in place, the probe is moved to the first point of check or reference position and both coordinate readouts are set to zero by pressing the X and Y axes buttons. The probe is then moved in turn to the various points of inspection with the X and Y coordinate dimensions being displayed simultaneously at each inspection point. Somewhat the same procedure is used when using the vertical or Z axis accessory. The probe movements, in this case, are measured by a four-pickup counting head as it travels over the steel grating, which has one thousand lines per inch. The corresponding grating setment mounted on the counting head creates a Moiré fringe pattern as it passes over the grating. As the fringe patterns are counted, output signals from the head provide a continuous readout of probe movement and position. Machines such as pictured in Fig. 4 can be obtained in several sizes. The smallest are bench mounted and are suitable for measuring small parts such as circuit boards. The larger models have more measuring range and work height capacity with the largest size designed for the measurement of heavy parts such as engine blocks and big machined castings and forgings (Fig. 6).

More Speed with Electronics

Coordinate measuring machines of more sophisticated design with many special features and optional accessories that speed up the inspection (or layout) process are being built. The basic gaging principle is always similar to the previously shown example where a mechanical measuring machine with travel dial indicator readout was used to make a 100% check on a batch of 50 drilled castings with considerable time saved over conventional gaging methods. With more elaborate electronic equipment, not only is greater accuracy obtained but also greater speed — up to 20 times faster than with the old reliable surface plate technique. Some of the accessories available are: an

Courtesy of Portage Double Quick, Inc.

Fig. 6. A coordinate measuring machine built to handle large workpieces and with added features which make the machine, in effect, a universal measuring machine.

optical viewing screen or optical comparator attachment as shown in Fig. 3, a microscope attachment for the inspection of thin, soft, or delicate workpieces, and automatic printout as shown in Fig. 7 which eliminates the inspector's pad and pencil.

New CMMs even offer the distinctions of noncontact gage heads that rely on an optical/electronic interface to sense and translate dimensional characterizations. The range and resolution of these machines makes them flexible enough for any measurement operation.

CMMs are generally fully or partially computerized or computer-directed. While these units can be manually directed, moved by motor, or numerically controlled, the essential principle involves the programming of an inspection sequence, and derived measurements or comparisons, based on numerical data plotted against the known coordinate axis of the machine.

Direct computer control "DCC" is a CMM that is entirely directed by computerization. The machine is entirely programmed in computer language for all measurement and movement that are accomplished by the machine. Servo drive motors on each of the machine's three axes are linked with the computer through the reference points or positions

Courtesy of Sheffield

Fig. 7. A coordinate measuring machine with automatic printout as well as digital
display of the measurements.

that are to be measured in the inspection. Once these points are
established, the preprogrammed activity of the machine is fully automatic — machine proceeds through the inspection sequence and offers
dimensions based on variation at those preset reference points.

When the inspection system calls for a part to be measured
repeatedly, the CMM and its program can carry out the entire routine
of measurement. When a particular subroutine is desired, or a unique
inspection setup called for, the machine operator or inspector may
choose a portion of the available programs in order to carry out the
specific task.

Computer systems, and program and information storage systems
are now available to accompany large CMMs. These systems not only
accommodate the direct work of the CMMs — the work of measuring
by programmed movement — but they also interface with data storage
and output systems that provide the proper summaries of inspection

data, and the use of that coordinate or measurement data directly on the manufacturing line. Computer networks are available to link CMMs with machines on-line that do the work that CMMs measure — that is, the measurements and the ascribed variation are fed instantly back to the production environment in an automatic format — one machine measuring another machine's work, and then "talking" directly to that production machine about the results — perhaps even on how to correct the variation that has been discerned.

Accessories for Combined Measurements

The versatility that can be built into a coordinate measuring machine is shown in Fig. 6. Here is a machine that is not only equipped for three-axis coordinate measurement with digital readout but is also designed to permit the checking of angularity, roundness, taper, and concentricity. This machine has an electronic measuring system accurate to ± .0002 inch and repeatability in the order of .00015 inch. A feature that makes the machine so versatile is the rotary table for reaching other areas of the part being checked without changing the setup, thus eliminating possible error that could occur in multiple setups. There is also an electronic indicator probe which is mounted on

Courtesy of Portage Double Quick, Inc.

Fig. 8. A closeup of the electronic indicator probe mounted on the end of the spindle of the coordinate measurement machine illustrated in Fig. 6.

the end of the spindle, Fig. 8, which can reach over and under the workpiece to check squareness in a single setup. Designed for heavy part inspection and layout, as can be seen in Fig. 6, the machine has a main base of granite 18 × 46 × 70 inches. Another measuring machine which is more than just a coordinate measuring machine is shown in Fig. 9. This machine makes use of linear air bearings on the horizontal slide motions to achieve finer slide position resolution. In addition to a printer or typewriter, readout devices such as paper tape punch, magnetic tape, card punch, and outputs for postprocessing can be provided.

Courtesy of Brown & Sharpe Mfg. Co.

Fig. 9. A universal three-axis measuring machine with bridge type construction. This machine has linear air bearings for the horizontal slide motions.

Linking Coordinate Measuring Machines to Computers

The introduction of the electronic coordinate measuring machine was indeed a major breakthrough in mechanical inspection technique, but computer processing of inspection data from the measuring machine may be a breakthrough of equal importance. Thus, it is possible to take the inspector's measurements of a non-aligned workpiece to a computer that has been programmed for alignment of the work and the resultant printout of dimensions from the computer will be the actual X and Y dimensions. The computer program compensates for the out-of-square condition of the workpiece as it is positioned on the machine table.

The writer has had considerable success with this technique in saving set-up time on a Moore Measuring Machine. Since the lead screws in the Moore must be run at very slow speeds in order to avoid excessive heat of friction and subsequent thermal expansion, it is possible to save from 20 minutes to 2 hours in the squaring-up of the part along a reference datum by letting the computer do the work in seconds with raw data taken with a non-aligned workpiece.

A coordinate measuring machine such as that shown in Fig. 10 can be tied directly to the computer and raw coordinate dimensions fed from the measuring machine to the computer through an interface converter. The computer will process the data received in the manner that the software program directs and the resultant output data can be furnished as printed records, punched on magnetic tape, or displayed on a cathode ray tube in a visual manner.

Courtesy of Sheffield

Fig. 10 Coordinate measuring machine interfaced with a general-purpose computer and teletype reader printer.

There are standard computer programs available from the manufacturers of the coordinate measuring machines who will also supply a computer interface converter or a computer with printer as an option. Automatic alignment computation is only one of the many computer programs available to speed up the inspection job. True-position deviation computation is a great source of mathematical error for the average inspector, but when the readings are fed directly into the computer, the computer can take the coordinate measurements, supplied by the measuring machine, and compare them to true-position tolerances punched into the master tape. The system then prints out the results as either in- or out-of-tolerance dimensions by comparing the differences between actual readings and nominal readings punched in from the blue print, squaring the differences, summing up the squares, taking the square root of the sums, and comparing these with the true position tolerances.

The X and Y coordinate dimensions, or Cartesian coordinates, can be converted to polar coordinates by a computer program. In other words the X and Y dimensions will be converted to polar coordinates — radius and angle — which are printed out. Once again, tolerance comparisons can be made if needed. Other uses of the computer tied directly to a coordinate measuring machine are: (1) automatic pattern duplication in which a part can be checked by the measuring machine and the resultant readings fed directly into a computer which will generate a NC tape for the machining of duplicate parts, (2) integration of the area under a curve as traced by the probe, (3) scaling dimensions as traced from a model, and (4) conversion of information into any tape format desired. The versatility of computerized inspection depends only on the imagination of the inspectors, quality engineers, and programmers.

The linking of the computer with the coordinate measuring machine provides data for the control of the machine tool. Thus, while the computer component is controlling the machining, the measuring machine can be used independently for measurement of parts. There is, therefore, complete utilization of the system with minimum downtime. Figure 11 shows (D) just such a complete manufacturing and measurement system. This system consists of a coordinate measuring machine (A) with air bearings and with a work capacity in the X, Y, and Z mode, and simultaneous measurement of diameters; a general purpose "minicomputer" (B); a teletypewriter (C); tape/punch-reader; and the option of adding 32K disk memory, CRT (cathode ray tube communicator), magnetic tape and transports, and interfacing to an IBM-360 computer. Figures 12, 13, 14, and 15 present several very sophisticated examples of coordinate measuring machines.

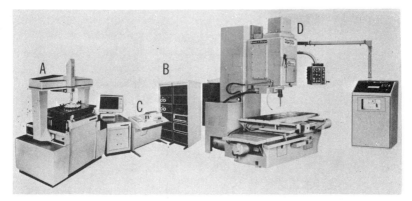

Courtesy of Brown & Sharpe Mfg. Co.

Fig. 11. A coordinate measuring machine inputting data through a mini-computer to direct the operation of the NC machine. (A) The coordinate measuring machine, (B) minicomputer, (C) teletypewriter, and (D) machining center.

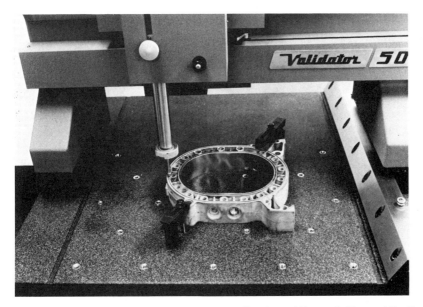

Courtesy of Brown & Sharpe Mfg. Co.

Fig. 12. An engine part under inspection on the table of a coordinate measuring machine capable of measuring hole location and diameter, simultaneously.

Courtesy of OPTO mechanisms Inc.

Fig. 13. A bridge-type coordinate measuring machine with air bearings on all three measuring axes.

When Masters are to be Measured

When the tolerances are in tenths of thousandths, such as in the case of precision masters, probably the most reliable method for accurately measuring hole locations is with a very precise universal measuring machine of the type illustrated in Fig. 16. This universal measuring machine has been described as the "ultimate in coordinate measuring" and it is hard to disagree with that assessment, although

Fig. 14. A large, bridge type coordinate measuring machine taking contour readings on the fly.

toolmaker's microscopes and optical gaging equipment of various sorts are now available which can be used with confidence to check to .0001 inch (see Chapter 12, Optical Measurement and Inspection Equipment). The universal measuring machine is built along the same lines as a jig borer or jig grinder. Indeed, if such a measuring machine is not available, a jig grinder can be a very useful substitute if the tolerances will allow it. The accuracy claims for the universal measuring machine shown in Fig. 16 are 35 millionths of an inch tolerance on the X axis and 35 millionths of an inch tolerance on the Y axis with spindle rotation accuracy of 5 millionths of an inch TIR. Where would one expect to find a machine like this and who would operate it? Well, certainly not the average mechanical inspector in a machine shop. The universal measuring machine should be located in a temperature controlled room or measurement laboratory. The machine is not normally used for ordinary production piece parts but is especially adapted to the checking of precision masters and very accurately machine special parts. The universal measuring machine is virtually a complete metrology laboratory in itself and can be used for measuring coordinate dimensions on small and large parts as well as other geometry of the part such as contour, taper, radii, roundness, squareness, and many other measurement tasks. For coordinate measurements, a piece is staged on the machine table and the reference datum, an edge or hole, is indicated out to zero using a high magnification electronic indicator built into the machine. Where applicable, the workpiece is squared up

Fig. 15. Coordinate measuring machine connected to a printer.

or leveled; then, using the leadscrews of the machine, the table is moved and the hole to be checked is indicated out to zero with the actual readings in X and Y axis being read on the verniers of the machine. When checking very close tolerances in "tenths" or fractions of a "tenth," the part must be brought to the same temperature as the measuring machine, preferably 68° F.

Possible Sources of Error in Coordinate Inspection

The inspector must always remember that the table and the probes of coordinate measurement machines are not in perfect alignment nor

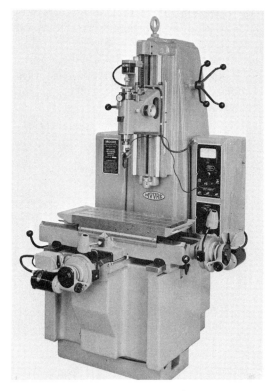

Courtesy of Moore Special Tool Co.

Fig. 16. This universal measuring machine is especially suitable for the checking out of special masters. Accuracy of the spindle is 5 millionths of an inch and the machine is accurate to 35 millionths of an inch on the X and Y axes.

will they always have perfect geometric form. The probes will have a degree of runout so it is advisable to locate a probe at the same rotational position; otherwise a "tenth" or two will be lost in extra tolerance — maybe even several tenths. Also, those probes that move up and down in the Z axis are liable to have some perpendicularity error. Speaking of perpendicularity, it is essential that the inspector always gets the primary datum reference surface of the part square to the machine probe. There is also the possibility in the optical readout of a digital system. The inspector should calibrate these machines before using them wherever possible. Some machines have a master plate to use as a master calibrator but in every case a master of known hole spacing should be available so that each coordinate measurement machine can be calibrated. Care should be taken not to try to get exact measurements from a coordinate measuring machine when the last digit in the digital readout is blinking between two adjacent values. Finally, the inspector would do well to have a thorough understanding of the operating instructions included with the coordinate measuring machine and should know the specified accuracies of the machine, both mechanical and electrical.

CHAPTER 18

Automatic Gaging

Measurement Techniques

The technique of measurement of workpieces on an automatic basis involves the application of mechanical handling and instrumentation to eliminate or minimize tedious, time-consuming, or otherwise impractical, manual inspection.

Once considered strictly for extremely high-volume manufacturing, automatic gaging is now being used more and more widely not only as a tool but as an integral part of the production line. It is, of course, utilized as "in-process" automatic gaging where the workpiece is monitored as it is being machined. When "zero" or a preset size is reached, the gage signals retraction of the tool, or stopping of the process. It will also provide signals for changes in speed of the process or for corrections to the positions of the tools.

As stations in transfer lines, automatic gaging provides extremely fast, efficient intelligence about the previous operation. This may involve signaling: (1) for adjustment of tools for trend away from size; (2) that an operation had not been performed; or (3) that a tool has broken. Generally, this type of automatic gaging has provision for rejecting a faulty workpiece off the line to avoid damage that might occur to tooling during machining in subsequent operations. This type of gaging can be considered as a combination of "in-process — post-process" measuring.

Specific postprocess automatic measurement is usually located immediately adjacent to a machine and, in addition to generally providing corrective information, will provide a sorting feature such as "good," "oversize," and "undersize," or possibly, into selected good and reject categories.

All of the above can or should provide other information in the measurement process. For example, if the tools have to be adjusted or replaced too often, percentage counters will provide visual or audible warning, The more sophisticated gages can provide additional data up to and including feeding continual information to central computers controlling an entire process.

Lastly, there is the final inspection automatic gage that processes workpieces after all machining has been accomplished. These gages are increasingly becoming very necessary as part of the total production. They provide for classifying the workpieces into discrete close-tolerance categories, thereby allowing more economical machining in a broader tolerance range. Again, in terms of economics, purchased components can be brought in with a broad tolerance and sorted into finer size categories.

Workpieces that are at too high a temperature for correct sizing during or after machining are brought to ambient temperature and then gaged.

Workpieces that must be matched with close-fit tolerance can be automatically gaged, sorted into discrete categories, compared with each other, and brought together in a "marriage," and even assembled; all as part of an automatic gage machine's functions.

In this chapter the various types of automatic gage philosophies outlined previously will be developed in detail. The inspector will find himself progressively more involved with gages located throughout the machining process with lesser emphasis on benchtop inspection.

In describing the many types of automatic gaging, the full range of instrumentation employed to accomplish the "automatic" process will be covered. This will range from electric, air–electric, and electronic types to the inclusion of micro- and minicomputers as dedicated components of the gages. It is important for the inspector to realize that though electric and air–electric instrumentation are utilized and will be, for a long time to come, the predominant trend is toward electronics. This is mainly because, with the advent of solid-state microelectronics in all phases of manufacturing, tolerances are getting closer and the application of digital electronics in measuring provides for extremely accurate, highly reliable gaging with a tremendous drop in maintenance requirements.

The use of automatic and continuous in-process gaging employs techniques of many different types of inspections — not only variables or dimensional measurements, but also go/no go characteristics. These setups and machines are also of course not limited to mechanical devices.

Infrared scanning systems are used to sort "good" from "bad," and other systems of video monitoring are very popular in continuous *and* automatic inspection operations. These systems, no matter what their specific configuration, do not necessarily require human intervention to carry out their direct function. Automation can be accomplished in many ways. The basic question drives the design of these and other inspection setups: What is critical to measure, and how can that be speedily, economically, and reliably accomplished?

In-Process Gaging

As already mentioned, this type of automatic measurement involves the continuous monitoring of the workpiece while it is being machined to a preset size and the stopping of the process or signalling for readjustment of the process to maintain an acceptable band of tolerance. Obviously, this philosophy of measuring, when it can be applied, is most attractive since it virtually eliminates making scrap work, increases output, and allows closer control of size.

Stopcycle Gaging

In all production grinding operations today — including external, internal, and centerless machines — some type of gage control is in use. These range from calipers equipped with electric switches, to solid-state electronic digital techniques for the most modern grinders.

Typical of the electric grinding gages in use for a number of years are those shown in Figs. 1 and 2. In the first, a caliper engaged to the work at the beginning of the grinding cycle employs a dial indicator with adjustable electrical limits. The caliper and electrical limits have been preset so that as the work is ground, the first limit "fires," and through relays connected to the grinder controls, will cause the grinding wheel to change from a fast feed to a fine, or "dwell," grind cycle. The second limit fires when the correct size is reached and a signal causes the grinding wheel to retract. Generally, this type of gage is used strictly on rough-grinding operations with a .0005-inch, or coarser, work tolerance.

In the second example, Fig. 2, an air–electric system is employed using the same type of caliper. Here more precision is obtained because of the ability to provide a higher magnification with the use of an air gage and because of the natural "damping" of the air providing a less vibration-prone operation. Again, adjustable electrical switch contacts are used to provide control of the grinding process. Although, generally, two electrical limits are employed to operate relays through a control box, two additional limits are part of the air–electric switch that can be connected in the circuit for further control.

Air–electric controls are very widely used for internal grinding where air fingers extend into the work bracketing and grinding wheel while in operation. (See Fig. 3.) As with external grinding, relays in the control box are tied into the grinder system to change or stop the cycle.

Courtesy of Federal Products Corp.

Fig. 1. Electric grinding gage.

Automatic Control in Centerless Grinding

In centerless-grinding operations, although work has and is being done to automatically control the relative position of the grinding wheel during the process, by continually readjusting the wheel position to control close sizing, the large mass of the wheel makes this very difficult. Therefore, control gaging is virtually limited to a kind of postprocess operation. Located immediately adjacent to the exit side of the wheel (Fig. 4A) a noncontact air caliper monitors the work passing through. Since the trend of the preset size away from "zero" is gradual, a four-limit air–electric control activates large signal lights. These lights

Courtesy of Federal Products Corp.

Fig. 2. Air–electric grinding gage.

Courtesy of Federal Products Corp.

Fig. 3. Internal grinding gage for microcentric grinder. (Left) Gage assembly and cabinet. (Right) Forks for in-process control of bore.

are set to go on bilaterally as "near under," "under," "near over," and "over." For example, if the total acceptable tolerance is +.0002 inch, the "near size" limits at either side of zero can be set at +.0001 inch. During the grinding cycle, if either the "near under" or "near over" light comes on, the operator knows that although the work is still being

Courtesy of Moore Products Co.

Fig. 4. Centerless grinding gages. (A) A noncontact air caliper work monitor.
(B) Another gaging system for centerless grinding.

Courtesy of Federal Products Corp.

Fig. 5. Analog electric grinding gage.

produced within tolerance, the machine should be readjusted. The outer limits — set at the tolerance limits — warn the operator that the process should be stopped and proper adjustments made. Figure 4B shows another gaging system for centerless grinding.

In the development of more accurate machine tools for closer tolerance grinding, the impact of solid-state electronics has resulted in the design of fully electronic gaging for in-process control. The reliability, precision, and lower and easier maintenance of such gaging equipment using total electronic techniques has led to its widespread adoption.

Whatever the grinding operation, typical examples of the solid-state electronic controls used today in conjunction with electronic transducers are shown in Figs. 5, 6, and 7. Illustrated in Fig. 5 is an analog electronic system. The amplifier is furnished with a meter with

Courtesy of Federal Products Corp.

Fig. 6. Digital electronic grinding gage.

logarithmic scale that enables the gage to operate over a long approach range and yet measure with high magnification as the work nears the critical size. As metal removal progresses the magnification becomes increasingly higher. Electronic steppers with minimum graduations of .0001 and .001 inch allow the operator to compensate for machine variables and wheel wear by resetting the "zero" size.

Digital Electronic Controls

In Figs. 6 and 7 digital electronic controls are illustrated. These systems employ a number of interesting features for grinder control:

1. Digital readout counting in one direction only toward size.
2. Out-of-round indicator utilizing a solid-state memory circuit

Fig. 7. All digital machine-control gage with digital setpoints, and BCD feedback
for adaptive control.

displaying a static reading that diminishes as the piece rounds
up.
3. Adjustable adaptive control rate to speed up grinding cycle to
   optimum rate.
4. Electronic steppers in .001- and .0001-inch steps, allowing the
   operator to offset the preset "zero" size. The coarser step is for
   use in case of tool or wheel wear; the finer steps are for the
   many machine variables that take place during the day.
5. Programmable plug-in cards with digital switches for adjustable
   approach control limits.
6. Isolated Binary Code Decimal (BDC) output available.

All of the above electronic in-process controls feature plug-in
interchangeable cards for ease in maintenance. They operate on very
low voltage levels; hence, have a cool, stabilized operation with a high
degree of repeatability and linearity in microinches.

Continuous Measurement in Process

The ability to signal a process to adjust its speed or position for
continuous operation is typified by gages in use in extrusion or
rolling-mill operations. These gages range through noncontact devices
such as X-ray, beta ray, capacitance transducers, and opposed air
orifices, up to contact roller types. All of these provide signals to adjust
the machine to control the size, within an acceptable band of tolerance,

for the end product. Usually, and it is recommended, four electrical limits are utilized — two inner limits are set well within the maximum and minimum tolerances, and the other two are set at outer limits for protection against gross changes in the process.

Also, two sets of time controls are usually incorporated since the gages are located at some point *after* the processing of the work begins to take place.

When a limit is fired in measuring, a relay is closed for an adjustable time and then opens. This relay is tied electrically to a motor to speed it up or slow it down (extrusion process, see Fig. 8), or to a motor for adjusting a lead screw for position (rolling mill, see Fig. 9). In either case, the relay is closed only long enough to provide a specific change of speed or position designed to cause the process to produce larger or smaller, aimed toward the desired size.

Courtesy of Federal Products Corp.

Fig. 8. Continuous measurement control of the diameter of plastic-coated electrical wire.

Fig. 9. Noncontact air continuous measurement at rolling mill.

The second time control is operated simultaneously with the first and shuts off the gage signal until the new-size work arrives at the gage. Obviously, it is not desirable for the gage to call for a new adjustment until the results of the previous adjustment have been monitored. The time of gage signal shut-off is adjustable, depending on the location of the gage relative to the process. On extruders such as those producing plastic or rubber-coated electrical wire, the gage could be located 100 feet away from the extruder die, after a water bath for cooling and hardening the insulation. On rolling mills the gage is usually located within a few feet of the working rolls of the machine.

Very often, with this type of gaging, a strip chart recorder provides a permanent record of the size of the product. Particularly in the case of stock produced on rolling mills, a copy of this record will stay with the coil for reference, as proof of the size in the entire coil when, for example, it is sent through expensive dies in a stamping press.

In examining a strip chart recording of such a process it is interesting to see that a definite size is not maintained. The curve continually changes or "hunts" toward "zero," both plus and minus. As the machine and its components drift or wear, the monitoring of the gage causes adjustments fairly continuously. Without a gage in the process, however, at the speeds involved — 100 feet per minute up to 8000 feet per minute — thousands of feet of out-of-tolerance stock could be produced before the machine could be corrected.

In the whole gamut of in-process measurement described here, the inspector's position in large companies is sometimes remote. Usually production setup personnel, the machine operator, and maintenance personnel are more intimately involved with the gages. Nevertheless, the inspector could be called in at any time or may actually have the responsibility of periodically certifying and assessing the performance of the gaging equipment. The gage manufacturing companies provide complete instruction books for the set-up, use, and recommended maintenance procedures of their products. In all cases, complete data should be available on what to look for in case of problems. Electrical and mechanical diagrams and/or schematics should be part of these manuals. The inspector should make sure to have copies of the instructions for all the equipment for which he will be in any way responsible and should become thoroughly familiar with their operation.

Some general hints to follow in maintaining such equipment really fall in the realm of good gaging practice, but may be more strongly emphasized because of the continual automatic use of the gages. On contact gages where the work is engaged physically, one should make certain that the gage contacts are replaced or readjusted when worn. The advance and retraction or engagement and disengagement of the calipers or rollers should be maintained under the recommended pressures or tensions and repeatable positions.

Although electric or air–electric gages usually operate with very low voltage and currents at the switch contacts to avoid arcing, dirt of all types can build up causing erroneous operation.

On air systems where noncontact orifices are located close to the work, and are thus continually flooded with coolant and fine debris, the air escape channels should be kept clear and care taken that the precision orifices are not worn or damaged. Usually, the high-pressure air escaping in operation tends to keep the system clean. The problem of dirt in this area occurs after the operation has shut down for the night, for example, as the combination of coolant or oil with accumulated debris solidifies around the orifices and chambers.

Although gage manufacturers consistently recommend that clean, dry air be used in the operation of air gaging, and provide filters and moisture traps at the entrance side of the systems, it is virtually impossible to maintain absolutely clean air; eventually, moisture and dirt will create a problem. Maintenance procedures are usually spelled out in detail for these cases. Most importantly, when internal dirt or moisture cause a problem, observing the action of the air meter and referring to maintenance data will reveal the source of the difficulty.

On modern, fully electronic gages, plug-in electronic component cards are easily replaceable, reducing downtime of the gage to practically nothing. Usually, these gages have very infrequent problems since they operate at very low power requirements and are sealed against dirt and moisture.

In-Line Postprocess Gaging

Many automatic transfer lines in use today incorporate gaging stations to monitor newly finished operations. These stations may simply probe to ensure that an operation has been performed, provide incremental signals for adjustment of tools, or monitor the production from multiple-tool stations — keeping track of from which station the parts came, and shutting down any or all stations, if necessary.

Typical of a simple in-line station is that shown in Fig. 10. Here the gage probes the part with a "Go" plug to ensure that the hole has been sufficiently rough-machined to be automatically processed in an internal grinder, thereby avoiding damage to the expensive grinding wheel and, of course, downtime to reset the machine. The mechanics are simple in that as the parts are processed successively, the gage probes one at a time. If the hole is grossly undersize, the gage will reject the part, off the line and into a tote pan. If the hole is satisfactory, the part will proceed to the next operation. A more sophisticated gage can measure the part and through the use of multiple electrical limits, electronic classifier limits, or digital pulses provide signals to incrementally adjust tools to keep production within acceptable parameters. Gross part changes cause immediate signals for the removal of worn or broken tools. (See Fig. 11.)

Because of the availability of solid-state electronics (particularly micro- and minicomputers), a single gaging system simultaneously monitoring several production lines producing the same part, is practical. The rapid response of these devices makes it possible to virtually receive data from several sources simultaneously, interpret them, and send signals to each separate source before the next workpieces are in position to be machined.

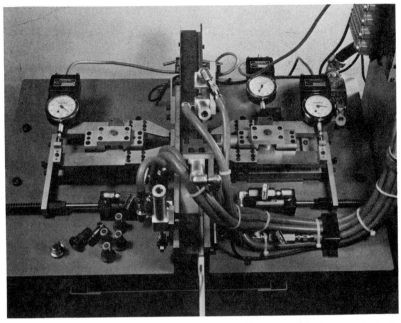

Courtesy of Federal Products Corp.

Fig. 10. In-line postprocess automatic gage probing hole for presence of rough machining.

Postprocess Gaging

In this philosophy of measurement we begin to approach the "automatic sorting" type of machine. These devices are usually located right after the exit of a machine and provide simple sorting for "good" and "scrap," "good," "salvage," and "scrap," or probably not more than two or three "good" categories, plus a "reject". The application generally involves measuring those parts where tight gage control is not practical in process.

In the measuring of the workpieces, adjustable percentage counters are incorporated to monitor the machine's ability to produce parts with a good size distribution. The counter is set so that if the ratio of sorted bad parts versus good parts exceeds a certain percentage, a warning is given to the operator by large dome lights, or audibly. A typical example is the chute gage shown in Fig. 12. A feature of this type of gage is that it has a fairly wide range of universality for measuring a range of sizes of like parts that the machine may be producing. Guide chutes, escapements, and disposals are all adjustable to fit a required size. Interchangeable tooling allows rapid changeover.

Fig. 11. In-line postprocess automatic gage with feedback signals. The system checks valve insert counterbores, valve guide bores, and dowel bores for engine cylinder heads.

In the use of postprocess and automatic inspection or sorting control systems, it is important to remember the basic precepts of measurement and control — understand the variation, as soon as possible, and fix it; don't just find the defects and scrap them and their costs. When these measurement systems are established, they should be set up to measure variation in terms of process or statistical limits, report the results, and derive action to limit the variation. Whatever the mode of measurement — numbers or dimensions or attributes, the

Courtesy of ITT Industrial and Automation Systems

Fig. 12. Automatic inspection system for roller bearings showing discharge chutes for classifying the gaged product.

purpose should be the same — to understand the inherent variation and to enact economical control of the system that produces the variation.

Automatic Sorting

The separate automatic gage usually devoted to final inspection of a finished workpiece customarily is isolated from the production area. The closer the required accuracy of measurement, the more isolated from all other areas it should be. Cleanliness and stabilization of the workpieces for precise measurements are the criteria for locating such machines in temperature and humidity controlled areas. The repeatability and linear accuracy of the sorting machine must be maintained hour after hour, day after day, and hence, demands stable conditions for successful operation. Frequent mastering of the gaging station(s) is always recommended to ensure that zero and classification settings are in order and that any contacts or tools used in the measuring area are kept clean and unworn.

In the design of automatic sorting gages, material handling to and from the actual measurement station can constitute the most complex part of the machine. From an unoriented position, or directly on a conveyor line, from a hopper, or even by manual means, successive workpieces must be aligned into a common position to arrive at the gaging station properly referenced for measuring. After measurment, workpieces often must be disposed of in an orderly stacked mode. However, just as often parts will arrive in disposal and merely drop into removable bins or tote boxes. Automatic gages use a variety of material-handling techniques depending on the requirements of speed of measurement, or the number of parts that are to be consistently measured per hour.

Semiautomatic Gages

The simplest automatic measuring is accomplished by an operator manually loading one workpiece at a time into the gage. This is generally called semiautomatic and is justified from several basic standpoints: (1) the workpiece cannot be practically oriented on an automatic basis; (2) the number of parts to be measured per hour is too great for a few inspectors to measure; and (3) the workpiece is too difficult to measure manually.

In the semiautomatic gage, the operator usually places the workpiece in a nested station, removes his or her hands, and presses a button. The gage is activated, performs the measurement and:

(a) In the simplest version, one of a number of trapdoors located on either side of the operator opens. The operator removes the gaged piece manually from the retracted gage, disposes of it into the category indicated by the open trap, and at the same time loads another workpiece. This version (shown in Fig. 13) relieves the inspector of tedious decision making and certainly eliminates the possibility of missorting.

Courtesy of Federal Products Corp.

Fig. 13. Semiautomatic gage measures thickness of discs. Manually loaded with automatically opened trapdoors for manual disposal.

(b) The gage completes the measurement and automatically shunts the workpiece into disposal. A typical gage of this type is shown in Fig. 14.

Manually loaded automatic gages are restricted only by the ability of the operator to feed parts to them and average consistent speeds of 1800 to 2400 parts per hour generally are attained.

The automatic gage next in simplicity is the "pusher" gage. In this concept parts are fed to the gage by a hopper and feed tube, or from manually loaded magazines attached to the entrance of the gage. From these tubes, or magazines, alternating escapement stops allow one workpiece at a time to align in front of a bar or blade which usually reciprocates back and forth by means of cam action. In retraction, the blade moves to a position clearing the escapement, a workpiece lands in position in front of the blade, which pushes the part to or through the gaging station where it is measured. Usually, the successive workpiece pushes the previously gaged piece into the disposal area, where a previously opened trap shunts the piece into its category. In the interest of speed, or in more complex situations, other pushers or mechanical

Courtesy of Moore Products Co.

Fig. 14. Semiautomatic gage (automatic disposal) measures half shell bearings for thickness and height and sorts them automatically.

devices shunt the gaged piece into disposal, while the receiver pusher is retracting to pick up another part. Speeds of up to an average of 3600 parts per hour are recommended for pusher gages.

All of the above, as with the more complex gages to be described, are controlled independently of the operator by electrical timing mechanisms and/or cam operation.

Indexing Belts and Walking Beam Transfer Gages

When more than one, or several, dimensions are to be measured on a workpiece, a walking beam transfer design or indexing belt or chain is employed. This allows the workpiece to be received, oriented into a loading station, and shunted progressively into a successive number of gaging stations. In this manner, many dimensions are measured separately, can be compared to each other, and if any dimension is out of tolerance, the workpiece can be ejected from the gage at that station without proceeding through the remainder of the gaging stations.

Very often this type of transfer is utilized when functions other than measurement are necessary. For example, after measurement identification as to category, a permanent marking for coding may be required; either metal stamping or paint color coding. Automatic weighing is another such function. An example of a walking beam transfer with automatic gages is shown in Fig. 15.

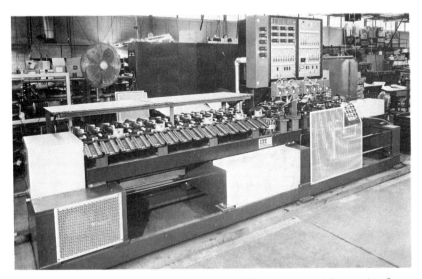

Courtesy of ITT Industrial and Automation Systems

Fig. 15. Indexing or walking beam transfer automatic gages.

The automatic gage system, operating at 2000 cycles per hour, stamps and classifies automotive pistons. It probes for the piston pin bore, gages the bore; piston skirt diameter; and taper, and automatically classifies acceptable pistons into nine categories then ink-stamps them.

Where protection of critical workpiece surfaces from marring or scratching in handling is required, this philosophy is also utilized. Individually transferred without bumping into each other, each part can be carefully inserted and removed from each gaging station on into controlled disposal. Indexing chains or belts transferring parts separately, are also used for this reason.

Another common type of automatic gage is the rotary wheel segregator where extremely high speeds of measurement, up to approximately 18,000 parts per hour, can be obtained. Roller or ball-bearing parts are typical of those that lend themselves to this type of measurement. In other words, cylindrical or spherical parts that can be passed through gaging stations without any retraction or advance out of the pass line of either the gages or parts can be handled in this way. Figure 16 illustrates this principle.

The parts are fed into a guide tube, or chute, above the rotating wheel. The wheel contains cutouts, or segments, to accept one part at a time. Parts are carried around the periphery of the wheel through gaging stations placed at intervals in fixed positions. The stations may include outside diameter measurement (two points in the case of tapered cylindrical parts for taper comparison and overall length).

Courtesy of Federal Products Corp.

Fig. 16. Rotary wheel automatic segreg. tor.

After the measurement (through electronic signals) and further along in the travel of the wheel, disposal tubes with trapdoors located in fixed positions will accept the parts into the various categories designated.

Matching Components

In the past few years, as a result of closer tolerances and the ability to measure accurately and consistently to "millionths" of an inch, automatic gages that can measure components and "match" them together are becoming increasingly popular. Fuel injection components or other precision valves are typical of such match requirements. Historically, because such components have high finish, lapped surfaces, and a fit where an ID to an OD may be on the order of .0001-inch clearance with plus or minus .000050-inch tolerance, the material handling to avoid damage and achieve accurate measurement to such close tolerances had to be on a strictly manual basis. Generally, the ID measurements were made carefully with visual readout air- or electronic comparators. The OD parts were then carefully lapped individually to fit each ID; a tedious and expensive procedure.

The state-of-the-art of solid-state electronics now, however, makes automatic matching practical. The components can be machined through their separate processes of grinding, honing, and/or lapping to tolerances of .0003, .0004, or .0006 inch. They can then be brought to a clean, stabilized condition in a controlled atmosphere. An automatic sorting and matching machine, such as that shown in Fig. 17, also located in a controlled atmosphere, will accept the components one by one, measure and sort them into categories, then from these categories select parts that will "match" as a pair and dispense them from the machine — or, in fact, even assemble the components.

In actual practice on a machine of this design, there is a kind of "lag effect" in the measurement of the components. Since it is known that controlling the final size of an outside diameter is easier, what actually takes effect is that a production of closely held ID parts is measured and categorized into storage. From the distribution of these parts a bell curve is discerned. The final size of the OD parts can be "aimed" at the center of the bell curve of stored ID part sizes; in other words, where the "bulk" of these parts is stored. Thus, the OD parts brought to the gage and matched with stored parts allows a continual flow output of "married" pairs. An efficient dispensation of stored ID parts is ensured by cutting down on the storage category space required and a minimum number of OD parts is required to get the maximum amount of matched pairs. Since efficient, careful materials-handing is mandatory to avoid marring or otherwise damaging the components, the speed of

Courtesy of ITT Industrial and Automation Systems

Fig. 17. Automatic gaging and matching gage.

such machines is generally restricted to from 200 to 300 matched pairs per hour.

Instrumentation

In all of the automatic gages described, various measuring systems are involved ranging from simple electric switching, through air and electronics. The criteria that dictate which system is to be applied involve accuracy, number of parts per hour, and number of categories required.

The simplest electric switching utilized would be in probing for the existence of a hole, for example. If the hole is missing, a microswitch is closed and a signal is used to reject. This is very inaccurate, simply a "Go"/"No Go" situation. The next higher level of switching would be to use a device such as that illustrated in Fig. 18. The "Electricator" shown is a dial indicator with minimum graduations of from .001 inch down to .0001 inch. As a continuation of the indicator spindle travel, levers trip electrical contact closure (adjustable limits) to activate electrical relays for disposal action. Recommended accuracies would be

Courtesy of Federal Products Corp.

Fig. 18. Dial indicator showing electrical contacts which are closed when indicator
spindle reaches preset positions.

no closer than .0005-inch per category, with a maximum of four
categories.

Air meters are available with electrical contacts; maximum number
of categories, four. Here the accuracy of sorting can be recommended
to .0001 inch per category.

In any of the above, speeds of operation are generally in the range of 1800 to 2500 parts per hour, with a maximum of 3600 per hour under most favorable conditions. When sorting into many categories, when high speed is necessary, when tolerances of .0001 inch or higher, or any combination of these conditions is required, solid-state electronic instrumentation is mandatory. The most important advantage, of course, is accuracy. Electronic transducers, amplifiers, and classifiers each offer linearity characteristic of 0.1% or better, and in combination will consistently perform with an overall accuracy of 0.25% or better.

In electronic gaging, a transducer reacting to the movement of a gaging device, contacts, air spindle, etc., because of change of size, outputs a voltage to an amplifier. The amplifier, through several stages of integrated circuits, will amplify the minute voltage to a value that can be split into increments. The output from the amplifier, either continuously variable (analog), or in pulses (digital), is then divided into discrete values. Each of these values, or voltage splits, is designated or adjusted to represent a category of sorting. These "trigger" points are tied electrically to operate solenoids, for example, in electrical relays which, in turn, activate disposal shunts, etc. This is a basic representation, of course, merely to show the train of philosophy generally used.

In recent years, the impact of minicomputers used in the production line as dedicated instruments has been outstanding. These compact 4000- to 12,000-word memory computers are called "dedicated" because they are designed usually with a single software program to perform one set of functions only. Their implementation as part of automatic gaging is a natural and valuable development.

One important advantage is that a single minicomputer can take the place of a cabinet that may have required many amplifiers taking up a great deal of floor space and certainly requiring a long setup time and maintenance. A good example is shown in Fig. 19, A and B, where 16 dimensions and 10 relationships, such as eccentricity and squareness, are measured — all in two full turns of a rotary table. It is probable in this case that up to ten full turns of the table would have been required without the minicomputer (the savings in time and wear are obvious). In the two turns, the speed of the computer input logs 960 readings for each of the 26 parameters; a total of 24,000 bits of information. Now, while the part just measured is unloaded and another is put into place; the computer, on command, will print out all data on the size of the dimensions, minimum and maximum. That is, the actual decimal dimensions such as: 11.5020 max, 11.5010 min; for the relationship data — for eccentricity, for example: .0025 TIR. Furthermore, if desired, rather than print out complete data on all parameters, the computer will print out on command only the data on dimensions or relationships that exceed the programmed tolerances. This is a further

Courtesy of Federal Products Corp.

Fig. 19. Automatic gage with minicomputer. (A) Close-up view of gage ready for inspection. (B) Full view of same gage system showing minicomputer.

time saver, since in all likelihood, the part being measured is produced with a very high percentage, as "all good," or certainly with but few dimensions out-of-tolerance, and the important thing here is to quickly-find out-what is going wrong, if anything, and correct the process. On the other hand, occasionally, a full print-out of all data reveals significant trends in the machining process.

Another distinct advantage of listing all the tolerances in software memory is that one can, through the teletypewriter or printer, type in, or address the computer with a simple code to focus in on a particular dimension and, in simple language, change a tolerance — all in a few seconds, as opposed to loading a master, zeroing — changing the voltage to the tolerance limits electrically for each limit, and manually adjusting each limit — subject to the operator's visual ability to align the limit point.

Keeping track of successive workpieces as they are processed is another approach to minicomputer use. Any machine that is producing an out-of-tolerance part, can be identified automatically and alerted or shut down. Further data can be sent to alert a machine operation of trends before out-of-tolerance work is produced. If desired, part of the memory can store dimensional data, to be outputted to a main data process computer that may be analyzing data from all production.

There are many other examples of minicomputer application to automatic gaging, suffice it to state that many "minis" have been put into use, and almost daily there is a gage being designed and built with these fine devices as part of the assembly. Naturally, this has brought new types of technicians to the inspection area; a computer pro-grammer and a maintenance man trained in the service of such equipment. Fortunately, in the purchase of these minicomputers an educational service generally is offered either in programming or service.

Rather than feeling overwhelmed by the thought of its sophistica-tion, the gage inspector will find that an automatic gage, minicomputer equipped, actually is operated with more ease. The print-out record usually is listed in actual blue print dimension, rather than as a deviation from a "zero." As mentioned, the tolerance limits are easily changed by typing in a code and the required change. The entire software program, or a new program on punched tape or cards is inserted with very little effort or training.

All the foregoing, then, covers a representation of the automatic gages and machines in use throughout industry today. Hardly new, the use of automatic measuring goes back well before World War II. It is only in the relatively last few years, however, that the accuracy of the instrumentation applied became so excellent, that wider use could be made of its advantages. Then, too, it should not be considered as being utilized strictly in the metal cutting industry; paper and plastics are measured continuously. Continuous automatic measurement of the intrusion of meat into sausage casings, under pressure, is monitored to ensure, first, that there is "enough" meat and, secondly, that there is not "too much" meat to cause the casing to split. Such products as vitamin and aspirin tablets are measured to a few thousandths of an inch tolerance for thickness to avoid damaging automatic bottling machines, where 200 tablets are loaded at a time, and if they were "too thick," would overflow and jam the machine.

Any process that involves somewhat high production, is often too difficult to monitor annually, costs too much to manually inspect, or requires too much time, thus inhibiting profitable overall production, will eventually be measured automatically. Automatic gaging is a necessary, and certainly complimentary, part of production and is rapidly becoming a dominant part of the industrial inspection procedure

CHAPTER 19

Nondestructive Testing

By definition, nondestructive testing is the name given to procedures and techniques which allow a product to be inspected for internal defects, or microscopic defects on the surface, and those defects identified without the product being destroyed. Despite the widespread use of nondestructive testing in industry today, it is a science often not well understood by those who must rely on it. This chapter will outline the various techniques and equipment employed in nondestructive testing and point to the many sources of information and education which are available in this expanding field. Throughout the chapter the abbreviation NDT will be used interchangeably with the words Nondestructive Testing.

Nondestructive testing is a wide and varied field of Quality, Reliability, or Inspection, and is growing and changing rapidly, for some very good reasons. NDT is really an extension of the testing process itself, and may or may not overlap some tangible realm of inspection.

First of all, this testing may be conducted on individual parts or components, or on completed subassemblies or assemblies of a given product. The testing or inspection may be intended to first of all examine the part beyond the dimensional inspection, into the realm of quantifying visual/mechanical defects that could directly or indirectly contribute to a failure of the part or assembly.

Beyond the visual mechanical inspection, NDT may also involve some limited functional testing of the assembly — some test combined with an inspection NDT designed to destroy the product or part, but perhaps intended to test a physical relationship by introducing an outside force, and evaluating or quantifying the impact. In fact, many of these inspections and tests are now made with the product in a functional or "loaded" condition — the operation of the unit is integral to the evaluation, since the evaluation is based on understanding how the requirement (or defect criteria) will effect performance.

Much progress has been made in this area of quantifying or examining product, based on visual/mechanical requirements, especially in the fields of optics, different types of light and sensing systems, and electronic sensing. The point is to definitively measure variables that are known to be important, in a way that can be easily compared and quantified for purposes of control. The extent, of measurable characteristics of a crack in a casting, for instance, is more important to understanding how to prevent the cracked castings in the process, than just recognizing and rejecting the crack.

Discovering Faults that are Not Visible

All flaws in weldments and castings cannot be observed in ordinary external visual inspection. If blowholes show up on the surface, there are liable to be voids under the surface. Cracks, inclusions, cold spots, crystallization, and other structural weakening flaws occur deep down in the centers of casting selections or in welding beads; conditions which are permanently sealed off from the sight of the naked eye. Modern technology does provide us with several ways of finding out what is taking place in the interiors of metal sections in much the same way as the doctor can examine human internal organs.

Radiographic Inspection

X-rays and gamma rays are used most commonly in industry for checking cracks, cavities, slag, and blowholes in welds and castings. The principle used in X-ray inspection is the ability of short wavelength radiations to penetrate opaque material. If there are holes or inclusions in the material being X-rayed there will be variations in the absorption of the rays and an image recorded on film will clearly show up these defects. Cracks show in the exposed film as dark lines on a light background while the presence of slag is indicated by a dark silhouette in the radiograph. Porosity shows up as dark spots. The source of radiation is the X-ray tube consisting of a glass tube containing a cathode and an anode. Heating the cathode filament in the vacuum of the tube forces it to give off electrons which by high voltage are sent at high velocity to the anode — the result is the emission of X-rays. Gamma rays are similar to X-rays but shorter in wavelength and are produced from the atomic disintegration of radioisotopes such as cobalt-60 or iridium-192. Generally, gamma rays require a much longer exposure time than do X-rays and there is some sacrifice of sharp definition. The big advantage of gamma ray radiography is portability, but because of the lack of definition gamma rays are used where the subject material has high contrast characteristics.

Portable X-ray System

Figures 1 and 2 show portable X-ray units set up and in use. Figure 1 shows a 100 kV X-ray unit and demonstrates the ease with which modern X-ray equipment can be transported to the job. A system like this one can be used for a wide range of radiographic applications in aircraft, construction, and industry. The tube head weighs only six pounds, and the control unit employs solid-state circuitry and a digital readout. Figure 2 shows another portable unit mounted on a hand truck. Notice the cassette containing the film to be exposed by the X-rays taped to the back of the weld inside the tank.

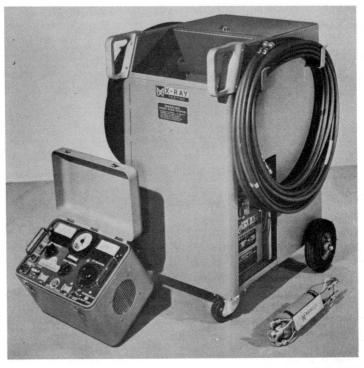

Courtesy of Magnaflux Corp.

Fig. 1. A compact portable X-ray machine which can be transported to the job.

Of course the operators of X-ray equipment must be qualified technicians well versed in the use of the equipment and in the interpretations of the results. Excellent courses in industrial radiography are available from several of the leading manufacturers of X-ray equipment. Special courses can be scheduled so that the entire class is

Courtesy of Philips Electric Institute

Fig. 2. An earlier model of a portable X-ray machine is shown here being used for
flaw detection in a welded seam.

made up of members of a single company and if the equipment is
available, instruction can take place on the company premises. A
typical 40-hour course might include basic metallurgy and principles of
radiography, making of a radiograph, photographic factors, techniques, diagnosis of radiographs, radiographic standards, and radiography by gamma rays.

Radiation Hazards

A special warning in regard to radiation from gamma rays or X-rays
and the exposure of the human body to them is certainly in order. The
effects of even a small amount of X-ray is cumulative and while the
body can absorb a certain minimum dosage on a regular basis, all
precautions should be taken against heavy exposure. Radiography
equipment should always be used only by thoroughly trained persons
and under controlled conditions. All industrial codes for use of
industrial X-ray should be known and observed by the operator. There
are radiation meters and film exposure methods that check for stray
radiation. The American National Standard Safety Code for the
Industrial Use of X-rays ought to be obtained and studied

Ultrasonic Inspection

Recently, there has been an upsurge in the use of high-frequency vibrations to detect all types of flaws in welds and castings. Actually, ultrasonic waves are the same as sound waves but are generated at high frequencies. Most inspectors have at one time or another tested a weldment or a casting by tapping it with a hammer and listening to the clearness and purity of the resulting ring. With ultrasonic equipment it is possible to do the same thing with electronics. The high-frequency waves are generated in a transducer coupled with a piezoelectric crystal and are directed toward the object to be inspected in cycles as high as 1000 times per second. The waves are reflected back to the transducer and displayed on an oscilloscope as "blips." The patterns thus displayed on the cathode-ray tube of the oscilloscope are interpreted by their height above the base line blip and the distance between two reflections. An observer of the "spikes" on the display tube can, by his knowledge of the characteristics of the patterns, determine whether there are flaws or inclusions in the part and precisely where they are located. The operator of an ultrasonic tester moves the transducer probe back and forth so that sonic waves can reach the entire area to be checked. In the case of a weld he must cover the width of the weld and movement parallel to the weld should not be more than the transducer width in any one pass. To make sure that the transducer is making proper contact with the test material, a wetting agent, such as a light oil, is applied to the surface.

It is important for proper evaluation of the results on the display tube and for a continuing check on the performance of the ultrasonic equipment that standard reference plates be used to calibrate and show up any deterioration in the electronic components. Standard ultrasonic reference plates are approved by ASME (American Society of Mechanical Engineers) and ASTM (American Society for Testing and Materials). Figure 3 shows a part being examined by portable ultrasonic equipment. The outstanding features of ultrasonic inspection are portability, the ability of ultrasonics to be used from one side of a test piece to be checked, and complete safety for the operator.

Ultrasonics has also been used in the measurement of dimensional thickness. An especially useful application is in the inspection of hollow casting wall-thickness or the thickness of the walls of hollow drawn tubing where a mechanical device cannot be employed because of the impossibility of referencing the inside of the wall. The principles for checking thickness with ultrasonics are the same as used for checking for flaws, i.e., a continuously varying frequency is generated by electrical means and by means of a piezoelectric transducer. The ultrasonic vibrations are sent into the part wall being checked through a

Courtesy of Magnaflux Corp.

Fig. 3. Use of pulse ultrasound for measuring wall thickness of hollow drawn tubing by applying transducer to the outside wall; the exact thickness is then shown on the digital device.

liquid coupler on the surface — a light oil, or more commonly, glycerin. When the constantly varying frequency matches the resonant frequency of the workpiece the actual thickness dimension is indicated on a calibrated meter dial.

Ultrasonic thickness gages have, through refined technology, gained acceptance as reliable indicators or thickness dimensions, in a wide range of materials applications. These units are available in small, portable, configurations, or they are fixtured into production environments, even used in continuous-monitoring applications. Ultrasonic principles have a variety of applications in industry — from oil and gas exploration to oceanography. Whatever the application, the notion of inspection and measurement is easily transferrable to this technology, and has been in many different areas of measurement.

Penetrant Inspection

Dye penetrant inspection is used to locate porosity, cracks, and other defects on the surfaces of nonmagnetic solids. The procedures employed in penetrant inspection are fairly simple and may be learned by the inspector with a minimum of instruction. Using penetrant dye, the inspector makes visible any defects on the surfaces of castings, bar stock, and forgings, and on such finished or semifinished components as electronic and machined parts. The defects might be caused by shrinkage, entrapped air, heat, metallurgical variables, fatigue, seams, leaks, "shuts," or improper bonding, in fact any type of microscopic flaw that has come to the surface of the part. Penetrant inspection usually is employed on nonmetallic materials which cannot be inspected by the magnetic particle method. The most extensive use is to check for surface defects and leaks in castings and weldments. The simplest procedure employed in the detection of cracks by dye penetrant is to wash and dry the part and then, with brush or spray, apply the penetrant liquid. After a short interval to allow for penetration of the liquid by capillary action into any of the extremely small surface openings — the waiting time may vary from a few minutes to several hours, however, depending on how fine and tight the cracks are — the penetrant is wiped off and a thin film of developer is sprayed on the surface. The developer colors the area being inspected white, and also serves to pull the penetrant from the defects and show up any cracks or flaws as dark or red lines or spots. Figure 4 shows the spraying of a workpiece with penetrant, and Fig. 5 is a picture of the crack now visibly displayed after being treated by the developer.

Courtesy of Magnaflux Corp.

Fig. 4. A penetrant is sprayed on the workpiece and by capillary action it penetrates surface defects or cracks.

Courtesy of Magnaflux Corp.

Fig. 5. The penetrant applied in Fig. 4 has been wiped off and the developer sprayed on and allowed to dry. The crack in the workpiece now becomes visible.

Fluorescent Dye Penetrant

A slightly more sophisticated variation of the ability of the penetrant dye process to discover cracks or leaks, is the use of fluorescent liquid and black light. In this method a highly fluorescent liquid is applied to the part surface; after being wiped off, the surface is then sprayed with a developer. A "black" (ultra-violet) light is employed and the defects then show up bright and sharp.

Magnetic Particle Inspection

The magnetic particle method for testing for flaws is probably the most widely used of all tests on magnetic materials. The process is widely known or described as "Magnaflux," which is actually a trade name for a process (developed and sold by the Magnaflux Corporation) where the principles of the electromagnet are used. An intense magnetic field is set up in the part being tested; any sudden interruptions in this magnetic field, caused by cracks or discontinuities, will crowd some of the magnetic flux outside the surface of the part. These leakage fields act as local magnets to attract and hold finely divided ferromagnetic particles applied to the surface of the part, showing the exact size and shape of the crack or other discontinuity. Actually, small magnetic fields form north and south magnetic poles at each juncture. The junctures may be cracks, nonmetallic inclusions, or even changes of density of the materials. The magnetic particles may be dusted on the part surface or suspended in a suitable liquid. The most widely used practice is the wet method (spray or dip), using a water or oil suspension. However, the dry method is generally used for inspection of welds, large forgings, and castings and other parts with extremely rough surfaces and where location of subsurface defects is needed.

There are two methods of magnetization used to find cracks or flaws in all directions in ferromagnetic parts. A part can be magnetized longitudinally, or lengthwise, by placing it inside a coil carrying electric current. A magnetic field is created running lengthwise in the part. When this field is across a crack, it attracts and holds iron powder to indicate the crack. Circular magnetization is produced by passing a current lengthwise through the part; this creates a magnetic field around and within the part at right angles to the direction of the current within the part. Where the circular magnetic field cuts across a crack it also attracts magnetic particles and thereby indicates a subsurface or microscopic crack or flaw in the material.

There are also several other different methods of magnetization two of which are most commonly used: residual and continuous magnetization. In the first of these methods the residual magnetism in the part attracts the ferromagnetic powder, while in the second, a continuous magnetizing current is employed as powder is applied. The continuous method is more sensitive and is commonly applied to ordinary soft steel which has low magnetic permeability and retentivity.

Eddy Current Testing

Eddy current testing came into use for checking tubing used for nuclear fuel element cladding. An eddy current is an induced electrical current within a mass of metal. A coil carrying alternating current induces an alternating current (eddy current) of the same frequency in the part being inspected. These currents are affected by variations of conductivity, permeability, mass, and homogeneity within the test part. The current changes are reflected as changes in a signal output from a probe coil placed on the surface with the coil axis perpendicular to that surface. An alternative method of detecting changes in electrical characteristics is to pass the test part through a circumferential coil. The latter method is well suited for checking shafts, bar stock, and tubing. The probe coil procedure does have the advantage of high resolution so that a very small section of the part can be checked at a time. The eddy current principle can be applied to flaw detection in metal parts, such as voids, inclusions, seams, and laps. It also can be used for sorting parts according to alloy, temper, conductivity, and other metallurgical factors and for dimensional gaging to size, shape, plating thickness, or thickness of insulation.

Figure 6 shows a probe coil being used for aircraft maintenance, to check for possible fatigue cracks around rivet holes. A portable unit

Courtesy of Magnaflux Corp.

Fig. 6. All wet: A billet is grasped by magnetizing heads and is bathed with fluorescent magnetic particles in suspension. Seams in the billet, viewed under black light, show up as yellow lines.

measuring conductivity of large copper coils is shown in Fig. 7; and Fig. 8 shows steel pins being automatically sorted for hardness. Because so many properties of metals are directly related to electrical conductivity, eddy current testing principles have been employed to: evaluate metal hardness, identify alloys, check uniformity of heat treatment; sort mixed nonmagnetic metals; measure absolute conductivity; test for surface discontinuities; detect fire damage to aircraft surfaces; measure thermal conductivity; and measure coating thicknesses. The list could probably be extended due to the variations in other properties of metals which also result in changes in conductivity.

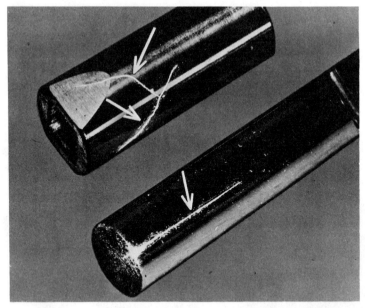

Courtesy of Magnaflux Corp.

Fig. 7. Magnetic particles inspection by the wet method using fluorescent particles
under black light reveals a deep seam opened in heat treating and heat treat cracks
in aircraft parts.

The advantages of eddy current testing are: (1) The probe does not
have to physically touch the test piece so it can be used for high-speed
testing; and (2) The test can be made in a poor optical environment and
in the presence of contaminants.

The disadvantages of eddy current testing are: (1) It can only be
used in conjunction with conductive materials; (2) It is dependent on
uniform hardness or alloy content unless specifically being used to sort
for hardness and alloy content; and (3) It is affected by variations in
temperature and magnetic field.

Hardness Testing

The most common of all nondestructive tests is for hardness, and
the oldest of these is the file test. It is rarely used at present because of
the availability of modern hardness-testing machines or gages. File
testing uses standard files with hardness numbers stamped on the
shanks. In using hardness files, calibrate yourself by filing on workpieces of known hardness; strive to file at a consistent angle, and at the

Courtesy of Magnaflux Corp.

Fig. 8. Here eddy current testing is used to detect core and case hardness of piston pins. Test piece placed in right coil must match master (left). Coils left and right produce curve on screen to match graph overlay.

same uniform speed and pressure. If the file cuts the workpiece the latter is softer than the file, and vice versa, if the file makes no impression. To measure the degree of hardness in a workpiece it is usually necessary to use a series of files whose hardnesses are graded about 10 hardness numbers apart.

The Rockwell Hardness Tester

Most often, the hardness of a material or part is tested by forcing or indenting a harder material into it and measuring the resultant penetration. The most widely used machine for this purpose is the Rockwell tester. In this machine, the test is made by driving a penetrator into the surface under a dead-weight load applied through a series of levers. An indicator gage shows the depth of penetration. The softer the metal being checked the deeper will be the penetration. The indicator dial does not read directly the depth that has been penetrated but instead has arbitrary scales showing "Rockwell Numbers". There are two types of penetrators; a diamond cone, called a "brale," is used for hard tool steel while a hardened steel ball is used for indenting softer material. The load normally used when loading the brale is 150 kilograms, and the scale read on the dial is designated with the letter "C". . . the resultant readings are specified as being from the Rockwell C scale. As a rule of thumb, the hardest tool steel to be checked with this method will usually not be over Rockwell C 70. So-called "file hardness" is approximately Rockwell C 65. To obtain the hardness of softer material a 1/6-inch-diameter ball is used with a load of 100 kilograms. The readings taken with the 1/16-inch ball and 100-kilogram load are read from the "B" scale of the indicator and are called a Rockwell B hardness.

Superficial Hardness

When checking very thin sections of soft material or extreme surfaces of hardened tools, a superficial hardness test is specified. The instrument used in this case is a "Superficial Hardness Tester". It is possible to have one machine which can be converted to either superficial hardness testing or regular hardness testing. A much lighter load is used and when the diamond brale is used for checking hardness of very hard tool steel the letter "N" scale is read from the dial of the instrument and the reading is specified with the load applied as on the "30 N" scale or the "15 N" scale. When the 1/16-inch ball is used to check thin or softer material, the letter "T" is used, which means the calibration from the T scale on the dial is used. The reading then might be designated as being on the "30 T" scale or the "15 T" scale.

Master test pieces of known hardness are supplied with the hardness tester so that its calibration accuracy, and repetition can be checked. Where the hardness of the material is unknown, first test with the diamond point and C scale. A glass-hard surface could damage a steel-ball penetrator.

The condition of the surface of the workpiece is important. The penetrator may register on ridges of surface roughness, on scale, or on dirt, and give false readings. For accurate readings it is best to polish smooth at least the small area which is to receive the test identation.

The Brinell Test

A much older test for hardness than the Rockwell is the Brinell test. This test operates in a similar manner to the Rockwell test but employs a much larger steel ball (2/5-inch or 10 mm in diameter) and is pressed into the material being checked with a load of 3000 kilograms (approximately 6500 lbs per square inch). The tested part is then removed from the machine and the diameter of the impression is measured under a microscope and converted to a table of Brinell hardness numbers. On some present-day Brinell machines the Brinell hardness numbers can be read directly on the dial. The problem with the Brinell test is that it is only useful on soft and partially hardened material. The diameter of the impression made by even a tungsten carbide ball on hard steel is so small that it is difficult to get a good reading and when the steel ball is used, it tends to flatten out under the very heavy load.

Vickers Hardness

The Vickers test uses a square pyramid diamond penetrator under constant load and the degree of penetration is measured under a microscope across the corners of the impression. The microscope is swing-mounted on the machine. The Vickers test gives hardness numbers close to the Brinell numbers up to 300, but as the materials get harder the Brinell numbers are progressively lower than the Vickers numbers. Above 600 the Brinell numbers are not valid. The Vickers numbers can be used as a measure of hardness up to over 1800 which is comparable to approximately 80 Rockwell C. In the Vickers test the area to be checked must be smooth ground, and round or curved surfaces should have flats ground on them for the test. Thin pieces can be checked by the Vickers pyramid point but the opposite side of the part must be smooth and making complete contact with the supporting surface.

Scleroscope Hardness Test

The Scleroscope test consists of striking a part with a diamond-tipped hammer then measuring the height of the rebound on a calibrated self-recording scale. In some cases the height of the rebound is signalled by photo-switches to an electronic readout. Some simple rules to follow when using the Scleroscope are: (1) the surface of the part must be flat and smooth; (2) the tube containing the striker must be in contact with the test part and perpendicular to the test surface; and (3) the striker must not be dropped on the same spot twice because not only will the measurement be incorrect but the metal which has been cold worked by the first strike may cause the diamond point to crack or shatter.

The Scleroscope can be used in continuous production and has in fact been used in conjunction with automatic and one-hundred per cent inspection. Parts as thin as 0.005 inch can be checked with the Scleroscope as long as they are clean and well supported. One advantage of the Scleroscope method is that superhard or brittle materials, which do not lend themselves to being indented or scratched by other tests, will allow the striker of the Scleroscope to penetrate and rebound but the marring effect of the impression is so small as to be barely visible.

Other NDT Procedures

While this chapter has dealt with the most common nondestructive testing processes, there are many new developments in the field of testing. Laminated and bonded material is being tested for flaws and improper bonding, by means of infrared rays. Eddy current testing has been perfected for high-speed checking of metal parts for hardness. Microhardness testing can be performed by use of ultrasonics. There are magnetic procedures for checking coating and plating thickness. Porosity can be checked by a patented X-ray technique called "Xero-radiography" and a process employing this technique is being developed to produce an X-ray image on paper for a permanent record.

An entirely new and exciting engineering development of recent years is called "Holography" and has been found to have great potential value in the field of NDT. In fact, an entirely new nondestructive testing discipline called HNDT is being developed which is already in wide use in industry. Holography is a method of recording a three-dimensional image on a photographic plate without the use of lenses; it became practical with the invention of the laser in 1960. Optical holography uses ruby, helium-neon, and argon lasers, while acoustical holography employs coherent beams of acoustic radiation. Holographic analysis has proven useful in disbond studies in tires, bonded panels, and bonded laminates. Holography is also being used for analyses of strain patterns in joints and welds and for vibration analysis.

Testing Laboratories

All of the NDT procedures and equipment which have been covered in this chapter are available on a contractual basis with private testing laboratories located in the major industrial areas. There are laboratories that specialize in one or several of the nondestructive testing methods and several that offer metallurgical, chemical, and mechanical tests as well. In selecting a laboratory it is desirable to pick one with up-to-date equipment and the capability of performing a variety of tests. Their personnel should be well trained by experience, by their participation in training programs and seminars, and by attendance at schools that specialize in teaching the various NDT techniques. The best laboratories are usually qualified to perform to military and NASA specifications and their personnel are participants in the work of professional societies such as the Society for Non-destructive Testing or the American Society for Metals. A testing laboratory should have available an operations manual which can be reviewed by the prospective customer who should look for evidence that there are written standard procedures and regular maintenance and calibration of equipment. The contract laboratory should also offer certification of tests when requested or required.

All the practices that a good contract NDT laboratory ought to do are identical to the things that anyone conducting tests in his own plant should do also. For example: material should be tagged for identification upon receipt, parts must be carefully cleaned, comparative test parts should be used when necessary, and all defective parts should be tagged and kept separate from good parts. Of course the laboratory does not indicate that the parts are rejected or acceptable; only whether there are defects present in the part or not. The test information is submitted to the customer who rejects or accepts the parts depending on whether or not the defects conform to his specifications. Sometimes, working with an independent laboratory is the first step into NDT before a company embarks on its own in-plant program. Many labs will help a customer design the proper tests, select the right equipment, and assist in training his operating personnel.

NDT — A Growing Field

This chapter has attempted to acquaint the reader with some knowledge of the most common nondestructive testing techniques and equipment. Obviously NDT is a big field and entire books can be written about some of its more technical and scientific aspects. There are books available which cover, in depth, the principles of Radio-graphic, Ultrasonic, Dye Penetrant, Magnetic Particle, and Eddy Current testing. One prominent manufacturer of NDT equipment also offers courses for Quality Control technicians and inspectors. These courses are usually 40 hours in length and consist of: laboratory work, lectures, film viewing, and class discussion. An inspector or technician who wishes to keep up with the continuing new developments in NDT should join one of the professional societies and attend its meetings and seminars. He should also read the current literature and publications.

All indications point to astounding growth for nondestructive testing during the next several years. Predictions are, that by 1980, expenditures for NDT equipment and material will reach $260 million a year, more than triple the amount now being spent. There should be rapid growth in the areas of acoustics, neutron radiography, and holography. The inspector or quality control technician who wants personal growth in this rapidly developing technology should take steps to upgrade his own knowledge of NDT theories, procedures, and techniques.

Index

Index

A

641

D

N